D0086776

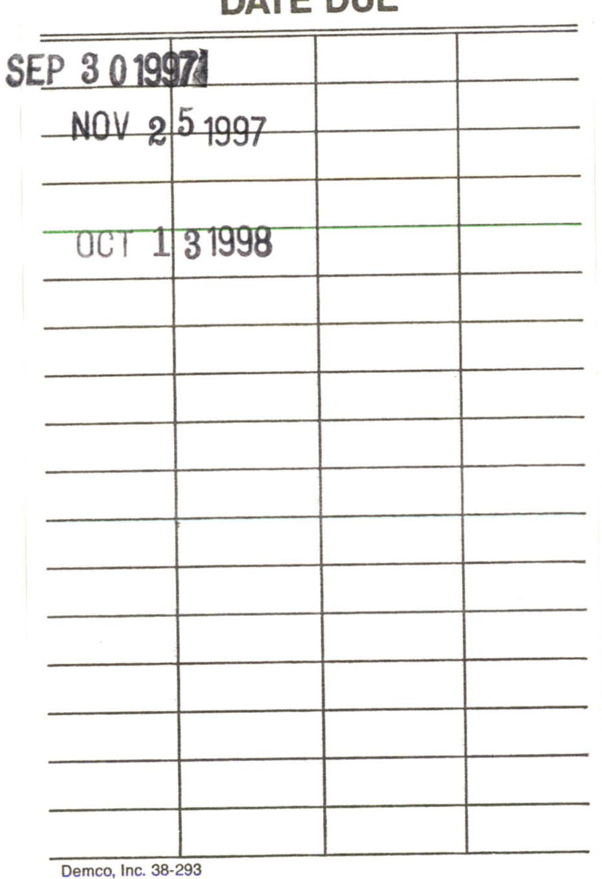
DATE DUE
SEP 30 1997
NOV 25 1997
OCT 13 1998
Demco, Inc. 38-293

WITHDRAWN

The Keys to Advanced Mathematics:

Recurrent Themes in Abstract Reasoning

Daniel Solow

Department of Operations Research
Weatherhead School of Management
Case Western Reserve University
Cleveland, OH 44106

Order from:

BookMasters Distribution Center
1444 U.S. Route 42, RD # 11
Mansfield, OH 44903
Fax: (419) 281-6883

For credit card, library, and bookstore orders call:

(800) 247-6553

Library of Congress Catalog Card Number: 94-96843

Solow, Daniel
The keys to advanced mathematics:
recurrent themes in abstract reasoning

ISBN 0-9644519-0-5

Printed in the United States of America

Ordering Information for

The Keys to Advanced Mathematics:

Recurrent Themes in Abstract Reasoning

by

Daniel Solow

ISBN number: 0-9644519-0-5

Order from: BookMasters Distribution Center
1444 U.S. Route 42, RD # 11
Mansfield, OH 44903
Fax: (419) 281-6883

For credit card, library, and bookstore orders call toll free (800) 247-6553.

A complimentary desk copy and Solutions Manual for all the exercises are available to instructors who adopt this book as a *required* text by sending the course name and number together with the instructors's name and address to

Professor Daniel Solow
Department of Operations Research
Weatherhead School of Management
Case Western Reserve University
Cleveland, OH 44106

The companion book, *How to Read and Do Proofs*, second ed., by Daniel Solow is available from John Wiley & Sons, Inc., New York, NY (ISBN number: 0-471-51004-1).

To my brother, Jon, who always wanted to know,

"What, exactly, does a mathematician do for a living, anyway?"

Foreword

Mathematics is a remarkable discipline. It is universal and it is unitary. It is uniquely suited to the study of the real world and it represents a highly sophisticated systematization of our process of rational thought. "Nature speaks to us in the language of mathematics," said Richard Feynman. Eugene Wigner noted "... the unreasonable effectiveness of mathematics."

But how can students learn to use that language? How can students become truly effective in mathematical reasoning? It is plainly not enough to learn a number of algorithms – there must be genuine understanding. The author of this imaginative text has long been concerned with the crucial pedagogical issues involved in conveying mathematical fluency, as well as with the nature of effective mathematics; and, in this text, he brings to bear his insights and experience in order to familiarize the reader not simply with important and frequently recurring mathematical techniques, but also with the types of reasoning, formal and informal, that characterize productive mathematical activity. For much of the work done by mathematicians in solving problems does not appear in their published research – or even in their expository articles and books. This is the informal, unrigorous thinking that determines the strategy of attack on a problem and the modifications of tried and tested strategies that are likely to succeed in the new situation under analysis.

Daniel Solow is trying to demystify mathematics, to show *why* it succeeds *when* it succeeds. He has done this before in demystifying mathematical proof[1]; here, he is much more ambitious in seeking to explain and clarify the methods of reasoning used by experienced practitioners of the art of mathematics. His initiative and enterprise are to be saluted, for certainly mathematical exposition cannot succeed without the kind of approach that he adopts. I wish him and his readers every success in this thoroughly worthwhile, indeed, essential endeavor.

Peter Hilton
Distinguished Professor Emeritus
State University of New York at Binghamton

---

[1] See *How to Read and Do Proofs*, second ed., by Daniel Solow, published by John Wiley & Sons, Inc., NY, 1990.

Preface to the Instructor

The Objective of this Book

The objective of this book is to simpify and to shorten the task of learning the underlying concepts involved in discrete math, linear algebra, abstract algebra, real analysis, and most other college-level advanced mathematics courses. These fundamental ideas include not only doing mathematical proofs, but also such topics as unification, generalization, abstraction, identifying similarities and differences, converting visual images to symbolic form and vice versa, understanding and creating definitions, and developing and working with axiomatic systems. This foundation reduces the amount of time and frustration it takes to learn these ideas and should enable undergraduate and graduate students in mathematics and the related fields of computer science, operations research, engineering, statistics, economics, and the physical sciences, to use mathematics more effectively as a problem-solving tool in their personal and professional lives.

The General Approach

There are two distinct philosophies toward teaching advanced mathematical ideas, such as the ones described above. One belief is that the only way to do so is in the context of teaching a specific body of knowledge, like abstract algebra or real analysis. The advantage of such an approach is that the student sees the application of general mathematical processes as they arise in "real" mathematics. Unfortunately, due to the amount of material that must be covered in a typical advanced math course, most of the class time is spent teaching the specific subject matter, leaving little, if any, time to discuss the underlying conceptual ideas (which are also unavailable in current textbooks). Even when an instructor *does* attempt to teach explicitly unification, generalization, and abstraction, the student is faced with the challenging task of trying to learn these ideas in addition to the actual subject matter.

The alternative philosophy – which is the one adopted here – is to separate *temporarily* the teaching of general mathematical processes from any one specific subject. The approach is to identify and to teach, with simple and easy-to-understand examples selected from a variety of mathematical areas, the foregoing conceptual ideas that are common to virtually all college-level mathematics courses. The disadvantage of this approach, it can be argued, is that the general mathematical processes may seem meaningless to the student when taught without the context of a specific subject. The advantage, however, is that this approach allows the student to focus attention exclusively on these important ideas without having to worry about learning specific subject matter as well. It is hoped that by presenting these ideas explicitly beforehand, the student will have an easier time understanding subsequent advanced math courses. Such an approach is analogous to teaching a student of chess the names of the pieces and how they move *before* letting them play a game that involves tactics and strategy as well.

Given these two separate and distinct philosophies, there seems to be only one way to teach the important conceptual ideas of advanced mathematics so as to satisfy both groups, and that is to write *two different books*. The current book is designed for those who believe in separating the teaching of conceptual ideas from specific subject matter. For those preferring the alternative approach, the *second* book will be a linear algebra text that covers not only the specific topics of a standard undergraduate course but also incorporates, whenever appropriate, a discussion of the underlying conceptual ideas of unification, generalization, abstraction, and so on, *as they arise in the context of linear algebra*.

Turning now to the specific approach of the current book, conceptual ideas from two types of mathematics are addressed. One is the "how-to" mathematics of solving specific problems: how to solve linear equations, how to differentiate functions, how to solve differential equations, and so on. The second is the "why" mathematics, which is the use of mathematics as a tool for organizing knowledge and for understanding *why* certain facts are *true*.

The "How-To" Mathematics

For the most part, the "how-to" mathematics is currently taught in applied linear algebra, calculus, and differential equations. In this book, students are given an overview of this use of mathematics as it spans *all* of these subjects so that, in each such course, they will have a general context in which to understand what they are learning. In particular, students are taught that a problem (algorithm, theorem, and so on) consists of some *known* information that is used to obtain some *unknown* information. Another objective is to make the student aware of the difference between a *closed-form solution* and a *numerical method* – two approaches they see in many of their courses. To this end, a specific problem involving a single linear equation is solved in closed form and then contrasted to an example of the traveling-salesperson problem that is solved by a numerical method.

The "Why" Mathematics

With regard to the "why" mathematics, the objective is to teach the techniques of unification, generalization, abstraction, and developing abstract and axiomatic systems. These terms are not always used consistently in mathematics, but have the following meanings in this book.

Unification is the process of combining two or more problems (theorems, ideas, and so on) into a *single* framework for studying all of the special cases. For example, to illustrate unification in this book, two specific numerical problems that each require solving a linear equation are unified into the single problem of solving a general linear equation.

Generalization is the process of creating, from a problem (theorem, definition, concept, and so on), a more *general* problem that includes not only the original one but also other *different* problems as special cases. These additional problems often necessitate creating a solution procedure that is significantly different from the solution procedure for the original problem that gives rise to the generalization in the first place. For example, to illustrate generalization in this book, the linear equation obtained by unification is generalized in two different ways: (1) to solving a quadratic equation, then to a polynomial equation, and then to a general nonlinear equation; and (2) to solving a (2×2) system of linear equations, then to an $(n \times n)$ system of linear equations, and finally to an $(n \times n)$ system of nonlinear equations.

Although both unification and generalization result in a single class of more general problems, what differentiates unification from generalization in this book is that unification comes about by looking at *two or more* special cases, with the objective of constructing a common framework to view those and all *similar* problems. In contrast, generalization is often applied to a *single* problem, with the objective of developing a framework that also enables the solution to problems of a *different* nature, as well.

Abstraction is the process of learning to work with *objects* rather than specific items. For example, instead of thinking of x and y as integers in the expression $x + y$, the student is taught that abstraction requires thinking of x and y as *objects*. Of course the student is also made aware of the resulting problems in doings so – namely, that when x and y are *objects*, the operation of addition does not make sense and is ultimately replaced by a binary operator.

This use of the term *abstraction* is different from that used in some areas of mathematical problem solving, such as the process of building an *abstract model* to represent a physical problem. The objective of teaching abstraction as done in this book is to prepare the student for the subsequent development of axiomatic systems, such as vector spaces, groups, metric spaces, and the like, that they will see in their courses.

An *abstract system* consists of one or more sets of objects together with various operations on those objects. An abstract system is the basis for a subsequent *axiomatic* system. For example, an abstract system consisting of a set G together with a closed binary operation $\oplus$ on G is the basis for a group.

An *axiomatic system* is an abstract system together with a list of axioms that are assumed to hold. This meaning includes, for example, the axiomatic system of Euclidean geometry as well as Peano's axiomatic system for the integers. Additionally, an axiomatic system, as used in this book, includes *any* abstract system that obeys cetain properties, such as the axiomatic system of a norm.

Developing Mathematical Skills

To understand and to use the techniques of unification, generalization, abstraction, and developing axiomatic systems, students need to acquire certain basic mathematical skills. These skills include doing mathematical proofs, learning to identify similarities and differences, converting visual images to symbolic form (and vice versa), and understanding and creating definitions.

The concept of a proof is fundamental to all advanced mathematics courses. The approach taught here is an adaptation of – but not a replacement for – *How to Read and Do Proofs*, Second Edition, Daniel Solow, copyright ©1990, John Wiley & Sons, Inc.

The ability to identify similarities and differences between various mathematical concepts is essential to the process of unification in that like properties from different problems must be isolated, identified, and subsequently unified into a single framework. This skill is also used in creating definitions, where it is necessary to identify a common property shared by all objects being defined. Section 1.3 of this book is devoted to presenting numerous examples of how to identify similarities and differences in various mathematical settings. This technique is then used throughout the book.

Much effort is also devoted to teaching the students how to convert mental and visual images to symbolic form. When students learn mathematics, they typically develop their own ways of imagining and picturing specific concepts, such as a bounded set of real numbers. However, it is one thing to *visualize* such sets; it is another thing entirely to translate such an image to symbolic form, especially when quantifiers are involved. The art of doing so is illustrated and explained with many examples. Students are also shown how to check for syntax and other logic errors that can arise in the translation process. The reverse technique of converting symbolic mathematics to visual form is also taught.

One area in which the student is given special help is in understanding mathematical definitions. Although a student may be able to *visualize* objects with a desirable property, such as linearly independent vectors, it is quite a challenge to *create* (or even to *understand*) the symbolic definition. Merely presenting the definition is inadequate because doing so fails to explain *how* the definition was arrived at and *why* the definition is correct. The approach taken here is to teach the student to identify similarities shared by all items *having* the property being defined, and then to translate those observations to symbolic form. The students are also taught to verify that the definition includes all objects having the desirable property while excluding all other objects.

Organization and Use of This Book

This book is designed for use by undergraduate and graduate students in mathematics, computer science, operations research, engineering, statistics, economics, physics, chemistry, biology, and related fields, in one of three ways:

1. As a text for a one-semester "transition"course to advanced mathematics for undergraduate math majors.
2. As a supplement to any standard undergraduate course in discrete mathematics, linear algebra, abstract algebra, and real analysis.
3. As a self-study guide for undergraduate and graduate students in the fields listed above who need to have a better understanding of how mathematics is used for problem solving and research.

To meet these needs, the first two chapters of the book present the fundamental ideas of unification, generalization, abstraction, developing and working with axiomatic systems, doing proofs, identifying similarities and differences, converting visual images to symbolic form, and understanding and creating definitions. Students should read and become familiar with all of these topics.

These first two chapters lay the ground work for understanding the underlying thinking processes used in advanced mathematics. To reinforce this material, these concepts are illustrated with selected topics from discrete mathematics, linear algebra, abstract algebra, and real analysis. One subsequent chapter of the book is devoted to each of these subjects. The primary objective is to show how the ideas from Chapters 1 and 2 are used in each of these areas, although the student also learns some *elementary* subject matter as well.

The material in Chapters 1 and 2 require some knowledge of sets and functions which the students should already have obtained from their previous mathematical studies. They can also find all the needed material pertaining to sets and functions in Chapter 3. That material is presented *after* Chapters 1 and 2 so as not to detract from the presentation in those first two chapters. The material on sets and functions in Chapter 3 therefore provides the students with elementary examples of the ideas in Chapters 1 and 2.

Pedagogical Features

The pedagogical features of this book include boxed and numbered examples with titles that describe the objective of each example. Where appropriate, solutions to the problems in these examples are also boxed. Definitions, propositions, and theorems are *double* boxed for easy reference. Extensive use of figures provides the student with visual images of mathematical concepts.

Each chapter, and most sections in Chapters 1 and 2, contain a summary of the key features as well as over 25 exercises designed to give the student practice in learning the concepts. Complete solutions (not just short answers) are provided in the back of the book for all *odd-numbered* problems. The instructor may wish to assign *even-numbered* problems for homework. A *Solutions Manual* for *all* the exercises is available to instructors who adopt this book as a required text for a course.

Appendix A contains a summary of the proof techniques, as presented in Chapter 13 of *How to Read and Do Proofs*, Second Edition, Daniel Solow, copyright ©1990, John Wiley & Sons, Inc., and is reprinted with permission from John Wiley & Sons, Inc. An alphabetized glossary of all terms used in the book is also included together with an extensive index.

Acknowledgments

This book benefited from my discussions with Peter Hilton, Paul Halmos, and Al Schoenfeld. However, no single person spent more time on this manuscript or had more influence than Bob Moore of Southern College. Many improvements are due to the comments he made after conducting a line-by-line review of the entire book. I appreciate not only his attention to mathematical detail but also his technical editing capabilities. The final product would not be as error-free without his efforts. A special thanks also goes to Roger Marty at Cleveland State University for having his students read a preliminary version of the first two chapters during the fall of 1994 and to Michael Grajek for suggesting the subtitle and for arranging for some of his colleagues to review preliminary versions of Chapters 4, 5, and 6. In fact, I thank all of the following people who read and gave me written comments on various parts of the manuscript:

Peter Hilton, State University of New York, Binghamton, NY,
Jean Pedersen, Santa Clara University, Santa Clara, CA,
Man Keung Siu, University of Hong Kong, Honk Kong,
Michael Grajek, Hiram College, OH,
Brad Grubser, Hiram College, OH,
Jan Green, Hiram College, OH,
Ed Smerek, Hiram College, OH,
Carolyn Cuff, Westminster College, OH,
Bob Kolesar, John Carrol University, OH.

I also thank Laurie Rosatone, the mathematics editor at Addison-Wesley, for obtaining numerous valuable reviews and for many beneficial discussions. I thank John Headlee for helping with the finer details of LATEX. It has also been a pleasure working with Amy Jenkins, who designed the cover. I also appreciate the help of my secretary, Tedda Nathan, with various miscellaneous aspects of this project.

I thank Professor Richard Karp for inviting me to the University of California at Berkeley where, as a Visiting Scholar in the Department of Computer Science, I was able to use the facilities of the International Computer Science Institute to work on this book. I am also grateful for the use of the excellent computing facilities at the Weatherhead School of Management at Case Western Reserve University.

Contents

Chapter 1

WHAT'S THE USE OF MATHEMATICS?

What is mathematics all about? What can you do with mathematics? Why should you learn proofs anyway?

One major goal of mathematicians is to discover methods for solving problems that arise in engineering, computer science, statistics, economics, business, medicine, and the physical sciences. This might be described as the "how-to" mathematics: how to solve linear equations, how to differentiate functions, how to solve differential equations, and so on. In general, however, it is necessary to understand not only *how* things work, but also *why* they work. This latter knowledge is attained through the use of theoretical mathematics and *proofs*. Although a single result in theoretical mathematics may not be so important – afterall, how many people have used the proof of the Pythagorean theorem – *the underlying thinking process* involved in doing theoretical mathematics *is* important. This thinking process provides a systematic approach to problem solving in general that you can use even outside the field of mathematics. The objective of this book is to enable you to use mathematics to solve specific problems, to understand why those solutions are correct, and to organize your knowledge in a simple, coherent, and logical manner – in other words, *to think mathematically*.

In the rest of this chapter, you will learn some of the different ways in which mathematics is used. When you learn mathematics, you will undoubtedly develop your own way of picturing and imagining specific concepts. To work effectively with those concepts, however, you need to convert those visual images to a more formal written form. The process for doing this is described in Chapter 2. Each of the remaining chapters of this book is then devoted to illustrating the ideas of Chapters 1 and 2 with topics from the specific mathematical subjects of discrete mathematics, linear algebra, abstract algebra, and real analysis.

1.1 Using Mathematics to Solve Specific Problems

One use of mathematics that you are already familiar with is to solve specific problems. In this section, you will see several typical ways in which this is done. Keep these general ideas in mind when you study individual courses in mathematics.

1.1.1 Closed-Form Solutions

Consider the following example of a specific problem together with its solution procedure.

Example 1.1 – A Problem Having a Closed-Form Solution

Suppose you are visiting Japan and want to buy a train ticket from Tokyo to Hiroshima. You are told that such a ticket on the bullet train costs 15,900 yen. If the current exchange rate is 106 yen to the dollar, how much does the ticket cost in dollars?

Solution to Example 1.1

Letting

x = the cost of the train ticket in dollars,

you are led to solving the following equation for x:

$$106x = 15,900.$$

Dividing both sides by 106 yields the following cost for this ticket:

$$x = \frac{15,900}{106} = \$150.$$

This is an example of using mathematics to solve a specific problem. You have undoubtedly seen many other such uses, for instance, deriving formulas for finding the area and volume of certain geometric figures, using trigonometry to find angles, using calculus to integrate and differentiate functions, and so on. One aspect that all these problems have in common is that you use some **inputs** – in the form of *known* information, called **data** – to determine some *unknown* information, called **outputs**. For instance, in Example 1.1 you use the known data consisting of the exchange rate of 106 yen to the dollar and the fare of 15,900 yen to compute the unknown cost of the train ticket in dollars (x).

When faced with the need to solve a specific problem, the ideal goal is to derive a **closed-form solution**, that is, a solution obtained from the given data by using a formula or simple rule requiring minimal computational effort. For Example 1.1, the closed-form solution using the given fare of 15,900 yen and the current exchange rate of 106 yen to the dollar is

$$x = \frac{15,900}{106}.$$

1.1.2 Numerical Methods

Unfortunately, a problem can be so complex that it is not possible to obtain a closed-form solution. Consider the following example.

Example 1.2 – A Problem Requiring a Numerical Method

Suppose you are living in New York (NY) and need to make a business trip by car to Cleveland (CL), Chicago (CH), Atlanta (AT), and Washington, D.C. (DC), before returning home. On the basis of the following mileages, you want to determine the order in which to visit those cities so as to drive the fewest miles:

From/To	NY	CL	CH	AT	DC
NY	–	475	800	850	225
CL	475	–	350	725	350
CH	800	350	–	700	700
AT	850	725	700	–	600
DC	225	350	700	600	–

No closed-form solution is known for Example 1.2. In such cases, an alternative approach is to use a **numerical method**, that is, a procedure or *algorithm* that uses the problem data to produce the solution. Sometimes, obtaining the exact solution requires too much computational time or effort. A practical alternative is to develop a numerical method that obtains an *approximate* solution efficiently. It is hoped that the solution thus obtained, though not optimal, is acceptable. For Example 1.2, you might use the following numerical method, in the form of an algorithm, to produce an approximate solution.

Approximate Solution to Example 1.2

Starting in New York, sequentially visit the closest remaining city until

Continued

returning to New York. Using this algorithm, you obtain the following route (which is an approximate solution):

$$\text{NY} \overset{225}{\longrightarrow} \text{DC} \overset{+\ 350}{\longrightarrow} \text{CL} \overset{+\ 350}{\longrightarrow}$$

$$\text{CH} \overset{+\ 700}{\longrightarrow} \text{AT} \overset{+\ 850}{\longrightarrow} \text{NY} = 2475 \text{ miles.}$$

In developing these numerical methods, it is generally the case that the solution is obtained by computer. Consequently, there are many *computational* concerns that must be taken into account, such as:

1. Developing procedures that are easy to implement and easy to use.
2. Insuring computational efficiency of the solution procedure so that large real-world problems can be solved in reasonable amounts of computer time.
3. The proper use of data structures to conserve computer storage space and increase efficiency.
4. Maintaining numerical accuracy by avoiding problems caused by round-off errors that occur when performing certain arithmetic operations.

These computational issues are beyond the scope of this book, as are the various different types of numerical methods, such as finite and infinite methods, those based on recursion, and so on.

1.1.3 Finding Solutions More Efficiently

Yet another use of mathematics in this regard is to find more efficient ways to solve a problem. Consider the following example.

Example 1.3 – Using Mathematics for Improving Efficiency

In tracking the performance of a particular stock, suppose you want to compute a 30-day "moving" average, that is, an average price based on the prices in the past 30 days. For instance, given the sales prices in the past 30 days, say, $x_1, x_2, \ldots, x_{30}$, you can compute:

$$\text{Average price} = \bar{x} = \frac{\sum_{i=1}^{30} x_i}{30}.$$

However, when you get the *next day's* sales price, say, x_{31}, how do you compute the new 30-day average?

Solution to Example 1.3

One approach to solving this problem is to replace the value of x_1 with x_{31} and compute the new average, as follows:

$$\text{New average} = \frac{\sum_{i=2}^{31} x_i}{30}.$$

This computation involves 29 additions and 1 division. You can obtain the same result more efficiently with only 1 addition, 1 subtraction, and 1 division by using the fact that you have already computed the average price, $\bar{x}$, for the first 30 days, as follows:

$$\begin{aligned}\text{New average} &= \frac{\sum_{i=2}^{31} x_i}{30} \\ &= \frac{\sum_{i=1}^{30} x_i + x_{31} - x_1}{30} \\ &= \frac{\sum_{i=1}^{30} x_i}{30} + \frac{x_{31} - x_1}{30} \\ &= \bar{x} + \frac{x_{31} - x_1}{30}.\end{aligned}$$

In this section, you have learned how mathematics is used for solving specific problems. You do this by using the inputs consisting of the given problem data to determine the desired outputs. When the solution to such a problem cannot be found in closed form, a numerical method is used to obtain the solution (or an approximate solution). Mathematics is also used to make these solution procedures as efficient as possible. In the next section, you will learn about another use of mathematics. In some places in this chapter, certain basic ideas pertaining to sets and functions are used. When necessary, you can review the appropriate material in Section 3.1 and Section 3.2.

1.2 Unification

One of the most important uses of mathematics is in grouping problems that have certain common features into one comprehensive problem. The advantage of doing so is that you can then use a single solution procedure developed for the common problem to solve each of the individual problems. Thus, you need create only *one* solution procedure rather than separate procedures for each specific problem in the class. In this section, you will learn the process of **unification** for combining two or more problems into a single, comprehensive problem.

1.2.1 Unification of Linear Equations

Recall Example 1.1, which is repeated here together with its solution procedure.

Example 1.4 – A Problem Having a Closed-Form Solution

Suppose you are visiting Japan and want to buy a train ticket from Tokyo to Hiroshima. You are told that such a ticket on the bullet train costs 15,900 yen. If the current exchange rate is 106 yen to the dollar, how much does the ticket cost in dollars?

Solution to Example 1.4

Letting

x = the cost of the train ticket in dollars,

you are led to solving the following equation for x:

$$106x = 15,900.$$

Dividing both sides by 106 yields the following cost for this ticket:

$$x = \frac{15,900}{106} = \$150.$$

Now consider another problem, described as follows.

Example 1.5 – Another Example of Solving a Specific Problem

While vacationing in Toronto, Canada, you heard that the temperature tomorrow is expected to reach 25° centigrade (25°C). You want to know what this temperature is in degrees Fahrenheit, using the formula:

$$C = \frac{5}{9}(F - 32). \qquad (1.1)$$

Solution to Example 1.5

The desired temperature is obtained by solving (1.1) for F, that is:

$$F = \frac{9}{5}C + 32.$$

Now, substitute $C = 25$ (because you know that the temperature in Toronto will be 25°C) to obtain

$$F = \frac{9}{5}C + 32 = \frac{9}{5}(25) + 32 = 45 + 32 = 77 \text{ degrees}.$$

The idea behind unification is to discover what is *common* between two or more problems, such as the ones in Example 1.4 and Example 1.5. Doing so may require some amount of manipulation and rewriting to make the problems look similar, and there is no unique way of doing so. (Sometimes, with great effort and creativity, two or more seemingly unrelated problems or theories can be unified into a single comprehensive problem or theory.) For instance, if you rewrite (1.1) in Example 1.5 as

$$5F = 9C + 160,$$

and then substitute $C = 25$, the result is

$$5F = 385. \tag{1.2}$$

Compare the form of the equation in (1.2) with that of $106x = 15,900$ in Example 1.4. You should now recognize the following *common* problem.

The Problem of Solving a Linear Equation

Given real numbers a and b, find a value for the variable x so that

$$ax = b.$$

This linear-equation problem is the *unification* of the two problems in Example 1.4 and Example 1.5, each of which is called a **special case** of the common problem. The common problem has general data (a and b, in this example of solving a linear equation). Each of the special cases is obtained by substituting appropriate values for the data of the general problem. For instance, the special case in Example 1.4 has $a = 106, b = 15,900$, and you want to find a value for x so that

$$ax = b \quad (106x = 15,900).$$

Similarly, from (1.2), the special case in Example 1.5 has $a = 5, b = 385$, and you want to find a value for x so that

$$ax = b \quad (5F = 385).$$

Having unified the two problems into a single common problem (that of solving a linear equation), your next step is to derive a solution for this general problem. In this case, you have the following closed-form solution.

Solution to the Linear Equation

The solution to the linear equation $ax = b$ is

$$x = \frac{b}{a} \text{ (provided that } a \neq 0).$$

Here you see the advantage of unification: You can use this one solution procedure to solve each of the specific problems in Example 1.4 and Example 1.5 (and many other problems, as well). For Example 1.4, $a = 106$ and $b = 15,900$, so the corresponding solution is

$$x = \frac{b}{a} = \frac{15,900}{106} = 150.$$

Similarly, for Example 1.5, $a = 5$ and $b = 385$, so the corresponding solution is

$$x = \frac{b}{a} = \frac{385}{5} = 77.$$

When you use unification to create a common problem and then develop an associated solution procedure, you often "lose sight" of the specific problems. For example, in developing the solution to the general linear equation, you do not think about the specific equation in Example 1.4, in Example 1.5, or elsewhere, but rather, you work only with the mathematics. You then apply the solution to the unified problem to each of the special cases. You should of course check that the solution thus obtained makes sense in the context of the original real-world problem.

1.2.2 Unification of the Traveling-Salesperson Problems

As another example of unification, recall Example 1.2, which is repeated here together with its approximate numerical-method solution procedure.

Example 1.6 – A Problem Requiring a Numerical Method

Suppose you are living in New York (NY) and need to make a business trip by car to Cleveland (CL), Chicago (CH), Atlanta (AT), and Washington, D.C. (DC), before returning home. On the basis of the

Continued

following mileages, you want to determine the order in which to visit those cities so as to drive the fewest miles:

From/To	NY	CL	CH	AT	DC
NY	–	475	800	850	225
CL	475	–	350	725	350
CH	800	350	–	700	700
AT	850	725	700	–	600
DC	225	350	700	600	–

Approximate Solution to Example 1.6

Starting in New York, sequentially visit the closest remaining city until returning to New York. Using this algorithm, you obtain the following route (which is an approximate solution):

$$\text{NY} \overset{225}{\longrightarrow} \text{DC} \overset{+\;350}{\longrightarrow} \text{CL} \overset{+\;350}{\longrightarrow} \text{CH} \overset{+\;700}{\longrightarrow} \text{AT} \overset{+\;850}{\longrightarrow} \text{NY} = 2475 \text{ miles.}$$

Now consider the following problem of the Philadelphia Paint Company.

Example 1.7 – Another Example of a Problem Requiring a Numerical Method

During each production cycle, the Philadelphia Paint Company must produce red, green, blue, white, and yellow paint. After finishing one color, the machine is cleaned and prepared for the next color. The amount of time required to do this depends on both the color that was just produced and the next color to be produced. Those preparation times (in minutes) are given in the following table:

From/To	Red	Green	Blue	White	Yellow
Red	–	120	150	180	100
Green	110	–	80	120	130
Blue	160	100	–	140	120
White	90	100	80	–	70
Yellow	120	140	110	90	–

In what order should these colors be produced so as to minimize the total preparation time?

If you recognize some similarities between Example 1.6 and Example 1.7, then you can use unification to avoid developing a new solution procedure for Example 1.7. To unify these two problems, identify what they have in *common*, such as:

1. In both problems, you have a number of "objects" you must "process" (cities to visit, in Example 1.6; colors to produce, in Example 1.7).

2. In both problems, the data include a nonnegative "cost" associated with each pair of different objects (distances, in Example 1.6; preparation times, in Example 1.7).

3. In both problems, the objective is to determine the *order* in which to process all of the objects exactly once so as to minimize the total "cost" of doing so (including the cost of returning to the starting object).

You can therefore unify Example 1.6 and Example 1.7 by creating the following *common* problem.

The Traveling-Salesperson Problem (TSP)

Given n objects and a nonnegative cost for each pair of different objects, find a sequence in which to process each of those objects exactly once so as to minimize the total cost of doing so (including the cost of returning to the starting object).

Having unified the two problems into a single, common problem (the TSP), your next step is to derive a solution procedure. By using the one for Example 1.6 as a guide, you might develop the following numerical-method solution procedure for the TSP (which is not necessarily the best one).

Approximate Solution Procedure for the TSP

Starting with any object, sequentially process the next remaining object of least additional cost, until returning to the starting object.

Here again, the advantage of unification is that you can use this one solution procedure to solve the two special cases in Example 1.6 and Example 1.7. Specifically, as shown in Section 1.1.2, the result of applying the foregoing solution procedure to the problem in Example 1.6 (starting in New York), is

$$\text{NY} \xrightarrow{225} \overset{+}{\text{DC}} \xrightarrow{350} \overset{+}{\text{CL}} \xrightarrow{350} \overset{+}{\text{CH}} \xrightarrow{700} \overset{+}{\text{AT}} \xrightarrow{850} \overset{=}{\text{NY}} \quad 2475 \text{ miles.}$$

Likewise, applying this *same* procedure to the problem in Example 1.7 (starting with Red) results in the following solution.

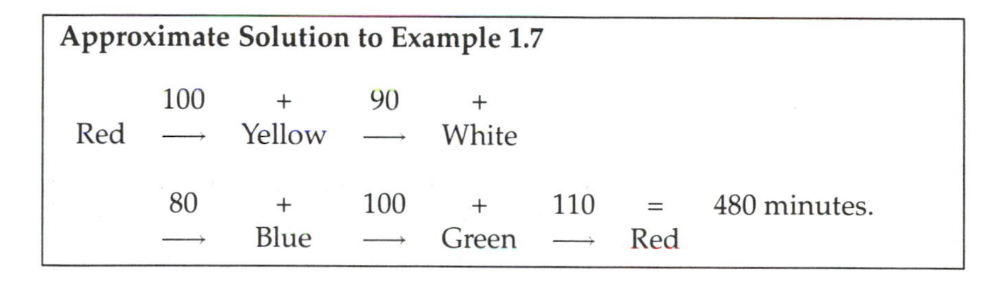
Approximate Solution to Example 1.7

$$\text{Red} \xrightarrow{100} \overset{+}{\text{Yellow}} \xrightarrow{90} \overset{+}{\text{White}} \xrightarrow{80} \overset{+}{\text{Blue}} \xrightarrow{100} \overset{+}{\text{Green}} \xrightarrow{110} \overset{=}{\text{Red}} \quad 480 \text{ minutes.}$$

Summary of Unification

In this section, you have learned how unification is used to identify a single class of problems whose solution procedure can be used to solve *any* specific problem in that class, that is, any special case. It is also possible to apply unification to definitions, theorems, theories, or other mathematical concepts. A summary of how to do so for problems (and the corresponding advantages and disadvantages) follows.

How to Apply Unification

Step 1. Identify two or more problems whose data and solution you know.

Step 2. By recognizing similarities between the problems in Step 1, create a single common problem with its data that includes all of the specific problems as special cases.

Step 3. By using the solution procedures for the special cases in Step 1 as a guide, develop a solution procedure for the common problem identified in Step 2.

Step 4. Verify that applying the solution procedure for the common problem to each of the specific problems in Step 1 results in the corresponding solution to those special cases.

Step 5. Apply the solution procedure for the common problem to *any* special case, including any new ones you encounter.

Advantages of Unification

1. Unifying many individual problems into one common problem allows you to develop a single solution procedure that you can then use to solve *all* problems in that class, even new ones you have not yet encountered.
2. Unifying problems into one common problem makes it easier to understand and to study the properties of all problems in that class.

Disadvantage of Unification

1. When you study the common problem, you lose sight of the individual problems in the class. Thus, for example, when trying to solve the common problem, you do not use any of the special properties that pertain to one specific problem in the class.

As you have just seen, the process of identifying similarities and differences is crucial to unification. Finding common properties is also important to many other areas of mathematics and even to other fields as well. For example, when a group of people suddenly develop the same illness, a medical researcher tries to identify what those people have in common in an attempt to discover the source of the illness. In the next section, you will see how to identify similarities and differences in various mathematical settings.

1.3 Identifying Similarities and Differences

In Section 1.2, you have seen that identifying common properties is the key to unification. Finding similarities and differences plays a central role throughout mathematics. Because of its importance, several examples of doing so are presented here. In each example, try to find as many similarities and differences as you can, *before* reading the solution. (The ones you identify may be different from the ones listed here.)

Example 1.8 – A First Example of Identifying Similarities and Differences

What are the similarities and differences between the following two problems?

Continued

1. Suppose you know the coordinates of n points in the plane, say, $(x_1, y_1), (x_2, y_2), \ldots, (x_n, y_n)$. You want to find the distance between the two closest points.

2. Suppose you have been keeping track of the price of a stock on each day of the previous year. You want to know the least amount you could have paid for the stock during that period.

Similarity in Example 1.8

1. Both problems require finding the smallest number in a finite list.

Differences in Example 1.8

1. In the first problem, the list consists of distances; in the second problem, the list consists of prices.

2. In the first problem, you must compute the numbers in the list (the distances); in the second problem, the numbers in the list (the prices) are given.

Example 1.9 – A Second Example of Identifying Similarities and Differences

What are the similarities and differences between the following two sequences of numbers?

1. $\frac{1}{2}, \frac{2}{3}, \frac{3}{4}, \frac{4}{5}, \ldots$.

2. $\frac{2}{1}, \frac{3}{2}, \frac{4}{3}, \frac{5}{4}, \ldots$.

Similarities in Example 1.9

1. Corresponding numbers in the two sequences are obtained from quotients of the same integers; in fact, these numbers are the reciprocals of each other.

2. The numbers in both sequences are getting closer and closer to 1.

Continued

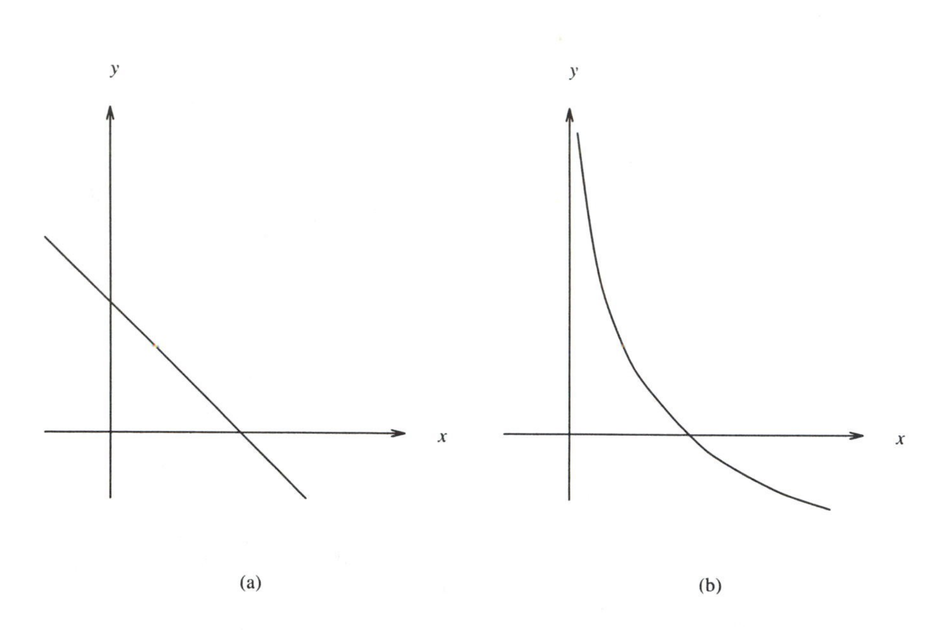

Figure 1.1: Two Different Graphs

Differences in Example 1.9

1. The numbers in the first sequence are increasing; the numbers in the second sequence are decreasing.
2. Every number in the first sequence is less than 1; every number in the second sequence is greater than 1.

Example 1.10 – A Third Example of Identifying Similarities and Differences

What are the similarities and differences between the two graphs in Figure 1.1?

Similarities in Example 1.10

1. As you move from left to right along the x-axis in both figures, the corresponding y-values get smaller and smaller.
2. Both graphs cross the x-axis exactly once.
3. You can think of both figures as the graph of a function.

Continued

Differences in Example 1.10

1. The graph in Figure 1.1(a) is linear; the graph in Figure 1.1(b) is not linear.
2. The graph in Figure 1.1(a) crosses the y-axis; the graph in Figure 1.1(b) does not.

Example 1.11 – A Fourth Example of Identifying Similarities and Differences

What are the similarities and differences between the operations indicated by the following underlined words?

1. x times y, which is the product of the real numbers x and y.
2. A intersect B, which is the intersection of the sets A and B.
3. i mod j, which is the remainder on dividing the integer i by the integer j.

Similarities in Example 1.11

1. All underlined words serve to combine two like objects (two real numbers; two sets; two integers).
2. The result of performing the underlined word is a new object of the same type as those being combined.

Differences in Example 1.11

1. Each underlined word combines *different* objects: times combines two *real numbers*; intersect combines two *sets*; mod combines two *integers*.
2. The specific *manner* in which the objects are combined is different in each case.

Mathematicians have devoted much effort to the study of the similarities and differences between various sets of numbers. Some of these results are presented in Chapters 5 and 6 of this book. To give you an idea of how this is done, consider the two sets of numbers in the following example.

Example 1.12 – A Fifth Example of Identifying Similarities and Differences

What are the similarities and differences between the following two sets of numbers?

1. $\mathbb{N} = \{1, 2, 3, \ldots\}$.
2. $\mathbb{R} = \{\text{real numbers}\}$.

Similarities in Example 1.12

1. Both sets contain an infinite number of elements.
2. You can write the numbers in both sets in decimal form.

Differences in Example 1.12

1. All numbers in $\mathbb{N}$ are positive, which is not the case for the numbers in $\mathbb{R}$.
2. For any number in $\mathbb{N}$, there is a "next" number – for example, for 5, the next number in $\mathbb{N}$ is 6. This is not the case for $\mathbb{R}$, that is, for a given real number, such as 1.2, there is no "next" real number.
3. Any nonempty subset of $\mathbb{N}$ has a smallest element – for example, the subset $\{5, 7, 9\}$ has 5 as the smallest element. This is not the case with $\mathbb{R}$ – for example, the subset $\{\text{real numbers } x : 2 < x < 4\}$ has no smallest element.
4. Although both sets contain an infinite number of elements, the set $\mathbb{R}$ contains *more* elements than $\mathbb{N}$, as shown in Chapter 6.

In this section, you have seen numerous examples of identifying similarities and differences in various mathematical settings. These ideas – including the next use of mathematics described in Section 1.4 – are prevalent throughout the rest of this book.

1.4 Generalization

Another fruitful use of mathematics is **generalization** in which, like unification, you also create a new, general class of problems on the basis of one or more specific problems. However, unlike unification, the new class, while containing all of the specific problems as special cases, *will also contain some significantly different problems as well.* These different problems necessitate developing a solution procedure for the general class that may be quite different from those of the specific problems. You therefore need to verify that this new solution procedure also solves the special cases.

As an example of the process of generalization, recall the problem and solution to a general linear equation presented in Section 1.2, which is repeated here in a slightly different form.

The Problem of Solving a Linear Equation

Given two real numbers a and b, find a value for the variable x so that

$$ax - b = 0.$$

Solution to the Linear Equation

The solution to the linear equation is:

$$x = \frac{b}{a} \text{ (provided that } a \neq 0\text{).}$$

You will now see how this problem is generalized in several different ways.

1.4.1 A Generalization to Quadratic Equations

One way to generalize the problem of solving the linear equation

$$ax - b = 0$$

is to realize that this *linear* equation is a special case of a *quadratic* equation.

Developing the Generalized Problem

Consider, therefore, the following quadratic equation.

The Problem of Solving a Quadratic Equation

Given values for the real numbers p, q, and r – each of which may be positive, 0, or negative – find a value for the variable x that satisfies

$$px^2 + qx + r = 0. \tag{1.3}$$

Technically, for (1.3) to be a *quadratic* equation, it should be the case that $p \neq 0$. However, in what follows, it may happen that $p = 0$ and, for the sake of discussion, (1.3) is still referred to as a quadratic equation.

This new class of quadratic equations includes all linear equations. To see that the linear equation $ax - b = 0$ is a special case of the quadratic equation, consider setting $p = 0, q = a$, and $r = -b$ in (1.3), so

$$px^2 + qx + r = 0x^2 + ax - b = ax - b = 0.$$

However, note that when $p \neq 0$, the class of quadratic equations contains problems that are *not* linear equations. For example, the quadratic equation

$$3x^2 - 6x + 5 = 0$$

is not a linear equation. The existence of these new and different problems in the general class requires developing a new solution procedure, as you will now see.

Developing the Generalized Solution Procedure

Turning to the solution procedure, you know from your previous studies that one closed-form solution to the quadratic equation in (1.3) is as follows.

Solution to the Quadratic Equation

Given the values of p, q, and r for which $p \neq 0$ and $q^2 - 4pr \geq 0$, the solutions to the quadratic equation in (1.3) are

$$x = \frac{-q \pm \sqrt{q^2 - 4pr}}{2p}. \tag{1.4}$$

Observe that the solutions in (1.4) are valid only when $p \neq 0$. Thus, you cannot apply the general solution in (1.4) to the special case of a linear equation because, in that case, $p = 0$.

When a solution procedure to a generalization does not lead to a solution of the special case, care is needed to determine why, that is, to see if either of the following has occurred:

1. A mistake has been made in the generalized problem or its solution.

2. A different type of generalized problem or solution is more appropriate.

In this case, you can dervie a different form of the solutions to the quadratic equation in (1.3) that you *can* apply to the special case of a linear equation. To see how, consider again equation (1.3), namely,

$$px^2 + qx + r = 0. \tag{1.5}$$

On dividing (1.5) through by x^2 (assuming that $x \neq 0$), you have

$$p + q\left(\frac{1}{x}\right) + r\left(\frac{1}{x^2}\right) = 0, \tag{1.6}$$

or, on letting $y = 1/x$,

$$p + qy + ry^2 = 0. \tag{1.7}$$

Thinking of (1.7) as a quadratic equation in y, you can apply the quadratic formula to obtain the following values for y:

$$y = \frac{1}{x} = \frac{-q \pm \sqrt{q^2 - 4pr}}{2r} \tag{1.8}$$

Taking the reciprocal of both sides of (1.8) results in the following alternative solutions to (1.3).

Alternative Solutions to the Quadratic Equation

Given the values of p, q, and r for which $q^2 - 4pr \geq 0$, the solutions to the quadratic equation in (1.3) are

$$x = \frac{2r}{-q \pm \sqrt{q^2 - 4pr}}. \tag{1.9}$$

The form in (1.9) allows you to obtain the solution to the linear problem in which $p = 0$, $q = a$, and $r = -b$, as follows:

$$\begin{aligned}
x &= \frac{2r}{-q \pm \sqrt{q^2 - 4pr}} \\
&= \frac{2(-b)}{-a \pm \sqrt{a^2 - 4(0)(-b)}} \\
&= \frac{-2b}{-a - a} \\
&= \frac{b}{a} \\
&= \text{the solution to the linear equation.}
\end{aligned}$$

Observe that the *negative* square root is used to avoid the division by 0.

Checking the Special Cases

Note that the problem obtained by generalization has *more data* than does the original problem. In the example here, the quadratic equation has *three* data values (p, q, and r) while the linear equation has only *two* data values (a and b). Because of this difference, when using generalization, be sure to *check the special cases*, that is:

1. Verify that the general problem includes the special case. You do this by using the data values from the special case (a and b, in this

example) to create appropriate data values for the general problem (p, q, and r, in this example) that, when substituted in the general problem, result in the specific problem. For the example presented here, when you substitute $p = 0, q = a$, and $r = -b$ in the quadratic equation, you obtain the linear equation.

2. Verify that the solution procedure for the general problem results in the solution to the specific problem when the corresponding data for the specific problem are substituted in the solution to the general problem. For the example presented here, when you substitute $p = 0$, $q = a$, and $r = -b$ in (1.9), you obtain the solution to the linear equation.

Note that using (1.9) to solve the linear equation requires more computational effort than solving the linear equation directly. This is usually the case because the solution procedure for a generalization requires using more data than does solving the special case directly. This is one of the major drawbacks of generalization.

1.4.2 A Generalization to Two Equations in Two Unknowns

The way in which you generalize is not necessarily unique. For example, as you just saw in Section 1.4.1, one generalization of the linear equation is the quadratic equation. Another generalization is also possible.

Developing the Generalized Problem and Checking the Special Cases

You can also generalize the problem of solving a linear equation to that of solving *two* linear equations in *two* unknowns, as follows.

The Problem of Solving (2 × 2) Linear Equations

Given real numbers p, q, r, s, t, and u, find values for the variables x and y so that

$$px + qy = t, \tag{1.10}$$

$$rx + sy = u. \tag{1.11}$$

Observe again that this more general problem has *six* data items (p, q, r, s, t, and u) whereas the linear equation has only *two* data items (a and b). As mentioned previously, you should make sure that this general problem includes the specific problem. In this case, there are many ways to do so. For example, you can use the values of a and b from the linear equation to create the following data for solving the (2 × 2) equations:

$$p = a, \ q = 0, \ t = b, \tag{1.12}$$

$$r = 0, \ s = 0, \ u = 0. \tag{1.13}$$

Substituting these values in (1.10) and (1.11) yields

$$px + qy = t \text{ or } ax + 0y = b \text{ or } ax = b, \tag{1.14}$$

$$rx + sy = u \text{ or } 0x + 0y = 0 \text{ or } 0 = 0. \tag{1.15}$$

Ignoring equation (1.15), you have obtained in (1.14) the linear equation as a special case of solving two equations in two unknowns.

Another way to verify that the linear equation is a special case is by setting

$$p = a, \ q = 0, \ t = b, \tag{1.16}$$

$$r = 0, \ s = 1, \ u = 0. \tag{1.17}$$

Substituting these values in (1.10) and (1.11) yields

$$px + qy = t \text{ or } ax + 0y = b \text{ or } ax = b, \tag{1.18}$$

$$rx + sy = u \text{ or } 0x + 1y = 0 \text{ or } y = 0. \tag{1.19}$$

Once again, ignoring the bottom equation (1.19), you have obtained the linear equation as a special case of solving two equations in two unknowns.

Developing the Generalized Solution Procedure and Checking the Special Cases

The next step is to develop a procedure for solving (2×2) linear equations which, from your previous studies, you know to be the following.

Solution to (2×2) Linear Equations

The solution to the two linear equations in (1.10) and (1.11), assuming that $ps - qr \neq 0$, is

$$x = \frac{st - qu}{ps - qr}, \tag{1.20}$$

$$y = \frac{pu - rt}{ps - qr}. \tag{1.21}$$

As mentioned previously, whenever you solve a more general problem, you want to make sure that you can use that method to solve the special cases (the linear equation, in this case). Here, the data for the (2×2) problem associated with the linear equation is given in (1.12) and (1.13), or, alternatively, in (1.16) and (1.17). Observe that you *cannot* use the values in (1.12) and (1.13) because, in that case,

$$ps - qr = a(0) - 0(0) = 0.$$

However, you *can* use the values in (1.16) and (1.17). Substituting those values in (1.20) and (1.21) yields

$$x = \frac{st - qu}{ps - qr} = \frac{1(b) - 0(0)}{a(1) - 0(0)} = \frac{b}{a},$$

$$y = \frac{pu - rt}{ps - qr} = \frac{a(0) - 0(b)}{a(1) - 0(0)} = 0.$$

Looking at the value for x, you have obtained the solution to the linear equation as a special case of solving (2×2) equations. Observe again that solving the more general problem is less efficient than solving the special case directly.

1.4.3 Additional Examples of Generalizations

In many problems, one generalization leads to further generalizations. For instance, the problem of solving a linear equation can be generalized successively in the following ways:

1. From solving a linear equation, to solving a quadratic equation, to solving a polynomial equation of degree n, to solving a general non-linear equation, and beyond.

2. From solving a linear equation, to solving two linear equations in two unknowns, to solving a system of n linear equations in n unknowns, to solving a system of n nonlinear equations in n unknowns, and beyond.

When creating such a **sequential generalization**, you should verify that each new generalization includes the previous one, which then becomes a special case. Also, whenever you create a new generalization and need to develop an associated solution procedure, check that you can use the new procedure to solve the previous special case.

In addition to problems, you can generalize theorems, proofs, and even mathematical concepts, as shown in the following examples.

Generalizing the Concept of Numbers

The following is a sequential generalization of sets of numbers:

1. The **natural numbers**: $\mathbb{N} = \{1, 2, 3, \ldots\}$.

2. The **integers**: $\mathbb{Z} = \{\ldots -2, -1, 0, 1, 2, \ldots\}$.

3. The **rationals**: $\mathbb{Q} = \left\{ \frac{p}{q} : p \text{ and } q \text{ are integers and } q \neq 0 \right\}$.

4. The **reals**: $\mathbb{R} = \{r : r \text{ is a number expressible in decimal form}\}$.

5. The **complex numbers**: $\mathbb{C} = \{a + bi : a \text{ and } b \text{ are reals and } i = \sqrt{-1}\}$.

Observe that each of these sets of numbers is a special case of the subsequent generalization. That is, the natural numbers are a special case of the integers, the integers are a special case of the rationals; the rationals are a special case of the reals; and the reals are a special case of the complex numbers.

Generalizing the Concept of Distance to the Origin

As a final example of generalizing a mathematical concept, suppose that x is a real number. You know that

$$\text{Distance to the origin} = |x|.$$

It is easy to generalize this concept to a point (x, y) in the plane, as follows:

$$\text{Distance to the origin} = \sqrt{x^2 + y^2} \text{ (recall that } \sqrt{x^2} = |x|\text{)},$$

and also to a point (x, y, z) in three dimensions:

$$\text{Distance to the origin} = \sqrt{x^2 + y^2 + z^2}.$$

You can even extend this generalization to a point (x, y, z, t), that is:

$$\text{Distance to the origin} = \sqrt{x^2 + y^2 + z^2 + t^2}.$$

Here, you can see one of the most powerful advantages of generalization – *the extension of concepts beyond your ability to visualize them.* In this example, you can easily picture a point (x, y) in the plane or even a point (x, y, z) in three dimensions. However, you *cannot* visualize the point (x, y, z, t) in four dimensions. Yet, through generalization, you can extend the concept of "distance to the origin" in a meaningful way to the point (x, y, z, t).

Throughout the history of mathematics, notation has played a significant role, as it can in generalizations. For example, instead of using the notation (x, y) to represent a point in the plane, consider using (x_1, x_2). Then,

$$\text{Distance to the origin } (0, 0) = ||(x_1, x_2)|| = \sqrt{x_1^2 + x_2^2}.$$

Here, the notation $||(x_1, x_2)||$ is also introduced. To see how useful the *subscript* notation is, consider now a point in three dimensions. Using the notation (x_1, x_2, x_3) instead of (x, y, z) results in the following:

$$\text{Distance to the origin } (0, 0, 0) = ||(x_1, x_2, x_3)|| = \sqrt{x_1^2 + x_2^2 + x_3^2}.$$

Now it is easy to generalize not only to a point in four dimensions, but in n dimensions. Specifically, representing such a point by $(x_1, x_2, \ldots, x_n)$,

$$\text{Distance to the origin } (0, \ldots, 0) = ||(x_1, \ldots, x_n)|| = \sqrt{x_1^2 + \cdots + x_n^2}.$$

The point $\mathbf{x} = (x_1, \ldots, x_n)$ is called an **n-vector** (or a **vector of dimension** n), meaning an ordered list of n real numbers. Each number x_i in the list is called a **component** of the vector.

1.4.4 Correcting Syntax Errors in Generalizations

Special care is needed when a direct generalization does not make sense. To illustrate, let x, y, and t be real numbers with $y \neq 0$. Suppose you want to generalize the statement

$$\left|\frac{x}{y}\right| \leq t \tag{1.22}$$

by thinking of x and y as *n-vectors* $\mathbf{x} = (x_1, \ldots, x_n)$ and $\mathbf{y} = (y_1, \ldots, y_n)$. However, simply substituting these n-vectors in (1.22) to obtain

$$\left|\frac{\mathbf{x}}{\mathbf{y}}\right| \leq t \tag{1.23}$$

makes no sense because the operation of dividing one n-vector by another n-vector is undefined. Such a mistake is called a **syntax error**, meaning that the symbols or operations in a mathematical expression make no sense or cannot be performed. How, then, can you generalize (1.22)?

There are two possible ways. One approach is to create a meaningful formula for dividing one n-vector by another. Another alternative – the one used here – is to rewrite (1.22) so that *the operation of division does not appear*. For instance, multiplying both sides of (1.22) by $|y|$ yields

$$|x| \leq t|y|. \tag{1.24}$$

Now that the operation of division has been eliminated, you can replace the real numbers x and y in (1.24) with the n-vectors $\mathbf{x} = (x_1, \ldots, x_n)$ and $\mathbf{y} = (y_1, \ldots, y_n)$ to obtain the following:

$$|\mathbf{x}| \leq t|\mathbf{y}|. \tag{1.25}$$

Unfortunately, (1.25) still contains a syntax error because you cannot take the absolute value of an n-vector (that is, $|\mathbf{x}|$ is undefined). So now what do you do?

One approach to resolving this syntax error is to realize that $|x|$ in (1.24) represents the distance from the real number x to the origin. As you learned in Section 1.4.3, the corresponding notion for an n-vector is $||\mathbf{x}||$. Rewriting (1.24) as

$$||\mathbf{x}|| \leq t||\mathbf{y}|| \tag{1.26}$$

results in a valid generalization of the original expression in (1.22).

Summary of Generalization

In this section, you have learned how generalization is used to create a new class of problems from a specific problem. The resulting class requires more data and also contains problems that are significantly different from the specific problem. You must therefore develop a new solution procedure for the generalized problem and make sure you can use that procedure to solve the special cases. You can also apply generalization to definitions, theorems, theories, or other mathematical concepts. A summary of how to do so for problems (and the corresponding advantages and disadvantages) follows.

How to Apply Generalization

Step 1. Identify a specific problem whose data and solution you know.

Step 2. Create a more general problem with its larger set of data.

Step 3. Verify that the specific problem is a special case of the general problem. You do this by using the data for the specific problem in Step 1 to create appropriate data for the general problem in Step 2 that, when substituted in the general problem, results in the specific problem.

Step 4. Develop a solution procedure for the general problem identified in Step 2.

Step 5. Verify that applying the solution procedure for the general problem in Step 2 to the specific problem in Step 1 results in the corresponding solution to that special case.

Step 6. Apply the solution procedure for the general problem to any special case of interest.

Advantages of Generalization

1. Generalization results in a class of problems and a single solution procedure that you can use to solve *any* problem in that class, including all the special cases, and even new problems you have not yet encountered.
2. Generalization allows you to work with concepts that you cannot visualize.

Disadvantages of Generalization

1. You must develop a solution for the general problem that may be quite different from the solutions for the special cases and thus requires significant effort to create.
2. The solution to the general problem may be less efficient computationally than solving the special cases directly. This is because the general problem has more data than the specific problem.

In the next section, you will see another use of mathematics that again allows you to solve more and more problems with less effort.

1.5 Abstraction

By now you realize that one of the primary objectives of mathematics is to create and solve larger and larger classes of problems. Unification and generalization are two methods for doing this. In this section, you will learn another such technique that is an extreme form of generalization.

1.5.1 Working with Objects

In mathematics, **abstraction** is the process of getting farther and farther away from specific items by working more and more with general *objects*. Thus, you become more abstract – hence the term "abstract mathematics." To illustrate the idea of abstraction, consider apples and oranges. You can unify these two items into the single comprehensive class of *fruits* (fruits include apples and oranges as special cases). You can then apply generalization by considering, instead of fruits, the more general class of *foods* (foods include fruits as a special case). With abstraction, you broaden the class even further by considering *objects* rather than specific items like food or fruits. By thinking of *objects*, you can now include such diverse items as foods, computers, houses, and much more. You will now see how abstraction arises in various mathematical settings.

An Abstract Concept of a Graph

You have already seen an example of abstraction in Section 1.2.2 where unification is used to create the traveling-salesperson problem (TSP), described as follows.

The Traveling-Salesperson Problem (TSP)

Given n objects and a nonnegative cost for each pair of different objects, find a sequence in which to process each of those objects exactly once so as to minimize the total cost of doing so (including the cost of returning to the starting object).

By working with *objects* in the TSP, rather than specifc items, you allow greater flexibility in the types of problems you can consider. For example, the objects in the TSP can be cities, colors, people, jobs, machines, workstations in a factory, or any other specific item.

You can create an even more general abstraction when you consider a pictorial representation of the TSP, such as the one in Figure 1.2 associated with the TSP in Example 1.6 in Section 1.2.2. Observe that there are a finite

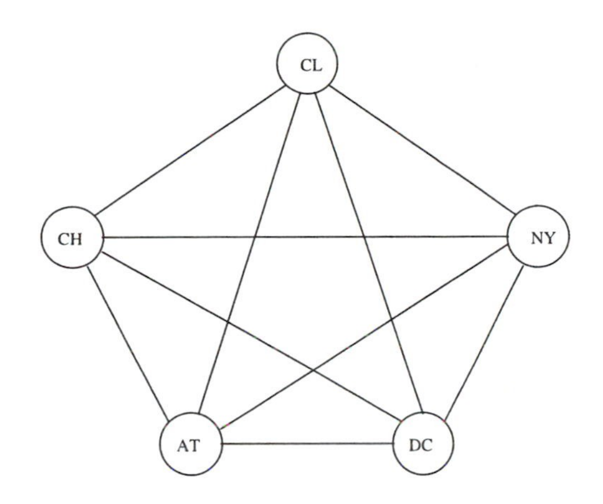

Figure 1.2: A Pictorial Representation of the Traveling-Salesperson Problem in Example 1.6 in Section 1.2.2

number of circles, called **vertices**, each of which stands for an object under consideration. Each pair of vertices is connected by a line, called an **edge,** that represents the possibility of going from the object represented by the vertex at one end of the edge to the object represented by the vertex at the other end of the edge.

In some problems, not all edges need be present. In other words, an edge is present if there is some specified relationship between the two objects corresponding to the two vertices associated with that edge. To illustrate, consider the following problem.

Example 1.13 – A Problem that Can Be Unified with the TSP

A company has two production facilities, one in San Francisco and one in Los Angeles. The products from these two plants are shipped to customers in Lake Tahoe, Barstow, and La Jolla. However, the plant in San Francisco ships its products only to the customers in Lake Tahoe and Barstow and the plant in Los Angeles sends its products only to the customers in Barstow and La Jolla. Given the amounts of products at the plants, the demands of the customers, and the unit shipping cost from each plant to each customer, you want to determine the amount to ship from each plant to each customer so as to incur the least total shipping cost.

To visualize this problem, create 5 vertices, one for each of the two production facilities and one for each of the three customers, as shown in Figure 1.3. In this case, an edge connects a vertex representing a plant to a vertex representing a customer only if that plant ships its product to the corresponding customer. The 4 appropriate edges are shown in Figure 1.3.

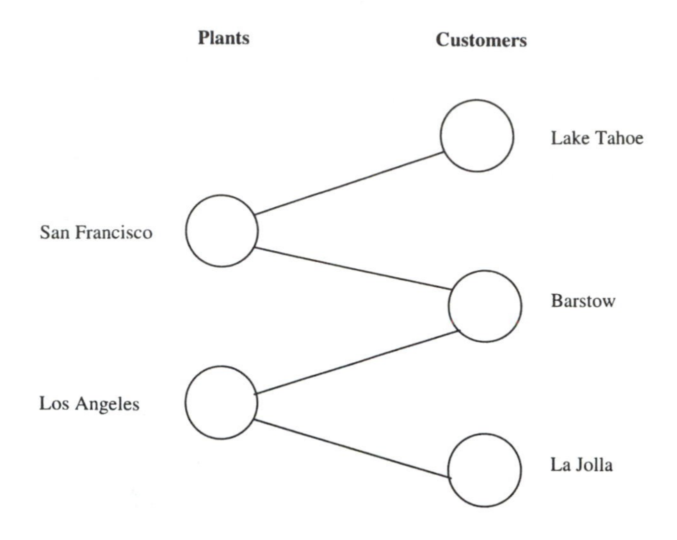

Figure 1.3: A Pictorial Representation of the Problem in Example 1.13

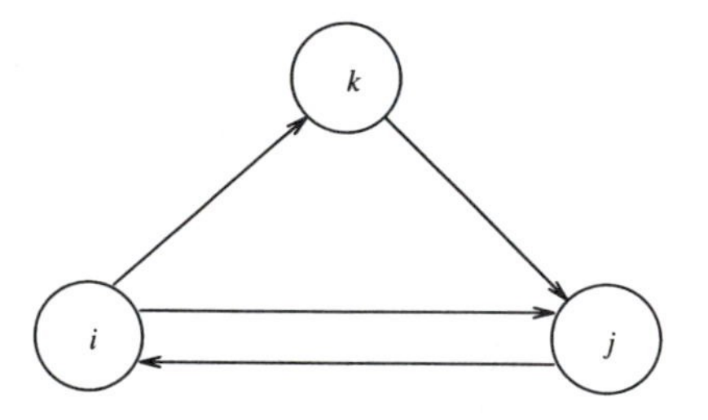

Figure 1.4: An Example of a Directed Graph

In summary, the abstract concept of a **graph** G consists of a nonempty finite collection of vertices, V, that represent objects, and a finite set of edges, E, each of which connects a pair of vertices in V. Each edge represents a relationship between the two objects associated with the corresponding connected vertices. (Do not confuse this concept of a graph with the graph of a *function*).

A further generalization of a graph $G = (V, E)$ is the concept of a **directed graph** $\overrightarrow{G} = (V, \overrightarrow{E})$, in which there is a *direction* associated with each edge of the graph. As seen in Figure 1.4, the edge from vertex i pointing to vertex j represents a relation between vertex i and vertex j. This edge is different from the one that points from vertex j to vertex i, which represents a relation between vertex j and vertex i. As an example of the difference between a directed and an undirected graph, you need a directed graph to represent the traveling-salesperson problem (TSP) of the Philadelphia Paint Company in Example 1.7 in Section 1.2.2 because the amount of time needed to clean the machine when changing from red to green, for example, is not the same as the amount of time needed to change from green to red. In contrast, an ordinary *undirected* graph suffices to represent the TSP in Example 1.6 because the distance is the same between two cities no matter in *which* direction you travel.

An Abstract Concept of a Problem

As another example of abstraction in mathematics, consider again the idea of a *problem*, as introduced in Section 1.1. So far, you have probably been thinking of the problem inputs and outputs as a group of numbers. However, consider what happens if instead you begin to think of those inputs and outputs as general *objects*. Thus, in addition to all of the examples presented so far, a problem might now be one of the following:

1. A problem in which the input is some text followed by a special word and the output is whether that special word appears in the text.
2. A problem in which the input is a real-valued function f of one variable and the output is a number (for example, the output might be the minimum value of f).
3. A problem in which the input is a function f and the output is a function (for example, the output might be a function whose derivative is f).
4. A problem in which the input is a differential equation and the output is a function that satisfies the differential equation.
5. A problem in which you want to prove a theorem. Here, the input is the hypothesis and conclusion of the theorem and the output is a proof that the desired conclusion follows logically from the hypothesis (see Section 1.6 for a detailed discussion of mathematical proofs).

From these examples, you see that by thinking of the inputs and outputs as objects rather than specific numbers, you extend significantly your concept of what a problem is. This is the primary advantage of abstraction.

1.5.2 Additional Examples of Abstraction

You will now see various other examples of how abstraction is used in mathematics. Observe that in each case, the key is to replace the specific items with general objects.

An Abstract Concept of Binary Operators and Relations

Where unification is used to combine items that are similar in nature, abstraction is a technique that you can sometimes use to combine items that are quite different from each other. For example, consider the operations indicated by the underlined words in each of the following mathematical statements:

1. x times y, which is the product of the real numbers x and y.
2. A intersect B, which is the intersection of the sets A and B.
3. i mod j, which is the remainder obtained on dividing the integer i by the integer j.
4. x less than y, which is the result of comparing the real numbers x and y.
5. A subset B, which is the result of comparing the two sets A and B.

To use abstraction, you must again look for similarities and differences while at the same time thinking about *objects*. The common feature between all of the operations indicated by the preceding underlined words is that they combine two *objects* of the same type: times combines two real numbers; intersect combines two sets; mod combines two integers; less than combines two real numbers; and subset combines two sets.

One difference, however, is that the *result* of performing the first three underlined words is *not* the same as performing the last two. In (1), (2), and (3), the result is an object of the *same* type (for example, the result of combining two real numbers with times is again a real number). In contrast, the result in (4) and (5) is *true* or *false* (for example, the result of combining x and y with less than is *true*, if $x < y$ and *false*, otherwise). In this case, abstraction results in two separate mathematical concepts: **binary operators,** that combine two like objects to produce a new object of the same type, and **binary relations**, that combine two like objects to produce a *true* or *false* result.

An Abstract Concept of a Function

Given real numbers a and b, you know that the linear function $ax + b$ associates to each real number x, the real number $ax + b$. First, think about *generalization*. You might therefore consider a more general function f that associates to each real number x, some arbitrary real number, $f(x)$. Continuing the process of generalization, consider a function that associates to an *ordered list* of real numbers, a single real number, or better yet, a function that associates to each ordered list of real numbers, an *ordered list* of real numbers.

Turning now to abstraction, think about *objects* rather than real numbers. For instance, consider a function that associates to each object in a set, some other object in another set. Here, abstraction allows you to expand greatly your concept of what a function is. For example, with this abstract concept, a *function* might now be one of the following:

1. A function that associates to each interval $[l, u] = \{x : l \leq x \leq u\}$, the length of that interval, namely, $u - l$.

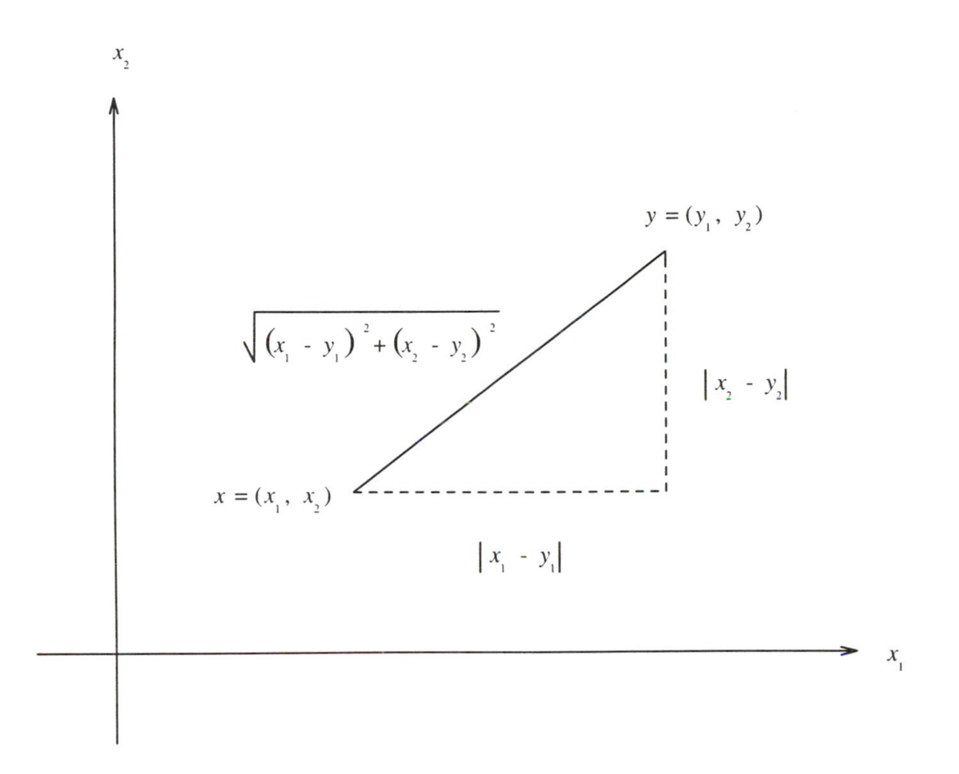

Figure 1.5: Two Measures of the Distance Between Two Points in the Plane

2. A function that associates to each positive integer, a *set* of positive integers. For example, a function that associates to each positive integer n, the set of all positive even integers that are less than n.

3. A function h that associates to each real-valued function, f, another real-valued function, g. For example, h might associate to a real-valued function f its derivative function, f'.

A detailed discussion of functions is given in Section 3.2.

An Abstract Concept of Distance

As another example of abstraction, suppose that $x = (x_1, x_2)$ and that $y = (y_1, y_2)$ are the coordinates of two points in the plane. Consider the following two measures of how far it is from one point to the other:

$$|x_1 - y_1| + |x_2 - y_2|, \tag{1.27}$$

$$\sqrt{(x_1 - y_1)^2 + (x_2 - y_2)^2}. \tag{1.28}$$

As seen in Figure 1.5, the expression in (1.27) represents the "driving" distance from x to y, as measured by a rectangular road system. The expression in (1.28) reflects the "flying" distance, that is, the straight-line distance from x to y.

First, think about generalization. You might therefore create the general concept of a distance *function*, d, that associates to two points x and y in the plane, a distance $d(x, y)$ between these two points (whose specific means of calculation is not specified). That is,

$$d(x, y) = \text{the distance between the points } x \text{ and } y \text{ in the plane.}$$

Continuing with generalization, think of $\mathbf{x}$ and $\mathbf{y}$ as two n-vectors. Then,

$$d(\mathbf{x}, \mathbf{y}) = \text{the distance between the } n\text{-vectors } \mathbf{x} \text{ and } \mathbf{y}.$$

Turning now to abstraction, think of x and y as two *objects* of the same type. Then, the distance function becomes

$$d(x, y) = \text{the distance between the objects } x \text{ and } y.$$

Abstraction greatly expands your ability to measure distance. For example, with this abstract concept, you can now consider measuring the *distance* between any of the following:

1. The distance between two points in n dimensions (rather than just in the plane).

2. The distance between two sets of real numbers.

3. The distance between two functions.

4. The distance between two computer programs.

Of course, in each of these cases, you would need to develop an appropriate method for defining and computing the distance between two specific objects. Observe that you cannot use the formulas in (1.27) and (1.28) because they apply only to points in the plane, and not to general objects. The distance between two objects would also have to satisfy certain properties, such as being nonnegative. Nevertheless, the concept of the distance between two objects of the same kind can be meaningful, even without specifying the formula for calculating those distances.

1.5.3 Correcting Syntax Errors When Developing Abstractions

As you have seen from the examples in Section 1.5.2, abstraction is an extreme form of generalization that allows you to work with systems involving general objects rather than specific items. The advantage of abstraction is that you can include many more problems in the one class. The disadvantage is that when you work with general objects, you lose sight of the specific items, even more so than with unification. Care is therefore needed when you attempt to develop an abstraction. For example, suppose that

x and y are real numbers and that you want to make an abstraction of the following operation:

$$x + y \text{ (where } x \text{ and } y \text{ are real numbers).} \qquad (1.29)$$

To perform this abstraction, think of x and y as objects rather than real numbers. However, now, the symbol + in (1.29) no longer makes sense because you cannot "add" two objects, so a syntax error arises. How, then, can you create an abstraction of (1.29)?

One approach is to replace the symbol + (that works only for real numbers) with the abstract concept of a *binary operator* (that combines two general objects), as described in Section 1.5.2. Using the symbol $\oplus$ for this binary operator, an abstraction of (1.29) is written as follows:

$$x \oplus y \text{ (where } x \text{ and } y \text{ are objects of the same kind).} \qquad (1.30)$$

In (1.30), x and y are objects of the same type that are combined by the binary operator $\oplus$ to create a *new* object of the same type, denoted by $x \oplus y$.

As another example of a syntax error, suppose that r is a real number and that you want to perform abstraction on the following absolute-value operation:

$$|r| \text{ (where } r \text{ is a real number).} \qquad (1.31)$$

Now think of r as a general object. A syntax error arises in (1.31) because the absolute value of an *object* makes no sense. One approach to overcoming this syntax error is to replace the absolute-value operation – that applies only to real numbers – by an arbitrary *function*, say, ν, that associates to each object r, some real number denoted by $\nu(r)$. Thus, the abstraction of (1.31) is written as

$$\nu(r) \text{ (where } r \text{ is a general object).} \qquad (1.32)$$

Putting the pieces together, an **abstract system** is a system that consists of a set of objects together with one or more operations on those objects.

Recall that the purpose of abstraction is to allow you to study more general problems and systems. How, then, do you *study* an abstract system? This and many other topics related to using mathematics are discussed in Chapter 2. To understand those topics properly, you need to be able to read and do proofs. A systematic method for doing so is presented in the final section of this chapter, following a summary of the process of abstraction.

Summary of Abstraction

In this section, you have learned how abstraction is used to create a broad class of problems by replacing specific items with general objects. A summary of how to do so (and the corresponding advantages and disadvantages) follows.

How to Apply Abstraction

Step 1. Identify one or more special cases you want to perform abstraction on.

Step 2. Create an abstract system by replacing specific items with general objects belonging to a set. Be sure to resolve syntax errors, that is, be sure you can perform the necessary operations on the general objects (using appropriate operators, if necessary).

Step 3. Interpret each of the special cases from Step 1 in the context of the abstract system developed in Step 2.

Step 4. Study and derive results about the abstract system. (You will learn how to do so in Chapter 2.)

Advantage of Abstraction

1. Abstraction allows you to expand greatly the corresponding problem to include not only the special cases, but a vast number of related problems obtained by interpreting the meaning of the general objects appropriately in the context of the special cases.

Disadvantages of Abstraction

1. The process of abstraction causes you to lose sight of the specific items to the point where the general objects may have little meaning you can relate to.
2. You can become so abstract that it is difficult to obtain meaningful results about the abstract system.

1.6 Mathematical Proofs

In Section 1.1, you learned that mathematics is often used to solve specific problems, that is, to determine *how* a solution is obtained. In this section, you will see that mathematics is also used to understand *why* a solution is correct. This is accomplished by using a **proof**, which is a logical argument for convincing someone that a particular mathematical statement is *true*. In this section, you will learn various **proof techniques**, which are methods for doing proofs. The approach is adapted from *How to Read and Do Proofs*, Second Edition, Daniel Solow, copyright ©1990 by John Wiley & Sons,

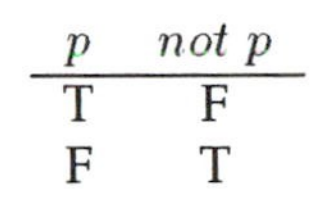

p	*not* p
T	F
F	T

Table 1.1: The Truth Table for *not* p

Inc., NY[1]. The reader will find a more detailed description in that book. You can also refer to the material on sets and functions in Section 3.1 and Section 3.2 of the current book, as needed.

1.6.1 Mathematical Statements and Implications

Proofs deal with **statements**, which, in mathematics, are expressions that are either *true* or *false*. For example, each of the following are statements:

1. $0 < 1$.
2. $3 > 4$.
3. The real number x is less than 2.
4. There are real numbers that are not rational.

You can see that (1) is *true* and (2) is *false*. In contrast, the truth of (3) depends on the specific value of x, which you do not know. Nevertheless, (3) is either *true* or *false* and, as such, is called a **conditional statement**. Finally, (4) is *true*, although this fact may not be so obvious. It is for this reason that a proof is used, namely, to convince someone that a particular statement is *true*. Throughout the rest of this section, p and q represent two statements.

You can use statements to create other statements. For example, the **negation of the statement** p, written *not* p or $\neg p$, is a statement that is *false*, if p is *true* and *true*, if p is *false*. This information is summarized in Table 1.1, called a **truth table**, that indicates the truth of a complex statement (*not* p, in this case) in terms of the truth of simpler statements (p, in this case).

As another example, the **conjunction** of the statements p and q is the statement p *and* q (also written $p \wedge q$) that is *true*, if both p and q are *true* and *false*, otherwise (see Table 1.2). The **disjunction** of the statements p and q is the statement p *or* q (also written $p \vee q$) that is *true* in all cases except when p is *false* and q is *false* (see Table 1.3).

Implications

Proofs apply to the following special kinds of statements:

If p is *true*, then q is *true*,

[1] Reprinted with permission from John Wiley & Sons, Inc.

p	q	p *and* q
T	T	T
T	F	F
F	T	F
F	F	F

Table 1.2: The Truth Table for p *and* q

p	q	p *or* q
T	T	T
T	F	T
F	T	T
F	F	F

Table 1.3: The Truth Table for p *or* q

or, more simply,

$$\text{If } p, \text{ then } q. \tag{1.33}$$

The statement in (1.33) is called an **implication** in which p is the **hypothesis** and q is the **conclusion**. You can write (1.33) in any of the following equivalent ways:

$$p \textit{ implies } q, \; p \Rightarrow q, \; p \rightarrow q.$$

In an implication, it is important to realize that p is a statement, q is a statement, and so is p *implies* q. The truth of p *implies* q depends on the truth of p and q themselves. Mathematicians have come to the agreement that the statement p *implies* q is *true* in all cases *except when* p *is true and* q *is false*, as summarized in Table 1.4. Three other statements related to p *implies* q are:

1. The **contrapositive statement**: (*not* q) *implies* (*not* p), whose truth table is given in Section 1.6.8.

2. The **converse statement**: q *implies* p, whose truth table you are asked to develop in the exercises.

3. The **inverse statement**: (*not* p) *implies* (*not* q), whose truth table you are asked to develop in the exercises.

The use of Table 1.4 as applied to the original statement p *implies* q is shown in the next example.

p	q	p *implies* q
T	T	T
T	F	F
F	T	T
F	F	T

Table 1.4: The Truth Table for p *implies* q

Example 1.14 – Using the Truth Table to Prove an Implication

In the implication

If $3 > 4$, then $0 < 1$,

the hypothesis $3 > 4$ is *false* and the conclusion $0 < 1$ is *true*. Thus, according to the third row of Table 1.4, the implication itself is *true*.

The use of Table 1.4 is not always so straightforward, as seen in the next example.

Example 1.15 – The Need for a New Proof Technique

In the implication:

If x and y are positive real numbers for which

$$x^2 \geq 2 \text{ and } y = \frac{1}{2}\left(x + \frac{2}{x}\right),$$

then $x \geq y$,

you do not know whether the hypothesis is *true* or *false* because you do not know the specific values of x and y. Likewise for the conclusion. Thus, you do not know which row of Table 1.4 is applicable. Nevertheless, as you will now see, you can still determine that this implication is *true*.

1.6.2 The Forward-Backward Method

The implication in Example 1.15 illustrates the need for some method of proving that p *implies* q is *true* other than using Table 1.4 directly. To understand this new approach, look again at Example 1.15. Although you do not know which line of Table 1.4 is applicable, you can reason as follows:

> "In the event that p is *false*, row 3 or row 4 of Table 1.4 is applicable and, in either case, p *implies* q is *true*. On the other hand, if p is *true*, then either row 1 or row 2 is applicable. If I want p *implies* q to be *true*, then I need to be sure that row 1 is applicable, that is, that q is *true*."

The result of this reasoning is that, if you want to prove p *implies* q is *true*,

> You can *assume that* p *is true* and your objective is to *show that* q *is true*.

The **forward-backward method** is a technique for using the *assumption* that p is *true* to reach the conclusion that q is *true*. As its name implies, the forward-backward method consists of two parts, each of which is described next.

The Backward Process

The objective of the **backward process** is to help you reach the conclusion that q is *true* by obtaining, from q, a new statement, q_1, with the property that if q_1 is *true*, then so is q. In the fortunate event that q_1 is p, you are done because you are assuming that p is *true* and so you would know that q_1 is *true* and thus so is q. On the other hand, if you do not know that q_1 is *true*, then you can apply the backward process again to q_1 to obtain a new statement, q_2, with the property that if q_2 is *true*, then so is q_1 (and hence q). This backward process is continued until you obtain a statement that you *know* is *true*.

One way to obtain these new statements is to look at the last statement you have in the backward process, say, q, and to ask the following **key question**:

> How can I show that q is *true*?

The specific way in which you ask this question often determines your answer. For example, associated with the conclusion

$$q : x \geq y$$

in the implication in Example 1.15 is the following key question:

> How can I show that a real number (namely, x) is greater than or equal to another real number (namely, y)?

Here are other examples of statements q and their associated key questions.

Statement q	**Key Questions**
The lines L_1 and L_2 are parallel.	How can I show that two lines are parallel?
$A \cup B = B$ (where A and B are sets).	How can I show that two sets are equal?
The right triangles ABC and RST are congruent.	How can I show that two triangles are congruent? or How can I show that two *right* triangles are congruent?
$f(x) = 0$ (where x is a given real number and f is a real-valued function of one variable).	How can I show that the value of a function at a given point is 0? or How can I show that a function crosses the x-axis at a given point? or How can I show that two real numbers are equal?

The following points are worth noting from these examples:

1. The key questions contain no symbols or notation from the statement under consideration. (Avoiding symbols allows you to focus on the more important aspects of the problem.)

2. There can be several different key questions associated with a statement q. (Choosing the correct one is an art that may require a trial-and-error process.)

After asking the key question, you must provide an answer. For instance, recall the key question associated with the conclusion

$$q : x \geq y$$

in the implication in Example 1.15, which is:

> How can I show that a real number (namely, x) is greater than another real number (namely, y)?

One answer to this question is to show that the difference of the two real numbers is nonnegative, that is, you must show that

$$q_1 : x - y \geq 0.$$

Answering the key question is done in the following two steps.

How to Answer a Key Question

1. First provide an answer that contains no symbols or notation from the specific problem under consideration.
2. Then write your answer in (1) using appropriate symbols and notation from the specific problem to obtain the new statement in the backward process.

There is often more than one answer to a key question. For the preceding example, another answer is to show that the first number (x) is equal to the second number (y) plus some nonnegative amount. Choosing the right answer to the key question is again an art that may require a trial-and-error process.

In summary, through asking and answering the key question, the objective of the backward process is to derive, from the last statement q, a new statement, q_1, with the property that if q_1 is *true*, then so is q. You continue from q_1 with the backward process until you create a statement that you *know* is *true*, at which time the proof is complete. If you have difficulties with the backward process, you can use the assumption that the hypothesis p is *true*, as described next.

The Forward Process

The **forward process** is the process of deriving, from the hypothesis p, some new statements that are necessarily *true* as a result of the assumption that p is *true*. For instance, from the following hypothesis in the implication in Example 1.15:

$$p : x^2 \geq 2,$$

you can create the new statement:

$$p_1 : x^2 - 2 \geq 0.$$

Observe that p_1 is *true* as a result of the assumption that p is true.

As another example of the forward process, from the following hypothesis in the implication in Example 1.15:

$$p : y = \frac{1}{2}\left(x + \frac{2}{x}\right),$$

you can create a new statement by multiplying both sides of the equality in p by -1, adding x, and then performing algebra to obtain:

$$p_2 : x - y = \frac{x^2 - 2}{2x}.$$

The objective of the forward process is to obtain the last statement in the backward process, at which time, the proof is complete. In this case, that last statement in the backward process is

$$q_1 : x - y \geq 0,$$

which is the motivation for obtaining p_2 in the forward process (compare p_2 and q_1). In fact, you can now conclude that q_1 is *true* because the numerator $x^2 - 2$ in p_2 is ≥ 0 (see p_1), and the denominator $2x$ is > 0 (because $x > 0$ from the hypothesis). It therefore follows that q_1 – and hence q – is *true*, thus completing the proof.

In the following written proof of the implication in Example 1.15, observe that there is little *explicit* reference to the forward and backward processes.

Proposition 1.1 *If x and y are positive real numbers for which*

$$x^2 \geq 2 \text{ and } y = \frac{1}{2}\left(x + \frac{2}{x}\right),$$

then $x \geq y$.

Proof. The conclusion is obtained by showing that $x - y \geq 0$. From the hypothesis, it follows by algebra that

$$x - y = \frac{x^2 - 2}{2x}.$$

Furthermore, from the hypothesis that x is positive and that $x^2 \geq 2$, the numerator $x^2 - 2 \geq 0$ and the denominator $2x > 0$, so $x - y \geq 0$ and hence $x \geq y$, completing the proof. QED

The letters QED indicate the end of the proof because QED stand for the words *quod erat demonstrandum*, meaning "which was to be demonstrated."

1.6.3 The Existential Quantifier and the Construction Method

The forward-backward method is only one of various proof techniques available. In many cases, you can choose a correct technique on the basis of certain key words appearing in the hypothesis or conclusion of the implication. One such set of key words involves the **existential quantifier** *there is*, (*there are*, *there exists*, and so on). Mathematicians use the symbol $\exists$ for the words *there is* and $\ni$ for the words *such that*, as illustrated in the following statements:

1. There is a real number $x > 0$ such that $x = 2^{-x}$, or
   $\exists$ a real number $x > 0 \ni x = 2^{-x}$, or
   $\exists\, x > 0 \ni x = 2^{-x}$.

2. There is an element x in the set A such that x is in the set B, or
   $\exists$ an element $x \in A \ni x \in B$ (where $\in$ stands for "in"), or
   $\exists\, x \in A \ni x \in B$.

3. The quadratic equation $ax^2 + bx + c = 0$ has two distinct real roots when a, b, and c are real numbers with $b^2 - 4ac > 0$ and $a \neq 0$.

From these examples, you can see that such statements have the following general form:

> There is an *object* with a *certain property* such that *something happens*.

In (1), the *object* is a real number x; the *certain property* is that of being > 0; and the *something that happens* is that $x = 2^{-x}$. In (2), the *object* is an element x; the *certain property* is that of being in the set A; and the *something that happens* is that x is in the set B. Even (3) has this form when rewritten as follows:

> There are real numbers y and z with $y \neq z$ such that $ay^2 + by + c = 0$ and $az^2 + bz + c = 0$.

When a statement containing the existential quantifier arises in the *forward process*, you can assume that *there is* an *object* with the *certain property* such that the *something happens*. You should use this *object* to arrive at the desired conclusion.

However, when a statement containing the existential quantifier arises in the *backward process*, it is your job to *show that* there is an *object* with the *certain property* such that the *something happens*. The **construction method** is a technique for accomplishing this goal. With the construction method, you must first *produce* the *object* – by guessing, by trial-and-error, by devising an algorithm whose output is the *object*, or by any other means. However, you must subsequently prove that the *object* you produced is the correct one in that the *object* has the *certain property* and that the *something happens*. The use of the construction method is demonstrated in the proof of the following proposition.

Definition 1.1 *An integer a* **divides** *an integer b if and only if there is an integer k such that $b = ka$.*

Proposition 1.2 *If an integer a divides an integer b, then a divides the integer b^2.*

Discussion of Proof. According to the forward-backward method, you can assume that the following hypothesis is *true*:

> p : The integer a divides the integer b.

You must use this assumption to conclude that

> q : a divides the integer b^2.

The backward process applied to q gives rise to the key question: How can I show that an integer (namely, a) divides another integer (namely, b^2)? By using the definition – one of the most common methods for answering a key question – you obtain the following new statement to prove:

q_1 : There is an integer k such that $b^2 = ka$.

The appearance of the existential quantifier *there is* in q_1 suggests that you now use the construction method to produce the integer k. How you find the value of k is not clear but, in general, you should use the information in the hypothesis, which you are assuming is *true*.

In this case, the hypothesis p is that a divides b. Working forward from Definition 1.1, this means that

p_1 : there is an integer m such that $b = ma$.

The idea is to use the *known* integer m in p_1 to construct the *unknown* integer k in q_1. This is accomplished by looking at what properties you want k to satisfy. In this case, you want $b^2 = ka$. To find such a value for k, you can work forward from p_1, as follows:

$p_2 : b^2 = (ma)^2 = m^2a^2 = (m^2a)a.$

From p_2, you can see that the desired value for k is $k = m^2a$.

It is important to note that, although the value of k has now been constructed as m^2a, you must still prove that this value of k is correct, that is, that $b^2 = ka$. This, however, is *true*, as seen in p_2.

Proof. By definition, it must be shown that there is an integer k such that $b^2 = ka$. However, from the hypothesis that a divides b, there is an integer m such that $b = ma$. Letting $k = m^2a$, it follows that

$$b^2 = (ma)^2 = (m^2a)a = ka.$$

Thus a divides b^2, completing the proof. QED

1.6.4 The Universal Quantifier and the Choose Method

Several other proof techniques are available when the hypothesis or conclusion of an implication contains the **universal quantifier** *for all* (*for each*, *for any*, and so on), written $\forall$. The way in which this quantifier arises is illustrated in the following statements:

1. For all real numbers $x > 1,\ x^2 > x$, or
   $\forall$ real numbers $x > 1,\ x^2 > x$, or
   $\forall\, x > 1,\ x^2 > x$.

2. For all elements x in the set A, x is in the set B, or
   $\forall$ elements x in A, x is in B, or
   $\forall x \in A,\ x \in B$.

From these examples, you can see that such statements have the following general form:

For all *objects* with a *certain property, something happens.*

In (1), the *objects* are real numbers x; the *certain property* is that of being > 1; and the *something that happens* is that $x^2 > x$. In (2), the *objects* are elements x; the *certain property* is that of being in the set A; and the *something that happens* is that x is in the set B.

When a statement containing the universal quantifier arises in the *backward* process, you should consider using the **choose method** to prove that for every *object* with the *certain property*, the *something happens*. This method works as follows. Rather than prove that the something happens *for every* object with the *certain property* (which you cannot do because there are usually too many of them), the idea is to show that, for *one representative object with the ceratin property, the something happens*. If you are successful for this one *chosen* object, then you could, in theory, repeat the same proof for each and every object with the *certain property*, thus showing that for all *objects* with the *certain property*, the *something happens*.

In the example that follows, observe that the act of choosing a generic *object* with the *certain property* provides a new statement in the forward process. The statement that, for this chosen *object*, the *something happens* becomes the new statement to be proved in the backward process.

Definition 1.2 *A set A is a* **subset** *of a set B if and only if for every element x in A, x is in B.*

Proposition 1.3 *If*

$$
\begin{aligned}
A &= \{\text{real numbers } x : -1 \leq x \leq 2\} \text{ and} \\
B &= \{\text{real numbers } x : x^2 - x - 2 \leq 0\},
\end{aligned}
$$

then A is a subset of B.

Discussion of Proof. You must conclude that

q : A is a subset of B.

The backward process gives rise to the key question: How can I show that a set (namely, A) is a subset of another set (namely, B)? According to Definition 1.2, the answer is to show that

q_1 : for every element x in A, x is in B.

Recognizing the quantifier *for every* in q_1, you should now proceed with the choose method, whereby, you choose one representative *object* with the *certain property* for which you must show that the *something happens*. In this case, that means you should choose

p_1 : an element $\bar{x}$ in A

for which you must show that

q_2 : $\bar{x}$ is in B,

or equivalently, from the fact that $B = \{\text{real numbers } x : x^2 - x - 2 \leq 0\}$, you must show that

$q_3 : \bar{x}^2 - \bar{x} - 2 \leq 0.$

Working forward from p_1, you know that $\bar{x}$ is in A so, from the hypothesis that $A = \{x : -1 \leq x \leq 2\}$, it follows that

$p_2 : -1 \leq \bar{x} \leq 2.$

Thus,

$p_3 : \bar{x} - 2 \leq 0$ and $\bar{x} + 1 \geq 0,$

so

$p_4 : \bar{x}^2 - \bar{x} - 2 = (\bar{x} - 2)(\bar{x} + 1) \leq 0.$

The proof is now complete because p_4 (in the forward process) is the same as q_3 (the last statement in the backward process).

In the written proof that follows, observe that the symbol x is used for the chosen *object* instead of $\bar{x}$. Also note that the word *let* indicates that the choose method is being used, as is often the case.

Proof. To see that A is a subset of B, *let* x be an element of A, that is, $-1 \leq x \leq 2$. But then $x - 2 \leq 0$ and $x + 1 \geq 0$, so

$$x^2 - x - 2 = (x - 2)(x + 1) \leq 0.$$

This means that x is an element of B and so it has been shown that every element x in A is also in B, thus completing the proof.

QED

1.6.5 Induction

You have just seen how the choose method is used when the last statement in the backward process contains the universal quantifier *for all* in the form:

For all *objects* with a *certain property, something happens*.

However, when the *objects* are *integers*, the *certain property* is that of being greater than or equal to some initial integer, and the *something that happens* is some statement $S(n)$ that depends on the integer n, a proof technique called **induction** is often the best method to use, even before the choose method. For example, you should consider using induction to prove each of the following statements:

1. For every integer $n \geq 1$,

$$S(n): \sum_{k=1}^{n} k = \frac{n(n+1)}{2}.$$

2. For every integer $n \geq 1$,

$$S(n): \frac{1}{n!} \leq \frac{1}{2^{n-1}}, \text{ [where } n! = 1(2)\cdots(n)\text{]}.$$

The idea of induction is to begin by proving that the statement $S(n)$ is *true* for $n = 1$. You could then use the fact that $S(1)$ is *true* to prove that $S(2)$ is *true*. Then you could use $S(2)$ to show that $S(3)$ is *true*, and so on. Because there are an *infinite* number of statements, you cannot actually prove *all* of them; however, you can accomplish the same goal by performing the following two steps:

1. Prove that the statement is *true* for $n = 1$.
2. *Assume* that $S(n)$ is *true* and prove that $S(n+1)$ is also *true*.

The first step of proving that $S(1)$ is *true* is usually straightforward, requiring little more than writing $S(1)$ and verifying that $S(1)$ is *true*. The second step is more challenging. Begin by writing $S(n)$ and assuming that this statement is *true*. You should then write $S(n+1)$ by replacing n everywhere with $n+1$ and try to prove that $S(n+1)$ is *true*. To do so, you want to use the assumption that $S(n)$ is *true* – often called the **induction hypothesis** – and this requires finding a relationship between $S(n+1)$ and $S(n)$. That is, you must express $S(n+1)$ in terms of $S(n)$ so that you can use the induction hypothesis that $S(n)$ is *true*. These two steps are illustrated in the following example.

Proposition 1.4 *For every integer $n \geq 1$,*

$$\frac{1}{n!} \leq \frac{1}{2^{n-1}}.$$

Proof. The first step of induction requires you to prove that $S(1)$ is *true*, that is,

$$S(1): \frac{1}{1!} \leq \frac{1}{2^{1-1}}.$$

You can see that $S(1)$ is *true* because the left side of the inequality is 1 and so is the right side.

Turning to the second step of induction, you should assume that $S(n)$ is *true*, which in this case means you should assume that

$$S(n): \frac{1}{n!} \leq \frac{1}{2^{n-1}}.$$

You must now use the assumption that $S(n)$ is *true* to prove that $S(n+1)$ is *true* which, in this case, means you must prove that

$$S(n+1): \frac{1}{(n+1)!} \leq \frac{1}{2^n}.$$

To use the induction hypothesis that $S(n)$ is *true*, relate $S(n+1)$ to $S(n)$. For example, from the left side of $S(n+1)$ and the definition of $(n+1)!$ you have that

$$\frac{1}{(n+1)!} = \frac{1}{n!\,(n+1)} = \left(\frac{1}{n!}\right)\left(\frac{1}{n+1}\right). \tag{1.34}$$

You can now apply the induction hypothesis to (1.34). Specifically, from $S(n)$ you know that

$$\frac{1}{n!} \leq \frac{1}{2^{n-1}},$$

and so (1.34) becomes

$$\frac{1}{(n+1)!} = \left(\frac{1}{n!}\right)\left(\frac{1}{n+1}\right) \leq \left(\frac{1}{2^{n-1}}\right)\left(\frac{1}{n+1}\right). \tag{1.35}$$

All that remains is to note that because $n \geq 1$,

$$\frac{1}{n+1} \leq \frac{1}{2},$$

so (1.35) becomes

$$\frac{1}{(n+1)!} \leq \left(\frac{1}{2^{n-1}}\right)\left(\frac{1}{n+1}\right) \leq \left(\frac{1}{2^{n-1}}\right)\left(\frac{1}{2}\right) = \frac{1}{2^n},$$

which is precisely $S(n+1)$, thus completing the proof.

QED

1.6.6 The Universal Quantifier and the Specialization Method

In Section 1.6.4 and Section 1.6.5, you learned to use the choose and induction methods when the universal quantifier *for all* arises in the backward process in the form:

For all *objects* with a *certain property*, *something happens*.

When this quantifier arises in the *forward* process, the **specialization method** is often used to obtain a new statement. For example, suppose you are doing a proof in which S and T are sets satisfying the following statement in the forward process:

p : For all elements $x \in S$, $x \in T$.

You can use specialization to create a new statement, p_1, by applying the general knowledge in p to *one specific element in* S. For instance, if, in the course of doing this proof, you come across a particular element y in S, then you can apply specialization to the statement p to conclude that, for this element y,

$p_1 : y \in T$.

In general, the specialization method allows you to obtain a new statement in the forward process from the following statement:

p : For all *objects* X with a *certain property, something happens,*

by performing these steps:

1. Identify one *particular* object, say, Y. (You may need to use trial-and-error to identify the *correct* object to use for specialization.)

2. Verify that the object Y satisfies the *certain property* in p.

3. Create a new statement, p_1, in the forward process by writing that the *something happens* for this specific object Y.

These steps are demonstrated in the following example.

Definition 1.3 *A real number a is a* **lower bound** *for a set T of real numbers if and only if for every element $x \in T$, $a \leq x$.*

Proposition 1.5 *If a is a lower bound for a set T of real numbers and $S \subseteq T$, then a is a lower bound for S.*

Discussion of Proof. You must show that

q : a is a lower bound for S.

The backward process of the forward-backward method leads to the key question:

How can I show that a real number (namely, a) is a lower bound for a set of real numbers (namely, S)?

From Definition 1.3, you must show that

q_1 : for every element $x \in S,\ a \leq x$.

Recognizing the universal quantifier *for every* in the *backward* process, you should use the choose method to choose an element

$p_1 : \bar{x} \in S$,

for which you must show that

$q_2 : a \leq \bar{x}$.

You will now see how specialization is used to obtain q_2.

Turning to the forward process, you know from the hypothesis p that $S \subseteq T$ which, by Definition 1.2, means that

p_2 : for every element $x \in S,\ x \in T$.

The universal quantifier *for every* now appears in the *forward* statement p_2, so you should consider applying specialization. To do so, follow the three steps in the discussion preceding this example. That is, first identify one particular object. In this case, that object is the element $\bar{x}$ in the statement p_1. Next, make sure that this special object satisfies the *certain property* in p_2 of being in the set S. But this is the case for $\bar{x}$, as stated in p_1. The final step of specialization is to create, from p_2, the following new statement in the forward process:

p_3 : In particular, for $\bar{x} \in S$, it follows that $\bar{x} \in T$.

The statement p_3 is the *something that happens* in p_2 applied to the particular *object* $\bar{x}$. Compare the statements p_3 and p_2.

To complete the proof, you must still obtain the last statement, q_2, in the backward process. You can use specialization again to accomplish this goal. Specifically, from the hypothesis that a is a lower bound for the set T of real numbers, by Definition 1.3 you know that

p_4 : for every element $x \in T,\ a \leq x$.

Recognizing the universal quantifier *for every* in the *forward* process, consider applying specialization. Once again, the particular object to use for specializing the general statement in p_4 is $\bar{x}$. Before doing so, however, make sure that $\bar{x}$ satisfies the *certain property* in p_4 of being in T. But you *do* know that $\bar{x} \in T$ from the statement p_3 in the forward process. The final step of specialization is to create, from p_4, the following new statement in the forward process:

p_5 : In particular, for $\bar{x} \in T,\ a \leq \bar{x}$.

The statement p_5 is the *something that happens* in p_4 applied to the particular *object* $\bar{x}$. Compare the statements p_5 and p_4.

The proof is now complete because the forward statement p_5 is the same as the last backward statement, q_2. In the written proof that follows, observe that no specific reference is made to the specialization or choose methods and that the symbol x is used throughout instead of $\bar{x}$.

Proof. To see that a is a lower bound for S, let $x \in S$. (The word *let* indicates that the choose method is used.) Because $S \subseteq T$, every element of S is an element of T. In particular, for $x \in S$, $x \in T$. (The words *in particular* indicate that specialization is used.) Also, from the hypothesis that a is a lower bound for T, by definition, for all $y \in T$, $a \leq y$. It now follows that for $x \in T$, $a \leq x$. (Here, the words *it now follows that* indicate that specialization is used.) It has therefore been shown that a is a lower bound for S, thus completing the proof.

QED

1.6.7 Nested Quantifiers

You have just learned the construction, choose, induction, and specialization methods for working with statements containing quantifiers. Some statements, however, contain **nested quantifiers**, that is, more than one quantifier. For example, you might encounter a *backward* statement of the form:

> For all *objects* X with a *certain property* P, there is an *object* Y with a *certain property* Q such that *something happens*.

When working with statements that have nested quantifiers, proceed as follows:

> **Rule for Working with Nested Quantifiers**
>
> Process the quantifiers one at a time as they appear *from left to right*, using the construction, choose, induction, and specialization methods, as appropriate.

For instance, in the foregoing example, the first quantifier from the left is *for all*, thus, you should use the choose method to choose

> p_1 : an *object* X with the *certain property* P,

for which you must show that

> q_1: there is an *object* Y with the *certain property* Q
> such that the *something happens*.

In trying to establish q_1, the appearance of the existential quantifier suggests that you now use the construction method to produce the object Y with the *certain property* Q and for which the *something happens*.

Alternatively, if a backward statement contains nested quantifiers in the form:

There is an *object* X with a *certain property* P such that for all *objects* Y with a *certain property* Q, *something happens*,

then you should proceed as follows. Because the first quantifier from the left is *there is*, you should use the construction method to produce an object X with the *certain property* P. After doing so, you will also have to show that, for this *object* X,

q_1 : for all *objects* Y with a *certain property* Q, *something happens*.

In trying to establish q_1, the appearance of the universal quantifier suggests that you now use the choose method to choose

$p1$: an *object* Y with the *certain property* Q,

for which you must show that

q_2 : the *something happens*.

This rule of processing nested quantifiers from left to right is demonstrated with the following example.

Proposition 1.6 *If f is the function defined by $f(x) = x^2$, then for every real number $y \geq 0$, there is a real number x such that $f(x) > y$.*

Discussion of Proof. The conclusion of this proposition contains nested quantifiers, the first of which is *for every*. Accordingly, you should first use the choose method to choose

p_1 : a real number $y \geq 0$,

for which you must show that

q_1 : there is a real number x such that $f(x) > y$.

The appearance of the existential quantifier in q_1 suggests that you now use the construction method to produce a real number x for which $f(x) = x^2 > y$. Thus, if you construct

p_2 : $x > \sqrt{y}$ (which you can do because $y \geq 0$),

it follows that

p_3 : $f(x) = x^2 > y$,

thus completing the proof.

Proof. To see that the conclusion is *true*, let $y \geq 0$ be a real number. (The word *let* indicates that the choose method is used.) By taking x to be a real number with $x > \sqrt{y}$, it follows that

$$f(x) = x^2 > y.$$

p	q	p *implies* q	*not* q	*not* p	(*not* q) *implies* (*not* p)
T	T	T	F	F	T
T	F	F	T	F	F
F	T	T	F	T	T
F	F	T	T	T	T

Table 1.5: The Truth Table for (*not* q) *implies* (*not* p)

Thus, it has been shown that there is a real number x for which $f(x) > y$, and so the proof is complete.

QED

Statements can contain any number of nested quantifiers. Just remember to process those quantifiers one at a time as they appear *from left to right* and to use the associated construction, choose, induction, and specialization methods.

1.6.8 The Contrapositive Method

Another proof technique for showing that

p *implies* q

arises by considering the following associated *contrapositive statement*:

$\neg q$ *implies* $\neg p$, that is, (*not* q) *implies* (*not* p).

Table 1.5 shows how to determine the truth of the contrapositive statement on the basis of the truth of the statements p and q. From Table 1.5, you can see that (*not* q) *implies* (*not* p) is *true* under the same conditions as p *implies* q, that is, in all cases except when p is *true* and q is *false*.

From this observation, you can conclude that to prove p *implies* q is *true*, you can just as well prove that the statement (*not* q) *implies* (*not* p) is *true*. The **contrapositive method**, then, is the forward-backward method applied to the the contrapositive statement. That is, to prove that

p *implies* q

with the contrapositive method, you

1. Assume that the statement *not* q is *true* – that is, that q is *false*.

2. Show that the statement *not* p is *true* – that is, that p is *false*.

You accomplish this objective by working forward from *not* q and backward from *not* p. The use of the contrapositve method is demonstrated in the following example.

Definition 1.4 *An integer n is* **even** *if and only if there is an integer k such that $n = 2k$.*

Definition 1.5 *An integer n is* **odd** *if and only if there is an integer k such that $n = 2k + 1$.*

Proposition 1.7 *If n is an integer for which n^2 is even, then n is even.*

Discussion of Proof. According to the contrapositive method, you should assume that the conclusion is *not true* which, in this case, means you should assume that

not q : n is not even,

or, equivalently,

p_1 : n is odd.

You must work forward from p_1 to show that the hypothesis is *not true* which, in this case, means you should show that

not p : n^2 is not even,

or, equivalently,

q_1 : n^2 is odd.

The remainder of this proof is to work forward from the statement p_1 and backward form the statement q_1.

Working backward from q_1, you have the key question:

How can I show that an integer (namely, n^2) is odd?

From Definition 1.5, you obtain the following new statement to prove:

q_2 : There is an integer k such that $n^2 = 2k + 1$.

The appearance of the existential quantifier in q_2 suggests that you now use the construction method (see Section 1.6.3) to produce the value for the integer k. You get the value for k from the forward process.

Turning to the forward process, you can apply Definition 1.5 to the statement that n is odd in p_1 to obtain the following:

p_2 : There is an integer m such that $n = 2m + 1$.

The idea is to use the *known* value of m in p_2 to construct the *unknown* value of k in q_2. Specifically, on squaring both sides of the equality in p_2 you obtain that

p_3 : $n^2 = (2m + 1)^2 = 4m^2 + 4m + 1 = 2(2m^2 + 2m) + 1$.

You can see from p_3 that the desired value of k in q_2 is $k = 2m^2 + 2m$. You must still show that this value for k is correct, but this follows from p_3.

Proof. By the contrapositive method, assume that n is not even, that is, that n is odd. It must be shown that n^2 is not even, that is, that n^2 is odd. However, because n is odd, by definition, there is an integer m such that $n = 2m + 1$. Letting $k = 2m^2 + 2m$, it follows that

$$n^2 = (2m + 1)^2 = 4m^2 + 4m + 1 = 2(2m^2 + 2m) + 1 = 2k + 1,$$

and so n^2 is odd, thus completing the proof.

QED

As a general rule, you will find the contrapositive method to be successful under either of the following circumstances:

1. When the conclusion – or the last statement in the backward process – is one of two alternatives. (This is the case in Proposition 1.7 where the conclusion that n is even is one of the two alternatives of n being odd or even.)

2. When the conclusion – or the last statement in the backward process – contains the key word *no* or *not*. For example, if x and y are real numbers and the conclusion of your proposition is to show that $x \neq y$, then you should consider the contrapositive method because, in so doing, you will assume that $x = y$, and this gives you some useful information.

1.6.9 The Contradiction Method

Another proof technique arises when you recall that the implication

p *implies* q

is *true* in all cases *except* when p is *true* and q is *false*. With the **contradiction method** (also called an **indirect proof**), you establish the truth of p *implies* q by ruling out this one unfavorable case, as follows:

1. Assume that the one bad case *does* happen, that is, that the statement p is *true* and the statement q is *false* (*not* q).

2. Work forward from p and *not* q to reach a contradiction to some fact that you absolutely know is *true*.

The specific contradiction you obtain depends on the problem under consideration, but some examples of valid contradictions are: (1) $1 < 0$, (2) $x \neq x$ (where x is a real number), and (3) some statement s is both *true* and *false* at the same time. Unfortunately, you will generally not know what the specific contradiction is beforehand, so you cannot work backward in this technique. The contradiction method is demonstrated in the following example.

Definition 1.6 *A real number r is* **rational** *if and only if there are integers a and b with $b \neq 0$ such that $r = a/b$.*

Proposition 1.8 *If r is a real number for which $r^2 = 2$, then r is not rational.*

Discussion of Proof. Proceeding by the contradiciton method, you should assume that the hypothesis is *true*, so, assume that

p : r is a real number for which $r^2 = 2$.

You should also assume that the conclusion is *false*, so, assume that

not q : r *is* rational.

The objective now is to work forward from the statements p and *not* q to reach some kind of contradiction. For example, working forward from the statement *not* q by using Definition 1.6, you can state that

p_1 : there are integers a and b with $b \neq 0$ such that $r = a/b$.

In this problem, the contradiction arises by noting that, in p_1, you can assume further that a and b have no common divisor, in other words,

p_2 : there is no integer that divides both a and b.

The reason p_2 is *true* is that, if there *were* an integer k that divides both a and b, then you could cancel that value from both the numerator a and the denominator b, ultimately writing $r = a/b$ in lowest terms.

You can now reach a contradiction by showing that 2 divides both a and b, that is, that both a and b are even. This is accomplished by working forward from *not* q and p_1, as shown in the following proof.

Proof. By contradiction, assume that r *is* rational, so, there are integers a and b with $b \neq 0$ such that

$$r = \frac{a}{b}. \tag{1.36}$$

It can be assumed further that a and b are in lowest terms and therefore have no common divisor. A contradiction is reached by showing that both a and b are even and thus they have 2 as a common divisor.

From the hypothesis that $r^2 = 2$ and (1.36), you have that

$$r^2 = \left(\frac{a}{b}\right)^2 = \frac{a^2}{b^2} = 2. \tag{1.37}$$

Multiplying (1.37) through by b^2 yields that

$$a^2 = 2b^2, \tag{1.38}$$

from which you conclude that

a^2 is even.

Now from Proposition 1.7, because a^2 is even, you can say that

a is even. (1.39)

It remains to show that b is even. However, from (1.39) you know by definition that there is an integer k such that

$$a = 2k.$$

By substituting this value of $a = 2k$ in (1.38), you have that

$$2b^2 = a^2 = (2k)^2 = 4k^2. \tag{1.40}$$

On dividing (1.40) through by 2, you obtain

$$b^2 = 2k^2,$$

from which it follows that b^2 is even. But now, because b^2 is even, by Proposition 1.7, b is also even. You now know that a is even [see (1.39)] and that b is even and these facts contradict the assumption that a and b have no common divisor, thus completing the proof.

QED

In general, you will find that a proof by contradiction is effective when the conclusion of a proposition contains the key words *no* or *not* (as is the case in Proposition 1.8).

1.6.10 Negations of Statements

To use either the contrapositive or contradiction method, you need to write the *negation of a statement* p, namely, *not* p. In some cases, finding the negation is straightforward. For example, if x is a real number, then the negation of the statement

$$p : x > 0$$

is

$$\textit{not } p : x \leq 0.$$

In other cases, however, there are certain rules you can follow to obtain the correct negation of the statement. For example, if the statement p contains the word *not* then, when you negate the statement, you remove the word *not*, as shown in the following example.

Example 1.16 – Negating the Word *Not*

Suppose that a and b are integers. The negation of the statement

$p : a$ does not divide b

is

not $p : a$ *does* divide b.

When the statement p contains the word *and* in the form

$p : p_1$ *and* p_2 (where p_1 and p_2 are statements),

the negation contains the word *or*, as follows:

not p : (*not* p_1) *or* (*not* p_2).

The following example illustrates this rule.

Example 1.17 – Negating the Word *And*

Suppose that x is a real number. The negation of the statement

$p : -1 \leq x$ *and* $x \leq 1$,

in which p_1 is "$-1 \leq x$" and p_2 is "$x \leq 1$," is

not p : (*not* $-1 \leq x$) *or* (*not* $x \leq 1$),

or equivalently,

not $p : x < -1$ *or* $x > 1$.

Analogously, the word *or* changes to *and*. That is, the negation of the statement

$p : p_1$ *or* p_2 (where p_1 and p_2 are statements),

is the statement

not p : (*not* p_1) *and* (*not* p_2),

as shown in the following example.

Example 1.18 – Negating the Word *Or*

Suppose that x is a real number. The negation of the statement

$$p : x > 4 \textit{ or } x < 2,$$

in which p_1 is "$x > 4$" and p_2 is "$x < 2$," is

$$\textit{not } p : (\textit{not } x > 4) \textit{ and } (\textit{not } x < 2),$$

or equivalently,

$$\textit{not } p : x \leq 4 \textit{ and } x \geq 2.$$

A more challenging situation arises when the statements contain quantifiers. For the existential quantifier, the negation of the statement that

> there is an *object* with a *certain property* such that *something happens*,

is the statement that

> for all *objects* with the *certain property*, the *something does not happen*,

as shown in the following example.

Example 1.19 – Negating the Words *There Is*

Suppose that S is set of real numbers. The negation of the statement

p : There is an element $x \in S$ such that $x > 2$

is

not p : For all elements $x \in S$, $x \leq 2$.

Observe in Example 1.19 that the negation of a statement containing the existential quantifier results in a statement containing the universal quantifier. Also note that the *something that happens* changes but the *certain property* does not.

These same ideas apply when negating a statement containing the universal quantifier. Specifically, the negation of the statement that

> for all *objects* with a *certain property*, *something happens*,

is the statement that

> there is an *object* with the *certain property* such that the *something does not happen*,

as shown in the following example.

Example 1.20 – Negating the Words *For All*

Suppose that S and T are sets. The negation of the statement that

p : for all elements $x \in S,\ x \in T$

is the statement that

not p : there is an element $x \in S$ such that $x \notin T$.

When a statement contains more than one quantifier, negate the quantifiers one at a time *from left to right* as they appear in the statement. For instance, the negation of the statement that

> there is an *object* X with a *certain property* P such that for all *objects* Y with *another property* Q, *something happens*,

is the statement that

> for all *objects* X with the *certain property* P, there is an *object* Y with the *property* Q such that the *something does not happen*.

This rule is demonstrated in the following example. Observe what happens as the word *not* is moved from left to right through the quantifiers.

Example 1.21 – Negating the Nested Quantifiers *There Is* **and** *For All*

Suppose that S and T are sets of real numbers. The negation of the statement that

there is an element $x \in S$ such that for all elements $y \in T,\ y \leq x$,

is the statement that

for all elements $x \in S$, it is *not true* that for all elements $y \in T$, $y \leq x$,

or, by applying the word *not* to the next quantifier to the right,

for all elements $x \in S$, there is an element $y \in T$ such that $y > x$.

Likewise, the negation of the statement that

> for all *objects* X with a *certain property* P, there is an *object* Y with *another property* Q such that *something happens*,

is the statement that

> there is an *object* X with the *certain property* P such that for all *objects* Y with the *property* Q, the *something does not happen*.

Statement	Negation
not p	p
p_1 *and* p_2	(*not* p_1) *or* (*not* p_2)
p_1 *or* p_2	(*not* p_1) *and* (*not* p_2)
There is an *object* with a *certain property* such that *something happens.*	For all *objects* with the *certain property,* the *something does not happen.*
For all *objects* with a *certain property, something happens.*	There is an *object* with the *certain property* such that the *something does not happen.*
Nested quantifiers	Process them from left to right.

Table 1.6: Summary of the Rules for Negating Statements

This rule is demonstrated in the following example.

Example 1.22 – Negating the Nested Quantifiers *For All* and *There Is*

Suppose that f is a real-valued function of one variable. The negation of the statement that

> for all real numbers $y \geq 0$, there is a real number $x \geq 0$ such that $f(x) \geq y$,

is the statement that

> there is a real number $y \geq 0$ such that it is *not true* that there is a real number $x \geq 0$ for which $f(x) \geq y$,

or, by applying the word *not* to the next quantifier to the right,

> there is a real number $y \geq 0$ such that for all real numbers $x \geq 0$, $f(x) < y$.

The various rules for negating statements are summarized in Table 1.6.

1.6.11 Either/Or Methods

Special proof techniques, referred to as the **either/or methods**, are available when the key words *either/or* arise in proving that

p *implies* q.

One technique is used when those key words appear in the *forward* process. The other technique is applicable when those key words appear in the *backward* process. Both techniques are described in what follows.

Proof by Cases

A **proof by cases** is used when a statement in the forward process contains the key words *either/or* in the form

p : either p_1 is *true* or p_2 is *true* (where p_1 and p_2 are statements).

Because p is a statement in the *forward* process, you *know that* at least one of p_1 and p_2 is *true* – the only question is, which one? To cover *both* cases, you must do *two* proofs:

1. First, assume that p_1 is *true* and prove that q is *true*. This is accomplished by working forward from p_1 and backward from q.
2. Next, assume that p_2 is *true* and again prove that q is *true*. This is accomplished by working forward from p_2 and backward from q.

A proof by cases is demonstrated in the following example where, for two sets A and B,

$$\begin{aligned} A \cup B &= \{x : x \in A \text{ or } x \in B\}, \\ A \cap B &= \{x : x \in A \text{ and } x \in B\}, \\ A^c &= \{x : x \notin A\}. \end{aligned}$$

Proposition 1.9 *If A and B are sets, then $(A \cap B)^c \subseteq A^c \cup B^c$.*

Discussion of Proof. The forward-backward method leads to the key question:

How can I show that a set (namely, $(A \cap B)^c$) is a subset of another set (namely, $A^c \cup B^c$)?

According to the definition, you must show that

q_1 : for every element $x \in (A \cap B)^c$, $x \in A^c \cup B^c$.

Recognizing the universal quantifier *for every* in the backward statement q_1, you should proceed by the *choose* method, whereby you choose

p_1 : an element $x \in (A \cap B)^c$,

for which you must show that

$q_2 : x \in A^c \cup B^c$.

Working forward from p_1 you know that

$p_2 : x \notin A \cap B$,

or, in other words, it is *not true* that $x \in A$ and $x \in B$. Applying the rules for negating a statement containing the word *and* (see Table 1.6), you can state from p_2 that

p_3 : either $x \notin A$ or $x \notin B$.

Recognizing the key words *either/or* in the forward statement p_3, you should now proceed using a proof by cases, as follows.

Case 1: Assume that $x \notin A$.

In this case, $x \in A^c$ and so $x \in A^c$ or $x \in B^c$. Thus, $x \in A^c \cup B^c$, which is precisely q_2, the last statement in the backward process.

Case 2: Assume that $x \notin B$.

In this case, $x \in B^c$ and so $x \in B^c$ or $x \in A^c$. Thus, $x \in A^c \cup B^c$, which again is q_2.

Thus, in either case, $x \in A^c \cup B^c$, which shows that $(A \cap B)^c \subseteq A^c \cup B^c$ and this completes the proof.

In the written proof that follows, the words *without loss of generality* mean that only one of the two foregoing cases is presented and the remaining case is left for the reader to verify.

Proof. To show that $(A \cap B)^c \subseteq A^c \cup B^c$, let $x \in (A \cap B)^c$. (The word *let* indicates that the choose method is used.) It follows that either $x \notin A$ or $x \notin B$. Assume, without loss of generality, that $x \notin A$. But then $x \in A^c$ and so $x \in A^c \cup B^c$. This means that $(A \cap B)^c \subseteq A^c \cup B^c$, completing the proof.

QED

Proof by Elimination

The other either/or method, referred to as a **proof by elimination**, is used when the key words *either/or* arise in the *backward* process in the form:

q : either q_1 or q_2 is *true*.

In this situation, you must reach the conclusion that one of q_1 or q_2 is *true*. One way to do so is to assume that q_1 is *not true*. You must therefore show that q_2 *is true*. In other words, you use a proof by elimination to prove that

p *implies* (q_1 *or* q_2),

as follows:

1. Assume that p is *true* and also that q_1 is *not true*.
2. Use the assumptions in (1) to reach the conclusion that q_2 is *true*. In particular, work forward from p and *not* q_1 and backward from q_2.

A proof by elimination is illustrated in the following example.

Proposition 1.10 *If x is a real number for which $x^2 + x - 6 \geq 0$, then* $|x| \geq 2$.

Discussion of Proof. Working backward from the conclusion, you can show that $|x| \geq 2$ by showing that

$$q_1 : \text{either } x \geq 2 \text{ or } x \leq -2.$$

Recognizing the key words *either/or* in the *backward* process, you should proceed with a proof by elimination. Accordingly, in addition to the hypothesis, you can assume that the statement $x \geq 2$ is *not true*, that is, that

$$p_2 : x < 2.$$

You must now show that

$$q_2 : x \leq -2.$$

Working forward from p_2, you have that

$$p_3 : x - 2 < 0,$$

and from the hypothesis, it follows that

$$p : x^2 + x - 6 = (x - 2)(x + 3) \geq 0.$$

On dividing both sides of p by the negative number $x - 2$ (see p_3), you have that

$$p_4 : x + 3 \leq 0,$$

from which it follows that

$$p_5 : x \leq -3,$$

and so q_2 is *true*, thus completing the proof.

Proof. To show that $|x| \geq 2$, it is shown that either $x \geq 2$ or $x \leq -2$, so assume that $x < 2$. It now follows from the hypothesis that

$$x^2 + x - 6 = (x - 2)(x + 3) \geq 0.$$

Because $x < 2$, you have that $x + 3 \leq 0$, that is, that $x \leq -3$ and hence $x \leq -2$, as desired. QED

When trying to prove that

p implies (q_1 *or* q_2),

you can equally well use a proof by elimination in which you assume that p and *not* q_2 are *true* and then show that q_1 is *true*. Try this approach for proving Proposition 1.10.

1.6.12 Uniqueness Methods

The final group of proof techniques presented here are the **uniqueness methods**, which are used when a statement has the following form:

> There is a *unique object* with a *certain property* such that *something happens*.

When the key word *unique* (or equivalent words like *one and only one*) arise in the *forward* process, you use this knowledge as follows. If you encounter two objects – say, X and Y – both of which satisfy the *certain property* and the *something happens*, then, by the uniqueness property, you can conclude that $X = Y$, that is, that X and Y are the same.

In contrast, when you encounter the key word *unique* in the *backward* process, you can use one of two proof techniques to establish that there is a unique such object. These methods are described in what follows.

The Direct Uniqueness Method

To prove that

> there is a *unique object* with a *certain property* such that *something happens*,

use the **direct uniqueness method**, as follows:

1. First establish that *there is* an *object* – say, X – with the *certain property* and for which the *something happens*. (You can use the construction method described in Section 1.6.3 to do so.)

2. Assume that Y is *also* an *object* with the *certain property* and for which the *something happens*.

3. Work forward from the properties that X and Y satisfy to show that $X = Y$, that is, that X and Y are the same. You can also work backward from the statement that $X = Y$; however, the key question depends on the specific objects X and Y. For example, showing that two *real numbers* are equal is different from showing that two *functions* are equal, which is different from showing that two *sets* are equal.

On completing these three steps, you can conclude that the object X is unique because you have shown that any other object Y with the same properties as X is, in fact, the same as X. The direct uniqueness method is demonstrated in the following example.

Proposition 1.11 *If*

$$\begin{aligned} C^1 &= \{\textit{points}\ (x, y) : x^2 + y^2 = 4\}\ \textit{and} \\ C^2 &= \{\textit{points}\ (x, y) : (x-3)^2 + y^2 = 1\}, \end{aligned}$$

then the circles C^1 and C^2 intersect in exactly one point.

Discussion of Proof. The appearance of the key word *exactly one* in the conclusion suggests using a uniqueness method. According to the direct uniqueness method, you must first construct a point (x, y) in the intersection of both circles. In this case, it is easy to see that (2, 0) is such a point.

Having constructed one such point, you should now assume that

$p_1 : (\bar{x}, \bar{y})$ is also a point in the intersection of C^1 and C^2,

from which it follows that

$$p_2 : \bar{x}^2 + \bar{y}^2 = 4$$

and

$$p_3 : (\bar{x} - 3)^2 + \bar{y}^2 = \bar{x}^2 - 6\bar{x} + 9 + \bar{y}^2 = 1.$$

Working forward from p_2 and p_3, you must show that

$$q_1 : (\bar{x}, \bar{y}) = (2, 0).$$

In this case, that means you must show that two pairs of real numbers are equal, in other words, that

$q_2 : \bar{x} = 2$ and $\bar{y} = 0$.

To establish q_2, replace the expression $\bar{x}^2 + \bar{y}^2$ in p_3 with the value of 4 from p_2 to obtain

$$p_4 : 13 - 6\bar{x} = 1,$$

from which it follows that

$$p_5 : \bar{x} = 2.$$

By replacing $\bar{x}$ in p_2 with its known value of 2, you then have that

$$p_6 : \bar{y} = 0.$$

Thus, from p_5 and p_6, you know that q_2 is *true* and so you have established the uniqueness of the point $(2, 0)$ in the intersection of the two circles C^1

and C^2. This is because, by the direct uniqueness method, you have shown that if $(\bar{x}, \bar{y})$ is *any* point in the intersection, then $(\bar{x}, \bar{y}) = (2, 0)$.

Proof. You can easily verify that the point $(2, 0)$ is in the intersection of the two circles C^1 and C^2. For the uniqueness, assume that $(\bar{x}, \bar{y})$ is also in the intersection, so

$$\bar{x}^2 + \bar{y}^2 = 4 \qquad (1.41)$$

and

$$(\bar{x} - 3)^2 + \bar{y}^2 = \bar{x}^2 - 6\bar{x} + 9 + \bar{y}^2 = 1. \qquad (1.42)$$

Substituting (1.41) in (1.42) yields that $13 - 6\bar{x} = 1$, or equivalently, that $\bar{x} = 2$. Replacing this value of $\bar{x}$ in (1.41) yields that $\bar{y} = 0$. Thus, if $(\bar{x}, \bar{y})$ is a point in the intersection of the two circles C^1 and C^2, then $(\bar{x}, \bar{y}) = (2, 0)$, and so the point $(2, 0)$ is unique.

QED

The Indirect Uniqueness Method

To prove that

> there is a *unique object* with a *certain property* such that *something happens*,

you can also use the **indirect uniqueness method**, as follows:

1. First establish that *there is* an *object* – say, X – with the *certain property* and for which the *something happens*. (You can use the construction method described in Section 1.6.3 to do so.)

2. Assume that Y is a *different object* with the *certain property* and for which the *something happens*.

3. You rule out the existence of the object Y by working forward from the properties that X and Y satisfy, especially the fact that $X \neq Y$, to reach a contradiction.

The indirect uniqueness method is demonstrated in the following example.

Proposition 1.12 *There is a unique real number $x > 0$ such that $x^2 = 2$.*

Discussion of Proof. According to the indirect uniqueness method, you must first construct a real number $x > 0$ such that $x^2 = 2$. Doing so involves quite a bit of work and is accomplished in Theorem 6.2 in Section 6.2.2. As a result, however, you now have a real number x such that

$p_1 : x > 0$ and $x^2 = 2$.

Having constructed one such real number, to use the indirect uniqueness method, you should now assume that y is also a real number for which

$$p_2 : y > 0 \text{ and } y^2 = 2,$$

and *also* that

$$p_3 : y \neq x.$$

Now work forward from p_1, p_2, and p_3 to reach a contradiction which, in this case, is that $x < 0$ (contradicting p_1). You can obtain this contradiction by combining p_1 and p_2 to claim that

$$p_4 : x^2 = y^2,$$

from which it follows that

$$p_5 : x^2 - y^2 = (x - y)(x + y) = 0.$$

From p_3, you know that $y \neq x$ and so $x - y \neq 0$. You can therefore divide both sides of p_5 by $x - y$ to obtain that

$$p_6 : x + y = 0,$$

or equivalently, that

$$p_7 : x = -y.$$

But from p_2, you know that $y > 0$ so $-y < 0$, hence, p_7 provides the contradiction that $x = -y < 0$.

Proof. The fact that there is a real number $x > 0$ such that $x^2 = 2$ is established in Theorem 6.2 in Section 6.2.2. To prove the uniqueness of x, suppose that $y > 0$ is also a real number for which $y^2 = 2$ and further that $y \neq x$. It then follows that $x^2 = y^2$, so

$$x^2 - y^2 = (x - y)(x + y) = 0. \tag{1.43}$$

Because $y \neq x$, you can divide (1.43) through by $x - y$ to conclude that $x + y = 0$, that is, that $x = -y < 0$. But this contradicts the fact that $x > 0$, hence showing that x is the unique positive real number for which $x^2 = 2$, thus completing the proof.

QED

In this section, you have learned various techniques for proving that an implication of the the form p *implies* q is *true*. You have also learned to select a proof technique on the basis of certain key words that appear in the hypothesis and/or conclusion of the implication. When no key words are present, you should probably try the forward-backward method. A summary of all the proof techniques is presented in Appendix A. Many additional details of these techniques are presented in the book *How to Read and Do Proofs*, Second Edition, Daniel Solow, John Wiley and Sons, Inc., NY, 1990. The final section of this chapter provides a complete summary of unification, generalization, and abstraction. Then, in Chapter 2, you will learn how to work with all of these ideas.

Chapter Summary

In this chapter, you have seen the following uses of mathematics:

1. To solve a specific problem preferably in closed form, or, alternatively, by developing an efficient numerical method.

2. To perform a unification that combines similar problems – each of which is called a *special case* – into a common problem whose single solution procedure is then used to solve each of the special cases as well as new problems in the class.

3. To perform a generalization that creates a more general problem from several individual special cases. The resulting general problem contains not only all of the special cases, but also some significantly different problems that necessitate developing a new solution procedure. You therefore need to verify that the general solution procedure can be used to solve each of the special cases. Keep in mind that this general solution procedure will not be as efficient as solving the special cases directly because solving the general problem involves using more data than does solving the special cases directly.

4. To create an abstract system by replacing specific items in one or more problems with general objects and by using operators to combine those objects in appropriate ways. The result is that the original problem is greatly expanded to contain not only the special cases, but a whole host of related problems that are obtained from the abstract system by appropriately interpreting the meaning of the objects.

5. To create a proof for establishing that a solution procedure and other mathematical statements are, in fact, correct.

You have also seen the role of identifying similarities and differences in unification, generalization, and abstraction, all of which are designed to allow you to understand many individual problems and concepts by having to study only one general class. You can solve many problems simultaneously (even some you have not yet encountered) by developing a single solution procedure for the generalized class of problems. You should make sure that the general problem and its solution apply to all of the special cases. This advantage of working with a single class far outweighs the disadvantages, which include the fact that as you work with more general and abstract concepts, you lose sight of the specific items that give rise to the abstraction in the first place.

Now it is time to see how to use mathematics creatively with the ideas you have learned in this chapter. That is the goal of Chapter 2.

Exercises

Exercise 1.1 Identify all of the inputs and outputs in the following problems. State your answer using appropriate symbols (that you may have to introduce). You need not solve the problems.

(a) Find the area of a given right triangle using the lengths of its legs.

(b) Find all n roots (that is, solutions) of a given polynomial equation of degree n.

(c) Florida State Airways will start providing service between the following cities: Miami (M), Daytona (D), Orlando (O), Tampa (T), and Jacksonville (J). The nonstop fares in dollars between certain pairs of these cities are given in the following table:

From/To	M	D	O	T	J
M	–	89	120	–	–
D	89	–	55	75	–
O	120	55	–	59	69
T	–	75	59	–	–
J	–	–	69	–	–

Management wants to use these nonstop fares as the basis for determining the fares between *all* pairs of cities. For example, they want to determine the fare from Miami to Tampa.

Exercise 1.2 Identify all of the inputs and outputs in the following problems. State your answer using appropriate symbols (that you may have to introduce). You need not solve the problems.

(a) Find the distance between the two closest of n given points in the plane.

(b) Find the area under the graph of a given function f between the numbers a and b on the x-axis and above that axis.

(c) The California Computer Company assembles microcomputers at two production facilities: one in San Francisco and one in Los Angeles. The computers from the San Francisco plant are shipped to retail stores in Lake Tahoe and Barstow. The computers assembled at the Los Angeles plant are shipped to retail stores in Barstow and La Jolla. The following shipping costs per computer (in dollars) are known:

From/To	Lake Tahoe	Barstow	La Jolla
San Francisco	5	3	–
Los Angeles	–	4	2

There are 1500 computers available at the plant in San Francisco and 2100 at the plant in Los Angeles. The retail store in Lake Tahoe has ordered 1000 computers, the one in Barstow wants 1800 computers, and the one in La Jolla needs 800. The company must determine how many computers to send from each plant to each retail store to minimize the total shipping cost while not exceeding the availability at any plant and also meeting the demand at each retail store.

Exercise 1.3 If possible, write a closed-form solution for each of the following problems by expressing the answer in terms of the given data. If unable to do so, indicate that a numerical method is needed to solve the problem.

(a) Find a point where the function $2x^2 - 3x + 1$ crosses the x-axis.

(b) Given the selling price of a stock on each day of the past year, identify the least amount you could have paid for that stock during that year.

(c) Suppose that $a < b$ are two positive real numbers. Find the area under the graph of the function $f(x) = 2x$ and above the x-axis between a and b. (Hint: Draw a picture.)

Exercise 1.4 If possible, write a closed-form solution for each of the following problems by expressing the answer in terms of the given data. If unable to do so, indicate that a numerical method is needed to solve the problem.

(a) Find a number x where the function $f(x) = 2^{-x} - x$ crosses the x-axis.

(b) Find the minimum value of the function $f(x) = ax^2 + bx + c$, where a, b, and c are given real numbers with $a > 0$.

(c) Find the area under the graph of the function

$$f(x) = \frac{1}{\sqrt{2\pi}} e^{-x^2} \text{ (where } e \approx 2.718)$$

above the x-axis, between 0 and some given positive number z.

Exercise 1.5 For each of the following problems, perform the given numerical-method solution procedure as indicated.

(a) Sort the list of numbers 10, 13, 8, 9, 12 in increasing order using the following approach. Interchange the smallest number with the number in the first position. Then interchange the next smallest number with the number in the second position. Continue in this manner until the list is sorted. (Indicate the new list each time.)

(b) Let $f(x) = 2^{-x} - x$. The *tangent line* to f at a point x is the line that just touches the graph of the function at the point $(x, f(x))$. By using a value of $t = 0.01$, compute an approximate value for the slope of the tangent line to f at the point $x = 1$ with the formula

$$\text{Slope} \approx \frac{f(x+t) - f(x)}{t}.$$

Exercise 1.6 For each of the following problems, perform the given numerical-method solution procedure as indicated.

(a) Find an approximate value for the area under the graph of the function

$$f(x) = \frac{1}{\sqrt{2\pi}} e^{-x^2} \text{ (where } e \approx 2.718)$$

above the x-axis, between 0 and 1, as follows.

(i) Compute the area of the rectangle whose base is on the x-axis between the points 0 and 1, and whose height is $f(0)$.

(ii) Compute the area of the rectangle whose base is on the x-axis between the points 0 and 1, and whose height is $f(1)$.

(iii) Compute the average of the two areas in (i) and (ii).

(b) Use the following approach to find values for the real numbers x and y that solve the problem

$$\begin{array}{lrcrl} \text{Maximize} & 3x + 2y & & & \\ \text{Subject to} & x + 2y & \leq & 10 & (1) \\ & x & \geq & 0 & (2) \\ & y & \geq & 0 & (3) \end{array}$$

(i) For each *pair* of numbered inequalities, replace the inequality signs with equality signs and solve the resulting two linear equations to find the values for x and y.

(ii) For each set of values for x and y found in part (i), compute the value of $3x + 2y$.

(iii) Use your results in part (ii) to identify the values for x and y that provide the largest value of $3x + 2y$.

Exercise 1.7 Suppose that $x_1, \ldots, x_n$ are positive real numbers and that you want to compute the value of

$$x = 2[\log(x_1) + \cdots + \log(x_n)].$$

Show how to do this more efficiently by taking the logarithm of only *one* number.

Exercise 1.8 Suppose that $x_1, \ldots, x_n, x_{n+1}$ are $n+1$ real numbers and that you have already computed

$$\bar{x} = \frac{\sum_{i=1}^{n} x_i}{n}.$$

Show how to use this value of $\bar{x}$ together with x_{n+1} to compute

$$\frac{\sum_{i=1}^{n+1} x_i}{n+1}.$$

Exercise 1.9 Use unification to create a single statement that, by appropriate substitution, includes each of the following as a special case:

(i) $\frac{1}{3} - \frac{1}{4} < \frac{1}{9}$.

(ii) $\frac{1}{5} - \frac{1}{6} < \frac{1}{25}$.

Exercise 1.10 Use unification to create a single differential equation that, by appropriate substitution, includes each of the following as a special case:

(i) The following differential equation describes the displacement, $u(t)$, at time t of an object of mass m on a spring whose damping constant is c and whose spring constant is k, when the mass is given an initial impressed force of $f(t)$:

$$mu''(t) + cu'(t) + ku(t) = f(t).$$

(ii) The following differential equation describes the charge, $Q(t)$, at time t of an electrical circuit having resistance R, inductance L, and elastance $1/C$, when given an initial impressed voltage of $E(t)$:

$$LQ''(t) + RQ'(t) + \frac{1}{C}Q(t) = E(t).$$

Exercise 1.11 Use unification to create a single statement that, by appropriate substitution, includes each of the following as a special case:

(i) $\sqrt{5(7)} \le \dfrac{5+7}{2}$.

(ii) $[6(8)(10)]^{1/3} \le \dfrac{6+8+10}{3}$.

Exercise 1.12 Use unification to create a single expression that, by appropriate substitution, includes each of the following as a special case (in which $x_1, \ldots, x_n$ are real numbers):

(i) $|x_1| + \cdots + |x_n|$.

(ii) $\sqrt{x_1^2 + \cdots + x_n^2}$.

(Hint: You may have to do some rewriting to identify a common underlying expression.)

Exercise 1.13 For the following list of numbers, (a) identify as many similarities and differences as you can and (b) on the basis of the differences you identify in part (a), separate these numbers into two groups so that the numbers in each group have similar properties:

$$\sqrt{2},\ \frac{1}{3},\ \pi, 0.090909\ldots.$$

Exercise 1.14 For the following list of equations, (a) identify as many similarities and differences in their *solutions* as you can and (b) on the basis of the differences you identify in part (a), separate the equations into two groups so that the solutions to the equations in each group have similar properties:

(i) $3x = 171$.

(ii) $9x = 109$.

(iii) $x^2 = 4$.

(iv) $x^2 - 3x + 1 = 0$.

(v) $7x = 273$.

Exercise 1.15 In the following list of sets, (a) identify as many similarities and differences as you can and (b) on the basis of the differences you identify in part (a), separate these sets into two groups so that the sets in each group have similar properties:

(i) $\{\text{real numbers } x : x^2 - x - 2 \le 0\}$.

(ii) {real numbers $x : x^2 - x - 2 \geq 0$}.

(iii) $\{\frac{1}{2}, \frac{1}{3}, \frac{1}{4}, \ldots\}$.

(iv) {real numbers $x : 3 \leq x \leq 5$}.

(v) {real numbers x : there is a real number y with $x^2 + y^2 \leq 9$}.

Exercise 1.16 Suppose that $\mathbf{x} = (x_1, \ldots, x_n)$ is an n-vector in which each x_i has a value of either 0 or 1. In the following list of mathematical operations, (a) identify as many similarities and differences as you can and (b) on the basis of the differences you identify in part (a), separate these operations into two groups so that the operations in each group have similar properties.

(i) $|\mathbf{x}|$ = the number of ones appearing in $\mathbf{x}$.

(ii) $\mathbf{x}' = (1 - x_1, \ldots, 1 - x_n)$.

(iii) left($\mathbf{x}$) = $(x_2, x_3, \ldots, x_n, x_1)$.

(iv) integer($\mathbf{x}$) = $\sum_{i=1}^{n} x_i \, 2^{n-i}$.

Exercise 1.17 Consider the following interval of real numbers:

{real numbers $x : -1 \leq x \leq 2$}.

(a) By introducing appropriate notation, generalize this set to allow for an interval between *any* two real numbers, instead of the specific values of -1 and 2.

(b) Generalize the set of real numbers in part (a) to a set of points (x, y) in the plane.

(c) Generalize the result in part (b) to a set in n dimensions.

Exercise 1.18 The real numbers p and q are *probabilities* for two events E and F provided that the following two conditions hold:

(i) $p \geq 0, q \geq 0$.

(ii) $p + q = 1$.

(a) By introducing appropriate notation, generalize this concept of the probabilities for *two* events so that it applies to n events (where n is any positive integer).

(b) Extend the generalization in part (a) so that the concept of probabilities applies to the infinite number of events $E_1, E_2, E_3, \ldots$.

(c) What difficulties arise in trying to generalize this concept of probabilities to an *arbitrary* collection of events, say, E_t, where t can be any real number?

Exercise 1.19 Perform a *sequential* generalization of the concept of a triangle in the plane. Generalize as far as you can but be sure that each subsequent generalization contains the previous one as a special case.

Exercise 1.20 Suppose that f is a real-valued function of one variable. Perform a *sequential* generalization of the concept of approximating the graph of f by a *straight line*. Generalize as far as you can but be sure that each subsequent generalization contains the previous one as a special case.

Exercise 1.21 Suppose that x and y are real numbers for which you want to compute

$$x^3 + \sqrt{y}.$$

What syntax errors arise when generalizing this computation to n-vectors $\mathbf{x}$ and $\mathbf{y}$?

Exercise 1.22 Let $A = \{x_1, \ldots, x_n\}$ (where each x_i is a real number) and suppose you want to compute

$$\text{center}(A) = \frac{\sum_{i=1}^{n} x_i}{n}.$$

Identify all syntax and other possible errors that might arise when generalizing the foregoing computation to an *arbitrary* set A of real numbers.

Exercise 1.23 By identifying similarities and working with objects instead of specific items, create an abstract system that includes each of the following as a special case:

(i) $|r|$, which is the absolute value of the real number r.

(ii) $\neg p$, which is the negation of the statement p.

(iii) $\mathbf{x}'$, which is the n-vector obtained from the n-vector $\mathbf{x}$ by replacing each 0 with a 1 and each 1 with a 0 (here, $\mathbf{x}$ is an n-vector in which each component is either 0 or 1).

Exercise 1.24 By identifying similarities and working with objects instead of specific items, create an abstract system that includes each of the following as a special case:

(i) The intersection of two intervals of real numbers is another interval.

(ii) The intersection of two polygons in the plane is another polygon.

(iii) The intersection of two sets in n dimensions is another set in n dimensions.

Exercise 1.25 Suppose that p and q are statements and consider the implication p *implies* q. Write truth tables for each of the following statements:

(a) The *converse* statement: q *implies* p.

(b) The *inverse* statement: (*not* p) *implies* (*not* q). How is this truth table related to the one in part (a)?

Exercise 1.26 Suppose that p and q are statements.

(a) Write the truth table for the statement (*not* p) *or* q. How does this truth table compare with the one for p *implies* q?

(b) Suppose you decide to use an either/or method (see Section 1.6.11) to prove that (*not* p) *or* q is *true*. Explain why this approach is the same as using either the forward-backward method or the contrapositive method to show that p *implies* q. (Hint: When you apply the either/or method to (*not* p) *or* q, what statements do you assume are *true* and what statements should you try to show are *true*?)

Exercise 1.27 Write two different key questions for the conclusions in each of the following implications. Then, for each key question, provide two answers in the form of new statements to be proved in the backward process.

(a) If a and b are nonnegative real numbers, then $\sqrt{ab} \leq (a+b)/2$.

(b) If

$$\begin{aligned} C^1 &= \{\text{points } (x,y) : x^2 + y^2 = 4\} \text{ and} \\ C^2 &= \{\text{points } (x,y) : (x-3)^2 + y^2 = 1\}, \end{aligned}$$

then $C^1 \cap C^2 \neq \emptyset$ (where $\emptyset$ is the empty set).

Exercise 1.28 Write two different key questions for the conclusions in each of the following implications. Then, for each key question, provide two answers in the form of new statements to be proved in the backward process.

(a) If ABC is an isosceles right triangle with legs of length a and a hypotenuse of length c, then the area of the triangle ABC is $c^2/4$.

(b) If $a, b, c, d,$ and e are real numbers for which $a(c-e) < 0$, then the line $y = dx + e$ crosses the parabola $y = ax^2 + bx + c$ at some point.

Exercise 1.29 Explain how you would use the construction method to prove each of the following statements. What object must you produce and what properties must you show that object has? (You need not actually produce the object.)

(a) There is an angle t between 0 and π for which $\sin(t) = \cos(t)$.

(b) There is a real number x such that $x^2 = y$ (where y is a given positive real number).

Exercise 1.30 Explain how you would use the construction method to prove each of the following statements. What object must you produce and what properties must you show that object has? (You need not actually produce the object.)

(a) Some point in the set $C = \{\text{real numbers } x : x^2 - 4x + 1 \leq 0\}$ is also in the interval $[3, 4] = \{\text{real numbers } x : 3 \leq x \leq 4\}$.

(b) There is a real-valued function f of one variable for which $f(0) = 2$ that satisfies the following differential equation:

$$f'(x) + 2f(x) = e^{-x}.$$

Exercise 1.31 Use the construction method to prove that if $a, b, c, d,$ and e are real numbers for which $a(c-e) < 0$, then there is a point on the graph of the line $y = dx + e$ that is also on the graph of the parabola $y = ax^2 + bx + c$.

Exercise 1.32 Use Definition 1.6 in Section 1.6.9 for a rational number to prove that if r_1 and r_2 are rational numbers for which $r_2 \neq 0$, then r_1/r_2 is also a rational number. Indicate clearly how and when you use the construction method.

Exercise 1.33 Prove each of the following statements by induction:

(a) For each integer $n \geq 1$, $\sum_{k=1}^{n} k = \dfrac{n(n+1)}{2}$.

(b) For each integer $n \geq 1$, $n! \leq n^n$.

Exercise 1.34 Prove each of the following statements by induction:

(a) A set S with $n \geq 0$ elements has 2^n subsets.

(b) For each integer $n \geq 1$, the derivative of the function x^n is nx^{n-1}.

Exercise 1.35 Explain how you would apply the choose method to prove each of the following statements. What should you assume? What should you try to conclude? (You need not prove the statements.)

(a) For every angle t, $\sin^2(t) + \cos^2(t) = 1$.

(b) For all real numbers t, x, and y with $0 \leq t \leq 1$, $f(tx + (1-t)y) \leq tf(x) + (1-t)f(y)$ (where f is a real-valued function of one variable).

Exercise 1.36 Explain how you would apply the choose method to prove each of the following statements. What should you assume? What should you try to conclude? (You need not prove the statements.)

(a) For every real number y, there is a real number x such that $f(x) = y$ (where f is a real-valued function of one variable).

(b) For every real number $\varepsilon > 0$, there is a real number $\delta > 0$ such that for all real numbers x and y with $|x - y| < \delta$, $|f(x) - f(y)| < \varepsilon$ (where f is a real-valued function of one variable).

Exercise 1.37 Prove that for all sets A and B, $A \cap B \subseteq A$. Indicate clearly when and how you use the choose method.

Exercise 1.38 Prove that for all real numbers m and b, the function $f(x) = mx + b$ satisfies the property that for all real numbers t, x, and y with $0 \leq t \leq 1$, $f(tx + (1-t)y) \leq tf(x) + (1-t)f(y)$. Indicate clearly when and how you use the choose method.

Exercise 1.39 Explain how you would apply the specialization method to each of the following statements using the given *specific* objects. What do you need to verify about those specific objects before you apply specialization and what is the resulting new statement in the forward process?

(a) Given statement: For every angle t, $\sin^2(t) + \cos^2(t) = 1$.
Specific object: Angle A of the triangle ABC.

(b) Given statement: For all real numbers t, x, and y with $0 \leq t \leq 1$, $f(tx + (1-t)y) \leq tf(x) + (1-t)f(y)$ (where f is a real-valued function of one variable).
Specific object: The real numbers λ, x, and z.

Exercise 1.40 Explain how you would apply the specialization method to each of the following statements using the given *specific* objects. What do you need to verify about those specific objects before you apply specialization and what is the resulting new statement in the forward process?

(a) Given statement: For every real number y, there is a real number x such that $f(x) = y$ (where f is a real-valued function of one variable). Specific object: The real number 3.

(b) Given statement: For all real numbers $\varepsilon > 0$, there is a real number $\delta > 0$ such that for all real numbers x and y with $|x - y| < \delta$, $|f(x) - f(y)| < \varepsilon$ (where f is a real-valued function of one variable). Specific object: The real number $\varepsilon/2$.

Exercise 1.41 For two real-valued functions f and g, $f \leq g$ means that for all real numbers x, $f(x) \leq g(x)$. Prove that if f, g, and h are real-valued functions of one variable for which $f \leq g$ and $g \leq h$, then $f \leq h$. Indicate clearly when and how you use the specialization method. Also, when applying the *choose* method, use symbols other than x for the *object* you choose.

Exercise 1.42 A set S of real numbers is **convex** if and only if for all elements $x, y \in S$ and for all real numbers t with $0 \leq t \leq 1$, $tx + (1 - t)y \in S$. Prove that if A and B are convex sets, then $A \cap B$ is a convex set. Indicate clearly when and how you use the specialization method. Also, when applying the *choose* method, use symbols other than x, y and t for the *objects* you choose.

Exercise 1.43 Write the negation of each of the following statements in such a way that the words *no* and *not* do not appear explicitly:

(a) $-1 \leq x \leq 1$ (where x is a real number).

(b) $x \in A$ or $x \in B$ (where A and B are sets).

(c) For all elements $x, y \in A$ with $x \neq y$, $f(x) \neq f(y)$ (where A is a set of real numbers and f is a real-valued function of one variable).

(d) There is no integer k with $1 < k < n$ such that k divides n (where n is a positive integer greater than 2).

Exercise 1.44 Write the negation of each of the following statements in such a way that the words *no* and *not* do not appear explicitly:

(a) p *implies* q (where p and q are statements). (Hint: Look at Table 1.4.)

(b) For every real number y, there is a real number x such that $f(x) = y$ (where f is a real-valued function of one variable).

(c) There is an element e in a set G such that for all elements $a \in G$, $a \odot e = e \odot a = a$ (where $\odot$ is a binary operator on G).

(d) For every real number $\varepsilon > 0$, there is a real number $\delta > 0$ such that for all real numbers x and y with $|x - y| < \delta$, $|f(x) - f(y)| < \varepsilon$ (where f is a real-valued function of one variable).

Exercise 1.45 Suppose that $x_1, \ldots, x_n$ are nonnegative real numbers. Use the contrapositive method to prove that if $x_1 + \cdots + x_n = 0$, then for each $i = 1, \ldots, n$, $x_i = 0$. Indicate clearly what statement you are assuming and what statement you are trying to prove.

Exercise 1.46 Suppose that m and n are integers with $m \neq 0$. Use the contrapositive method to prove that if m does not divide n, then $mx^2 + nx + (n - m)$ has no positive integer root. Indicate clearly what statement you are assuming and what statement you are trying to prove.

Exercise 1.47 Use the contradiction method to prove that if m and n are odd integers, then there is no even integer x for which $x^2 + 2mx + 2n = 0$.

Exercise 1.48 Use the contradiction method to prove that if x and r are positive real numbers for which $x^3 = r$, then there is no real number y with $y \neq x$ such that $y^3 = r$.

Exercise 1.49 Use the direct uniqueness method to prove that if n is a positive integer, then there is a unique positive integer m such that m divides n and n divides m.

Exercise 1.50 Use the indirect uniqueness method to prove that there is a unique positive integer n such that $n^3 - n - 6 = 0$.

Exercise 1.51 Prove Proposition 1.10 in Section 1.6.11 by assuming that $x^2 + x - 6 \geq 0$ and also that $x > -2$. Then show that $x \geq 2$.

Exercise 1.52 Use a proof by cases to show that if k, m, and n are integers for which k divides m or k divides n, then k divides mn.

Chapter 2

WORKING WITH MATHEMATICAL CONCEPTS

In Chapter 1, you saw how unification, generalization, and abstraction enable you to analyze a great number of problems and concepts by having to study only one single class. To use these techniques effectively, you must learn certain mathematical skills. One of the most important ones is to translate *informal* mathematics – consisting of the images you create – to a more *formal* mathematics – consisting of written symbols. A symbolic form has the following advantages:

1. You can manipulate, rewrite, and apply the proof techniques in Section 1.6 to the symbolic form that you cannot do easily with images.

2. Symbols allow you to work with abstract concepts that you cannot visualize (such as a vector in four dimensions).

3. Symbols are easier to communicate. For example, by using the notation $S = \{\text{real numbers } x : 3 \leq x \leq 5\}$, it is easier to refer to this set by the symbol S rather than $\{\text{real numbers } x : 3 \leq x \leq 5\}$.

You must learn to translate images from the "mental" language in which you picture these concepts to the "symbolic" language of mathematics, and vice versa.

One reason this translation process is challenging is that the language of mathematics has a specialized syntax and an extremely limited vocabulary, but you do not. To draw an analogy, imagine trying to explain complicated directions to a foreigner whose vocabulary consists of only 20 words in your language. So it is with translating images to symbolic mathematics. Yet, to use mathematics successfully, this is a necessary skill you must acquire. The techniques in this chapter are designed to help you achieve this goal.

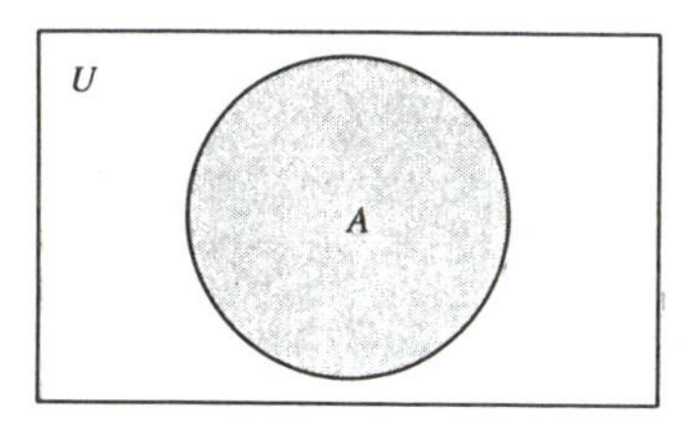

Figure 2.1: A Visual Image of a Set

2.1 Creating and Working with Images

The objective of this section is to illustrate how to create and work with visual images of mathematical concepts.

2.1.1 Creating Images

Whenever you learn a mathematical concept, you should develop an associated image, if possible. The objective of an image is to capture the essential features, thus making it easier for you to work with that concept. The process of creating images is illustrated in the following examples.

Example 2.1 – Visualizing a Set

Consider a set A consisting of selected elements from a universal set U. One of the most common images associated with the set A is the **Venn diagram** depicted in Figure 2.1. Everything inside the shaded region is in the set; everything outside the shaded region is *not* in the set.

There are no known rules for creating visual images. Moreover, you can create more than one image for the same concept, as shown in the next two examples.

Example 2.2 – Two Ways to Visualize a Function

Consider a real-valued function of one variable, say, f. One associated visual image is its *graph*, which is shown in Figure 2.2(a). However, as you learned in Section 1.5.2, through abstraction you can think of a function as a rule that associates to each object in a set A, some object in another set B. An appropriate picture might be the one in Figure 2.2(b).

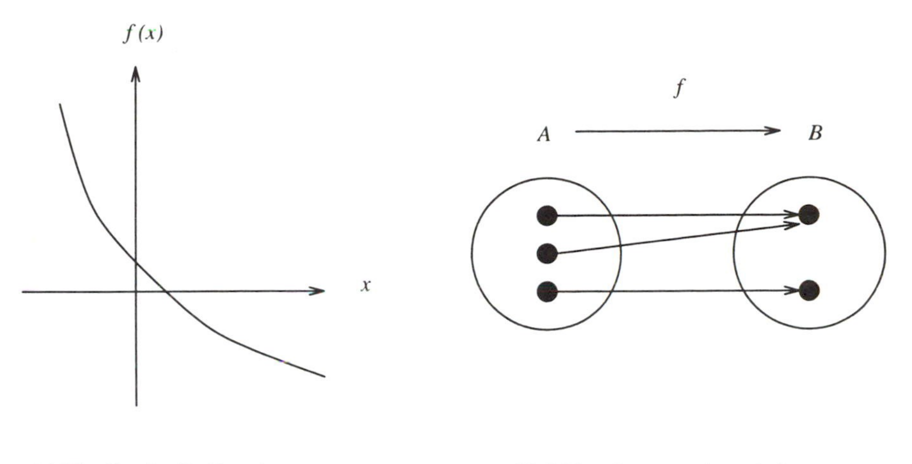

(a) The Graph of a Function

(b) A Function as an Association

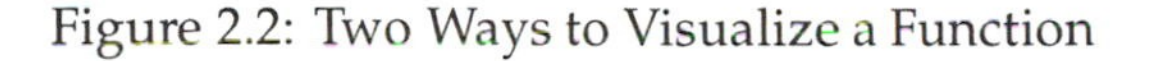

Figure 2.2: Two Ways to Visualize a Function

Example 2.3 – Two Ways to Visualize a Vector in the Plane

As you learned in Section 1.4.3, an n-vector is an ordered list of n real numbers. Consider, then, a vector $\mathbf{x} = (x_1, x_2)$ having the two components x_1 and x_2. One way to picture this vector $\mathbf{x}$ is as a *point* in the plane – whose coordinates are (x_1, x_2) – as shown in Figure 2.3(a). An alternative, but equally valid, visualization is that of an *arrow* – whose tail is at the origin and whose head is at the coordinates (x_1, x_2) – as shown in Figure 2.3(b).

The specific visualization often influences how you think about that concept. Consider the following example.

Example 2.4 – Visualizing a Graph in Two Different Ways

The graphs in Figure 2.4 are the same except for the way they are drawn. The *cycle* in Figure 2.4(a) might make you think of the traveling-salesperson problem (see Section 1.2.2) because you can start at a vertex and, by following the cycle, visit every other vertex exactly once, returning to the starting vertex. In contrast, the graph in Figure 2.4(b) conjures up a different image – one in which the vertices are split into two groups, with each edge connecting a vertex in one group to a vertex in the other group.

A mathematical concept is often presented in a symbolic form that contains quantifiers. In the event that no associated visual image is presented, you need to create one of your own, as shown in the following examples.

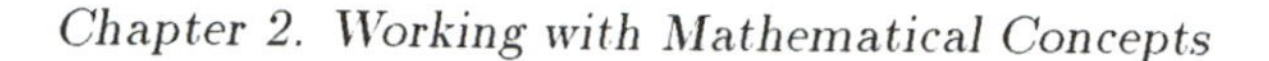

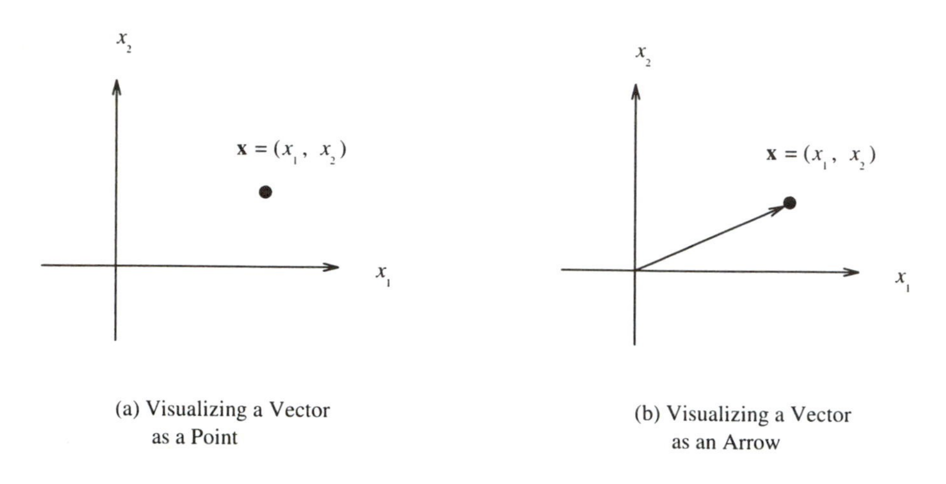

(a) Visualizing a Vector as a Point

(b) Visualizing a Vector as an Arrow

Figure 2.3: Two Ways to Visualize a Vector in the Plane

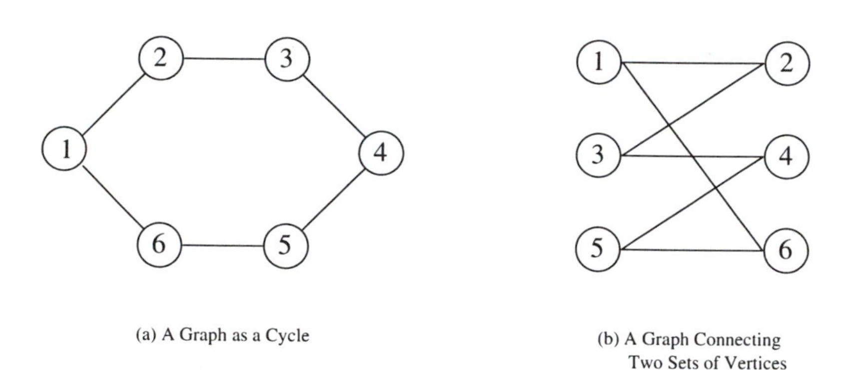

(a) A Graph as a Cycle

(b) A Graph Connecting Two Sets of Vertices

Figure 2.4: Two Ways to Visualize the Same Graph

Example 2.5 – Converting Symbolic Mathematics Involving One Quantifier to Visual Form

Consider the following definition:

> A real number x^* is a **minimum of a real-valued function** f if and only if for all real numbers x, $f(x^*) \leq f(x)$.

This definition describes a point x^* at which $f(x^*)$ is less than or equal to the value of the function at *any* other point, so, $f(x^*)$ is the *smallest* value of the function. By using the graph of f (see Example 2.2), an appropriate visual image for the foregoing definition is a point x^* on the x-axis at which the graph of f achieves its lowest value, as shown in Figure 2.5.

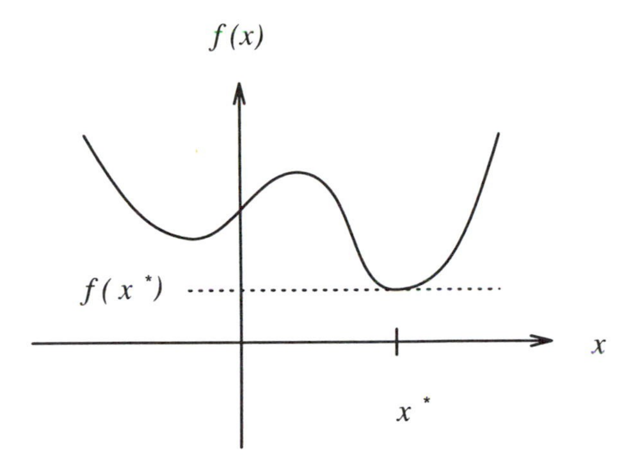

Figure 2.5: A Visual Image of the Minimum of a Function

Figure 2.6: A Visual Image of a Set of Real Numbers with No Upper Bound

Example 2.6 – Converting Symbolic Mathematics Involving Two Quantifiers to Visual Form

Consider the following definition:

> A set S of real numbers **has no uppper bound** if and only if for all real numbers x, there is an element $y \in S$ such that $y > x$.

This definition describes a set S of real numbers with a certain property, so you should visualize S as a subset of the real line, denoted by a horizontal straight line. To understand the property that S satisfies, specialize the *for-all* statement in the foregoing definition to one particular value of x, say, $x = 1$. So, for $x = 1$, you know that there is an element $y \in S$ such that $y > 1$ (in other words, there is a $y \in S$ somewhere to the right of 1). Likewise, for $x = 2$, you know that there is an element $y \in S$ for which y lies to the right of 2, and so on. This means that, as you proceed farther to the right along the real line, there are always elements of S. A visual image of a set with this property is shown in Figure 2.6.

Now that you know the process of creating a visual image from a symbolic form, you will learn about the reverse process of creating a symbolic form from a visual image.

2.1.2 Translating Images to Symbolic Form

The purpose of an image is to enable you to visualize and to think about a mathematical concept. For example, you might create your own image of how to solve a particular problem. You need to translate this image to a symbolic form that you (or a computer) can work with.

As one example of this translation process, consider the following quadratic equation:

$$ax^2 + bx + c = 0. \tag{2.1}$$

Visually, you can see that this quadratic equation contains a linear equation as a special case by "covering up the term ax^2 in (2.1)," for that would result in the linear equation:

$$bx + c = 0. \tag{2.2}$$

How, then, do you translate the visual image of "covering up the term ax^2 in (2.1)" to symbolic form?

The answer, in this case, is to set the value of a equal to 0 in (2.1), which results in the linear equation in (2.2). Thus, the visual image of "covering up the term ax^2 in (2.1)" translates to the symbolic mathematical action of "setting $a = 0$ in (2.1)."

Several other examples of this translation process are now presented. As you read these examples, keep in mind the *vocabulary* of mathematics that includes symbols, operators, relations, and special words, such as *no, not, and, or,* and the quantifiers *there is* and *for all,* amongst others.

Example 2.7 – Moving a Line Parallel to Itself

How do you translate the visusal image in Figure 2.7 of "moving the line $y = mx + b$ parallel to itself a certain amount" to symbolic form?

Compare the two lines in Figure 2.7 – what similarities and differences do you see? Because the lines are parallel to each other, their *slopes* are the same. What is different, however, is their y-intercepts. Thus, "moving the line $y = mx + b$ parallel to itself a certain amount" corresponds to the symbolic operation of changing the value of the y-intercept from b to b', resulting in the following equation:

$$y = mx + b'.$$

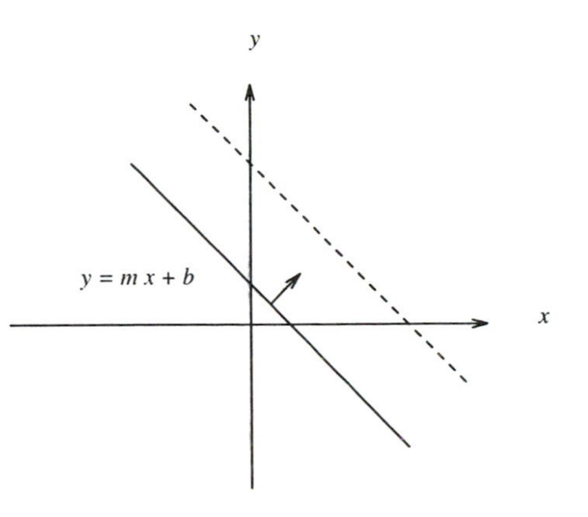

Figure 2.7: Moving a Line Parallel to Itself

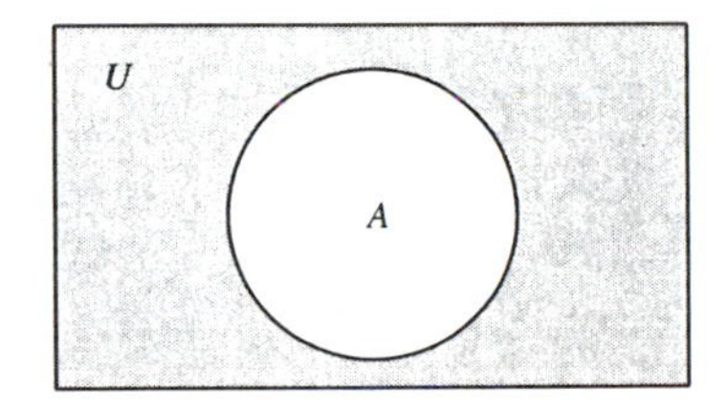

Figure 2.8: The Complement of a Set

Example 2.8 – Finding the Complement of a Set

Look again at the sets A and U in Figure 2.1. You can easily visualize "everything in U that is outside of A," as shown in Figure 2.8. How do you translate this image of the **complement of a set A in U** to symbolic form?

One way to do so is to use the word *not*, as follows:

$$\begin{aligned} \text{Complement of } A \text{ in } U &= A^c \\ &= \{\text{elements } x \text{ in } U : x \text{ is } \textit{not} \text{ in } A\} \\ &= \{x \in U : x \notin A\}. \end{aligned}$$

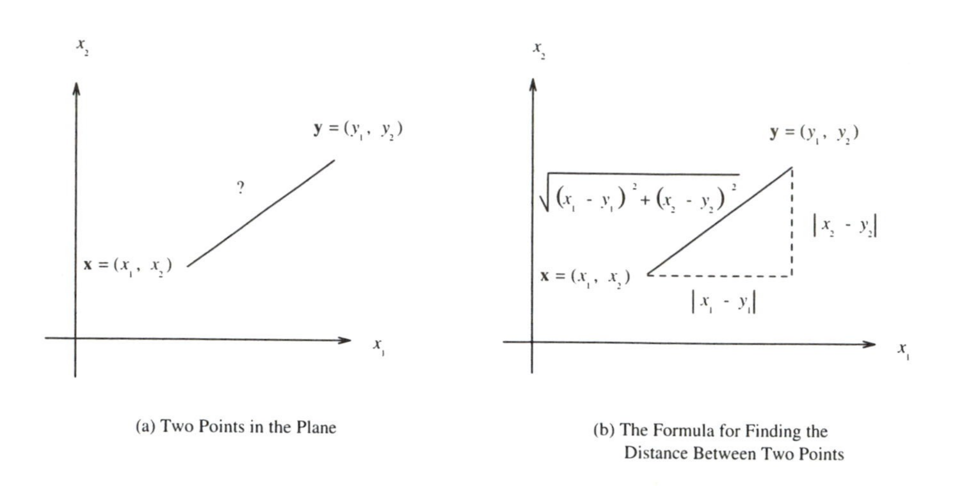

Figure 2.9: Finding the Distance Between Two Points in the Plane

Example 2.9 – Finding the Distance Between Two Points

Suppose that $\mathbf{x} = (x_1, x_2)$ and $\mathbf{y} = (y_1, y_2)$ are two points in the plane, as shown in Figure 2.9(a). How do you translate the idea of computing "the distance between **x** and **y**" to symbolic form?

From geometry, you know that this distance is the length of the hypotenuse of a right triangle whose two sides have lengths $|x_1 - y_1|$ and $|x_2 - y_2|$, respectively, as shown in Figure 2.9(b). From the Pythagorean theorem, then, the symbolic form is as follows:

$$\text{Distance from } \mathbf{x} \text{ to } \mathbf{y} = \sqrt{(x_1 - y_1)^2 + (x_2 - y_2)^2}.$$

Applying generalization, and recalling the notation from Section 1.4.3, you can define the distance from an n-vector $\mathbf{x} = (x_1, \ldots, x_n)$ to an n-vector $\mathbf{y} = (y_1, \ldots, y_n)$ by the following formula:

$$\text{Distance from } \mathbf{x} \text{ to } \mathbf{y} = ||\mathbf{x} - \mathbf{y}|| = \sqrt{(x_1 - y_1)^2 + \cdots + (x_n - y_n)^2}.$$

Example 2.10 – Reversing the Direction of a Vector

Look again at the vector **x** in Figure 2.3(b). Consider *reversing* this vector, that is, drawing the new vector $-\mathbf{x}$ that is the same as **x** but points in the opposite direction, as shown in Figure 2.10. How do you translate this visual image to symbolic mathematics?

Continued

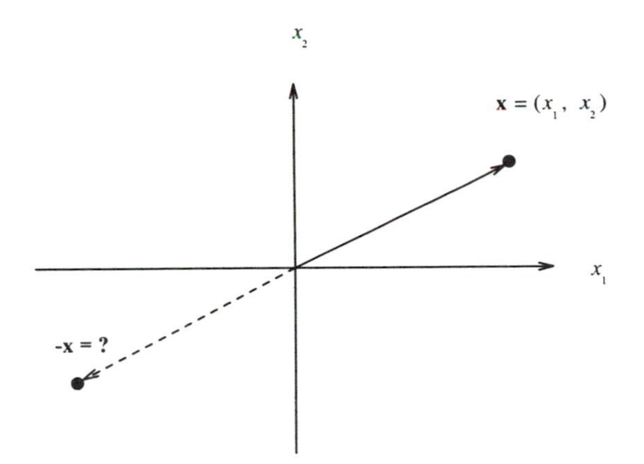

Figure 2.10: Reversing a Vector in the Plane

Compare $\mathbf{x}$ and $-\mathbf{x}$. What similarities and differences do you see? You might observe that the coordinates of the new vector $-\mathbf{x}$ are the *negatives* of those of the original vector. Thus, the visual process of "reversing a vector $\mathbf{x} = (x_1, x_2)$" translates to the mathematical operation of "multiplying each component of the vector $\mathbf{x}$ by -1," that is:

$$-\mathbf{x} = (-x_1, -x_2).$$

Applying generalization, you can reverse the n-vector $\mathbf{x} = (x_1, \ldots, x_n)$ by creating the following n-vector:

$$-\mathbf{x} = (-x_1, \ldots, -x_n).$$

Example 2.11 – Adding Two Vectors

Consider the two vectors $\mathbf{x} = (x_1, x_2)$ and $\mathbf{y} = (y_1, y_2)$ in Figure 2.11(a). You can add these two vectors to obtain the new vector $\mathbf{x} + \mathbf{y}$, as shown in Figure 2.11(b). How do you translate this visual image of "adding two vectors" to symbolic mathematics?

In this case, you probably need to work through several numerical examples to determine how the coordinates of the new vector $\mathbf{x}+\mathbf{y}$ are related to those of $\mathbf{x}$ and $\mathbf{y}$. You should come to the following conclusion:

$$\mathbf{x} + \mathbf{y} = (x_1, x_2) + (y_1, y_2) = (x_1 + y_1, x_2 + y_2).$$

Applying generalization, you can add two n-vectors $\mathbf{x} = (x_1, \ldots, x_n)$ and $\mathbf{y} = (y_1, \ldots, y_n)$ as follows:

$$\mathbf{x} + \mathbf{y} = (x_1 + y_1, \ldots, x_n + y_n).$$

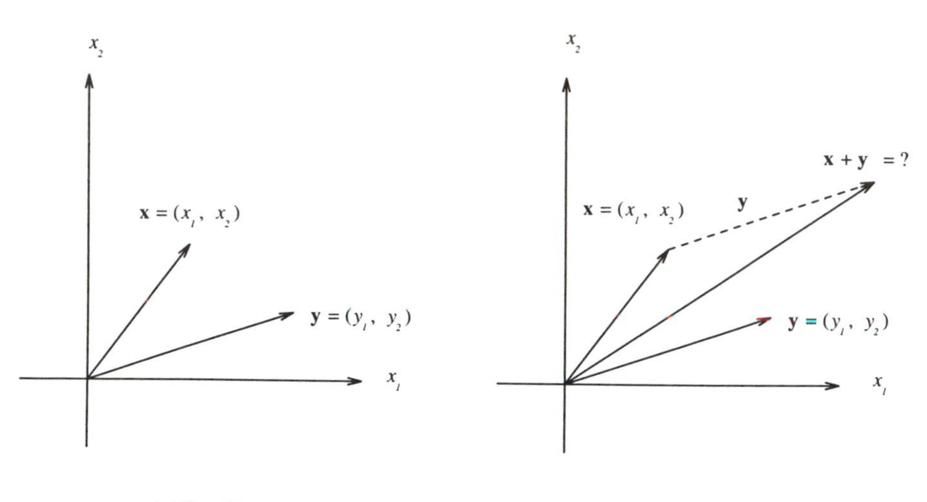

(a) Two Vectors

(b) Adding the Two Vectors in (a)

Figure 2.11: Adding Two Vectors in the Plane

Example 2.12 – A Function Crossing the x-axis

Look again at Figure 2.2(a). You will notice that this function crosses the x-axis. How do you translate this image of "a function f crosses the x-axis" to a symbolic statement?

One way to do so is to use the words *there is*, as follows:

There is a real number x such that $f(x) = 0$.

In this section, you have seen the process of creating images associated with mathematical concepts. You have also learned the reverse process of translating images to symbolic form. Doing so involves expressing your thoughts in the language of mathematics. This language has a limited syntax and a vocabulary consisting of symbols and special words, such as *no, not, and, or, there is, for all,* and several others. Like learning any language, the only way to become fluent is to practice. In the next section, you will see how this translation process is used to make mathematical definitions.

2.2 Creating Definitions

One of the most important steps toward using (and even understanding) mathematics is to learn to create and work with *definitions*. As you will see in this section, doing so involves looking for similarities and differences and translating images to symbolic form, as you learned to do in Section 2.1.

2.2.1 What is a Mathematical Definition?

In mathematics, a **definition** is a meaning given to a word (or group of words) that describes certain *desirable* objects – that is, objects satisfying certain desirable properties. The general form of a definition might appear as follows:

> An "object of a certain type" is defined to be some "name" if and only if the object has the following "desirable property."

The words in quotation marks depend on the specific definition, as shown in the following examples.

The Definition of a Parabola

Consider the following definition of a *parabola*.

> **Definition 2.1** *A real-valued function of one variable, say, f, is a* **parabola** *if and only if there are real numbers a, b, and c with $a \neq 0$ such that for all real numbers x,*
>
> $$f(x) = ax^2 + bx + c.$$

In this example,

1. The "object of a certain type" – that is, the kind of object being defined – is a real-valued function of one variable, f.

2. The "name" given to a function that satisfies the desirable property is a "parabola."

3. The "desirable property" is that there are real numbers a, b, and c with $a \neq 0$ so that for all real numbers x,

$$f(x) = ax^2 + bx + c.$$

Your image of this definition probably coincides with the one in Figure 2.12. Definition 2.1 is designed to include all the desirable objects, namely, functions that *are* parabolas (such as the one in Figure 2.12) *while at the same time, excluding all other objects* (such as the function in Figure 2.13). In this sense, a definition works both ways: (1) any desirable object must satisfy the property stated in the definition and (2) any object satisfying the stated property must be one of these desirable objects. For this reason, a definition is often written using the words "if and only if."

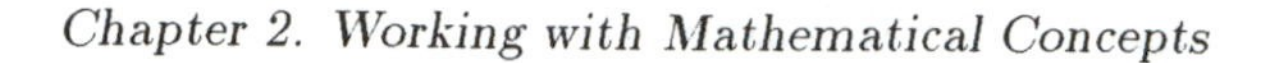

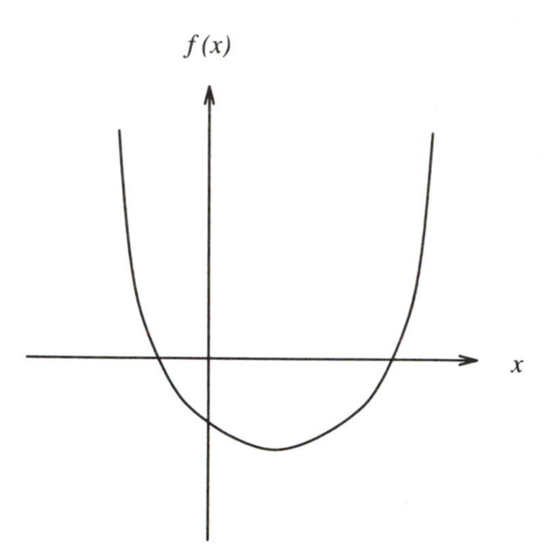

Figure 2.12: A Function that Is a Parabola

The Definition of a Closed Interval

As another example, consider the following definition of a certain kind of *interval* on the real line.

> **Definition 2.2** *Given real numbers l and u with $l \leq u$, the following set of real numbers, denoted by $[l, u]$, is the* **closed interval from l to** u*:*
>
> $$[l, u] = \{\text{real numbers } x : l \leq x \leq u\}.$$

Even though the words *if and only if* are omitted in Definition 2.2, you can still recognize the following items:

1. The "object of a certain type" – that is, the kind of object being defined – is a set of real numbers.

2. The "name" given to the set that satisfies the desirable property is the "closed interval from l to u."

3. The "desirable property" is that you can write the set in the form

$$[l, u] = \{\text{real numbers } x : l \leq x \leq u\}.$$

Undoubtedly, you have already created a visual image of this definition, such as the one in Figure 2.14. Once again, Definition 2.2 is designed to include all the desirable objects that *are* sets representing closed intervals (such as the one in Figure 2.14), *while at the same time, excluding all other objects* (such as the set in Figure 2.15). A definition captures a property

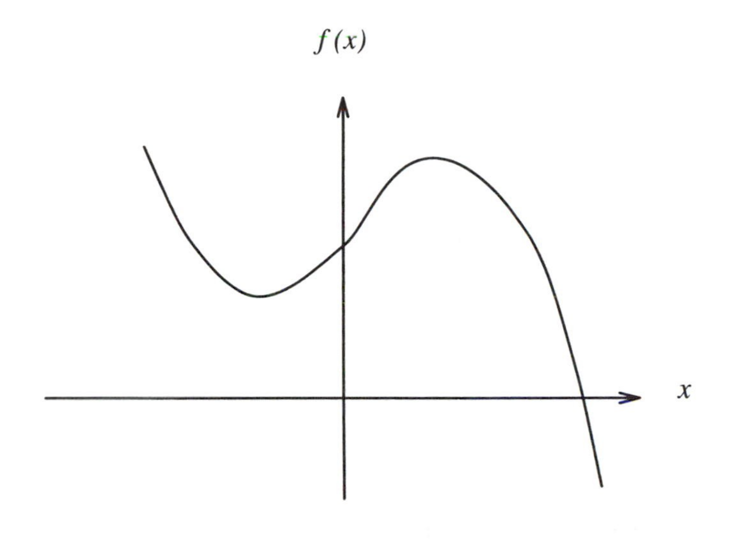

Figure 2.13: A Function that Is Not a Parabola

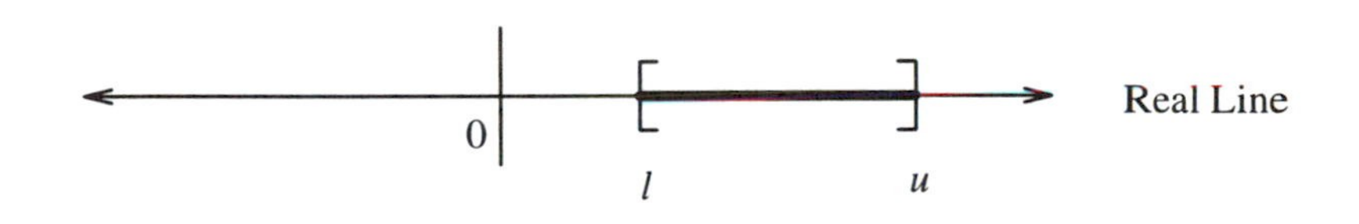

Figure 2.14: A Set that Is a Closed Interval

shared by all the desirable objects (closed intervals, in this case), while insuring that all the undesirable objects (such as the set in Figure 2.15) do not satisfy the property.

There are many different forms in which you can state a definition. Consider the following examples for the definition of a closed interval, each of which is, for all practical purposes, the same as Definition 2.2:

1. Suppose that l and u are real numbers with $l \leq u$. The set $[l, u]$ of real numbers is a *closed interval from l to u* if

   $$[l, u] = \{\text{real numbers } x : l \leq x \leq u\}.$$

   Notice that the explanation of the symbols l and u *precedes* the actual definition in which they are used. Also, the word "if" is used instead of "if and only if," but the meaning is the same because a definition must work both ways.

2. A *closed interval from l to u* is a set of real numbers, denoted by $[l, u]$, of the form

   $$[l, u] = \{\text{real numbers } x : l \leq x \leq u\},$$

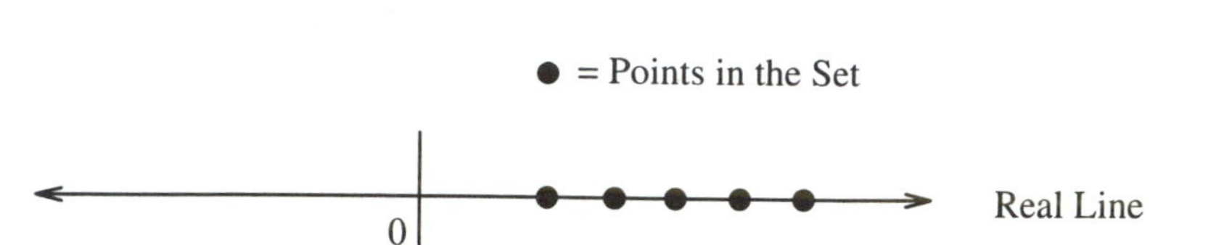

Figure 2.15: A Set that Is Not a Closed Interval

where l and u are real numbers with $l \leq u$.

Here, the explanation of l and u is given at the *end* of the definition. Also, notice that the words "if and only if" are not used at all.

The Definition of a Partition

As a final example of a definition, consider the concept of "splitting a set into two sets." For example, you can split the set $S = \{1, 2, 3, 4, 5\}$ into the two sets $S_1 = \{1, 3, 5\}$ and $S_2 = \{2, 4\}$. A formal definition of this concept follows.

> **Definition 2.3** *The sets S_1 and S_2 constitute a* **partition of a set S into two sets** *if and only*
>
> 1. *$S_1 \neq \emptyset$ and $S_2 \neq \emptyset$ (where $\emptyset$ is the empty set containing no elements).*
> 2. *$S_1 \cap S_2 = \emptyset$ (where $\cap$ is the intersection operator).*
> 3. *$S_1 \cup S_2 = S$ (where $\cup$ is the union operator).*

In this example, you can identify the following items:

1. The "object of a certain type" – that is, the kind of object being defined – is a pair of sets, S_1 and S_2.

2. The "name" given to a pair of sets that satisfies the desirable property is a "partition."

3. The "desirable property" is that the pair of sets satisfies conditions (1), (2), and (3) in Definition 2.3.

A visual image of a partition is shown in Figure 2.16 where you can see that the set S is split into the two nonempty sets S_1 and S_2. Definition 2.3 includes the sets in Figure 2.16 that *do* constitute a partition of S while excluding all other pairs of sets that are *not* a partition of S, such as the two sets shown in Figure 2.17.

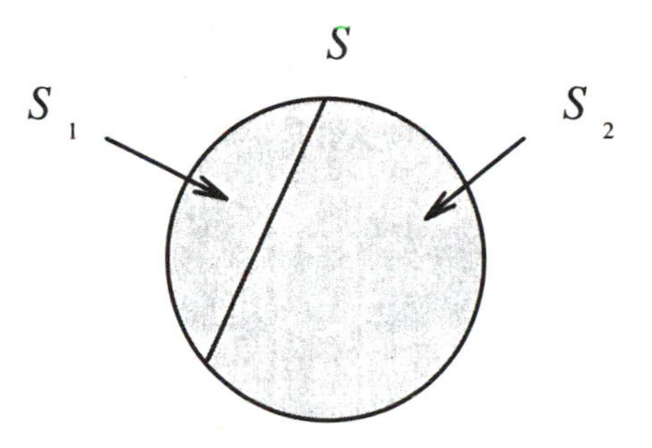

Figure 2.16: A Pair of Sets that Is a Partition of a Set S

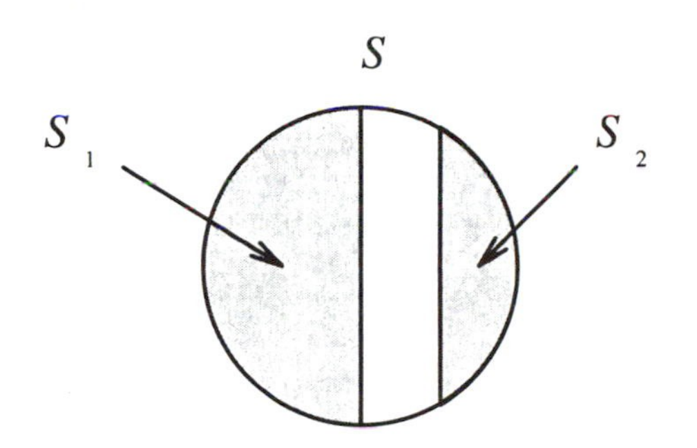

Figure 2.17: A Pair of Sets that Is Not a Partition of a Set S

A generalization of Definition 2.3 arises when you consider splitting a set into n sets, rather than just two, as shown in the following definition.

Definition 2.4 *Let* $n \geq 2$ *be an integer. The sets* $S_1, \ldots, S_n$ *constitute a* **partition of a set** S **into** n **sets** *if and only if*

1. *For each* $i = 1, \ldots, n$, $S_i \neq \emptyset$.
2. *For each* $i, j = 1, \ldots, n$ *with* $i \neq j$, $S_i \cap S_j = \emptyset$.
3. $S_1 \cup S_2 \cup \cdots \cup S_n = S$.

Compare Definition 2.3 and Definition 2.4. Observe the necessity of stating that $i \neq j$ in condition (2) of Definition 2.4.

A further generalization of Definition 2.4 arises when you consider splitting a set into an *arbitrary* number of sets, as shown in the following definition.

Definition 2.5 *Let X be a collection of subsets of a set S. Then X constitutes a* **partition of the set S into an arbitrary collection of sets** *if and only if*

1. *For each set $T \in X$, $T \neq \emptyset$.*
2. *For each pair of sets $T, U \in X$ with $T \neq U$, $T \cap U = \emptyset$.*
3. $\bigcup\{T : T \in X\} = S$.

Compare the three conditions in Definition 2.5 with the corresponding conditions in Definition 2.4 to see how this generalization is derived.

Each time you read a definition you should identify the "objects" being considered, the "name" given to those objects having the property being defined, and the "property" itself. Then try to create an image of the definition. The Glossary at the end of this book contains a complete list of the definitions used in this book.

2.2.2 Creating Your Own Definitions

Eventually you may need to create your own definitions to solve a particular problem. Doing so is an art that involves identifying and translating to symbolic form a desirable property of interest for an object you are thinking about. To develop your own definitions, follow these steps:

How to Create Your Own Definitions

Step 1. Identify a common property shared by all desirable objects you are considering.

In performing this step, it usually helps to look at numerous examples of desirable objects while comparing and contrasting them to numerous examples of undesirable objects. In this way, you can isolate and identify the common property that all the desirable items have.

Step 2. Choose a name for those objects having the desirable property.

This name should reflect the associated property to whatever degree possible.

Step 3. Translate your visual image of this property to a corresponding symbolic form that then becomes the definition.

You might find it helpful to translate the visual image first to a verbal form before trying to write the symbolic form. Then, after you write the final definition, make sure that there are no *syntax errors* that arise from undefined symbols or other items whose meaning are not properly understood and described. This includes not only the variables, but operations as well. For example, the operation

$$\frac{x}{y}$$

makes sense when x and y are real numbers, but *not* when $\mathbf{x}$ and $\mathbf{y}$ represent n*-vectors*. The process of looking for syntax errors in mathematical statements is similar to that same process when writing a computer program.

Step 4. Verify that the resulting definition works both ways: (1) all desirable objects should satisfy the stated property and (2) any object satisfying the stated property should be a desirable object.

You should check your definition against the examples used in Step 1. If you discover a mistake, then you need to make appropriate modifications to the definition and repeat this step. This process of testing and correcting a definition is similar to looking for run-time and logic errors in a computer program that must be corrected on a trial-and-error basis before the program is deemed to be correct.

The remainder of this section is devoted to showing various examples of how this process of creating definitions is done.

Defining the Ball of Radius r in n Dimensions

In this example, you will see how generalization is applied to create new definitions that include the previous ones as special cases. To begin with, consider the following definition that captures the concept of a circle of radius 1 centered at the origin in the plane, together with all points inside that circle (see Figure 2.18).

Definition 2.6 *A set B of points in the plane is the* **ball of radius 1 centered at the origin** *if and only if*

$$B = \{(x, y) : x^2 + y^2 \leq 1\}.$$

Defining the ball of radius r centered at the origin of a plane. Now apply generalization to Definition 2.6. Thus, for example, rather than restrict the definition to a ball of radius 1, why not consider a ball of radius r? To create the appropriate definition, follow the four steps.

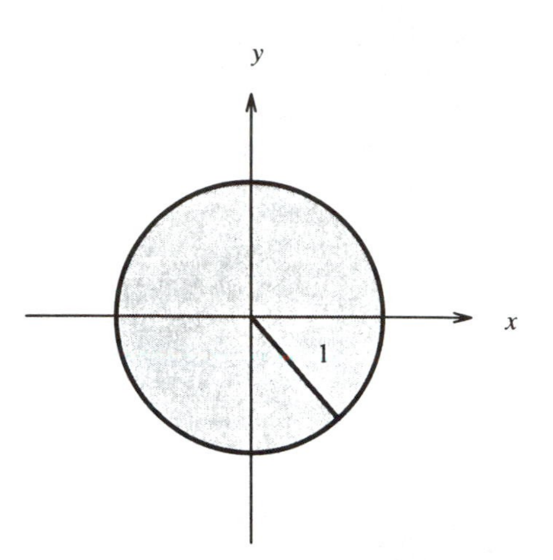

Figure 2.18: The Ball of Radius 1 Centered at the Origin

Step 1. Identify a common property shared by all the desirable objects. Look at the desirable set in Figure 2.19, as opposed to the undesirable set in Figure 2.20. The property is that the desirable set is a ball of some radius r whose center is at the origin.

Step 2. Choose a name for those objects having the desirable property. In this case, that name is the "ball of radius r centered at the origin."

Step 3. Translate your visual image to symbolic form. You might at first attempt to obtain the symbolic form by replacing the number 1 in Definition 2.6 everywhere with the symbol r, resulting in the following definition:

A set B_r of points in the plane is a *ball of radius r centered at the origin* if and only if

$$B_r = \{(x, y) : x^2 + y^2 \leq r\}.$$

Note how the notation B_r shows the dependence of this definition on the radius r. Unfortunately, however, this definition is *not* correct, and this is why you must perform the next step.

Step 4. Verify that the resulting definition works. The definition derived in Step 3 is incorrect because it is *not* true that all points (x, y) inside a ball of radius r centered at the origin satisfy the equation

$$x^2 + y^2 \leq r.$$

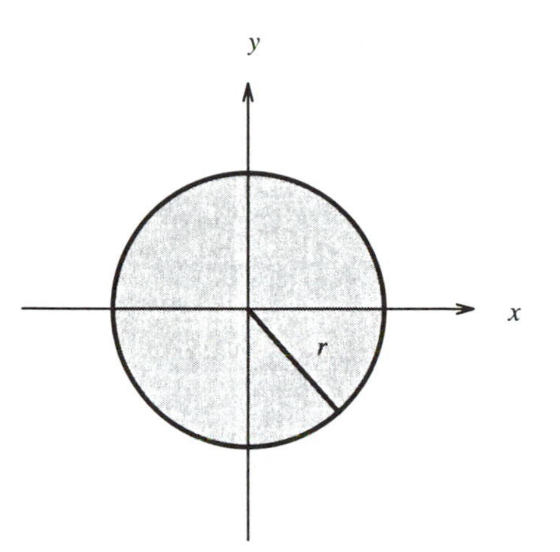

Figure 2.19: The Ball of Radius r Centered at the Origin

For example, according to this definition, every point (x, y) in the ball of radius 2 would satisfy

$$x^2 + y^2 \leq 2.$$

However, the point $x = 2$ and $y = 0$ – which *is* in the ball of radius 2 – does *not* satisfy the equation $x^2 + y^2 \leq 2$. When you discover an error in a definition, you must go back to Step 3 to correct the mistake.

Step 3 (again). Translate your visual image to symbolic form. To derive the correct definition, look again at Figure 2.19 and attempt to translate the picture to symbolic form. What common property is shared by all points (x, y) inside a ball of radius r centered at the origin? One answer – in verbal form – is the following:

The *distance* from (x, y) to the origin is $\leq r$.

Recalling how you measure distance from a point to the origin, you can create the following corresponding symbolic form:

$$\sqrt{x^2 + y^2} \leq r. \tag{2.3}$$

Squaring both sides of (2.3) now results in the following correct definition.

Definition 2.7 *For a real number $r \geq 0$, a set B_r of points in the plane is the* **ball of radius r centered at the origin** *if and only if*

$$B_r = \{(x, y) : x^2 + y^2 \leq r^2\}.$$

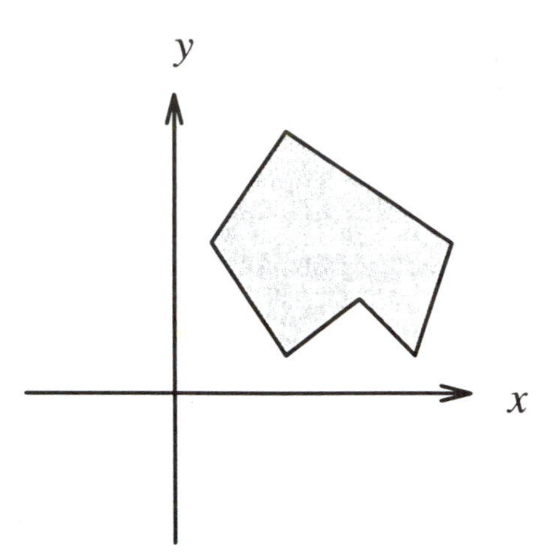

Figure 2.20: A Set that Is Not the Ball of Radius r Centered at the Origin

Observe that the generalization in Definition 2.7 reduces to the special case of a ball of radius 1 centered at the origin. That is, when you replace the radius r everywhere in Definition 2.7 with the value of 1, you obtain Definition 2.6.

Defining the ball of radius r centered at the point $(\bar{x}, \bar{y})$ in the plane. Continuing the process of generalization, you can create a more general definition that allows the ball to be centered at some *arbitrary* point in the plane – whose coordinates, say, are $(\bar{x}, \bar{y})$ – as seen in Figure 2.21. In this case, the property shared by all points (x, y) inside a ball of radius r centered at the point $(\bar{x}, \bar{y})$ – in verbal form – is the following:

The distance from (x, y) to $(\bar{x}, \bar{y})$ is $\leq r$.

The corresponding symbolic form is

$$\sqrt{(x - \bar{x})^2 + (y - \bar{y})^2} \leq r. \tag{2.4}$$

Squaring both sides of (2.4) now results in the following correct definition.

> **Definition 2.8** *For a real number $r \geq 0$, a set $B_r(\bar{x}, \bar{y})$ of points in the plane is the* **ball of radius r centered at the point** $(\bar{x}, \bar{y})$ *if and only if*
>
> $$B_r(\bar{x}, \bar{y}) = \{(x, y) : (x - \bar{x})^2 + (y - \bar{y})^2 \leq r^2\}.$$

Notice again how the notation $B_r(\bar{x}, \bar{y})$ shows the dependence of the definition on the radius r and the center $(\bar{x}, \bar{y})$. Also observe that the generalization in Definition 2.8 reduces to the special case of a ball of

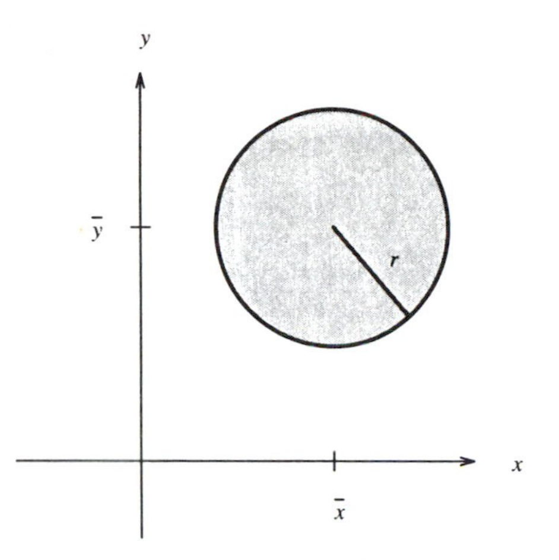

Figure 2.21: The Ball of Radius r Centered at the Point $(\bar{x}, \bar{y})$

radius r centered at the origin. That is, when you replace both $\bar{x}$ and $\bar{y}$ in Definition 2.8 with 0 (corresponding to the center being at the origin), you obtain Definition 2.7.

Defining the ball of radius r in n dimensions. As a final extension, consider a generalization to n dimensions. To define such an n-dimensional ball, a subscript notation is more appropriate. Thus, represent the center of the ball by the n-vector $(c_1, \ldots, c_n)$ instead of by $(\bar{x}, \bar{y})$. Similarly, instead of representing a point inside the ball as (x, y), denote such a point by the n-vector $(x_1, \ldots, x_n)$. With these notations, Definition 2.8 generalizes as follows.

Definition 2.9 *For a real number $r \geq 0$, a set $B_r^n(c_1, \ldots, c_n)$ of points in n dimensions is the* **ball in n dimensions of radius r centered at the point** $(c_1, \ldots, c_n)$ *if and only if*

$$B_r^n(c_1, \ldots, c_n) = \{(x_1, \ldots, x_n) : (x_1 - c_1)^2 + \cdots + (x_n - c_n)^2 \leq r^2\}.$$

Once again, the generalization in Definition 2.9 reduces to the special case of a ball in the plane. That is, when you replace the dimension n in Definition 2.9 with 2 (corresponding to a plane), you obtain Definition 2.8.

Appropriate notation can simplify the way in which Definition 2.9 is written. For example, letting

$$\mathbf{x} = (x_1, \ldots, x_n) \text{ and } \mathbf{c} = (c_1, \ldots, c_n)$$

Verbal Description	Mathematical Description
The 2-dimensional ball of radius 1 centered at the origin.	$\{(x_1, x_2) : x_1^2 + x_2^2 \leq 1\}$.
The 2-dimensional ball of radius r centered at the origin.	$\{(x_1, x_2) : x_1^2 + x_2^2 \leq r^2\}$.
The 2-dimensional ball of radius r centered at the point (c_1, c_2).	$\{(x_1, x_2) : (x_1 - c_1)^2 + (x_2 - c_2)^2 \leq r^2\}$.
The n-dimensional ball of radius r centered at the point **c**.	$\{(x_1, \ldots, x_n) :$ $(x_1 - c_1)^2 + \cdots + (x_n - c_n)^2 \leq r^2\}$.

Table 2.1: Summary of the Definitions of the n-Dimensional Ball

and recalling from Section 1.4.3 that

$$||\mathbf{x} - \mathbf{c}|| = \sqrt{(x_1 - c_1)^2 + \cdots + (x_n - c_n)^2},$$

you can rewrite Definition 2.9 as follows.

Alternative Way To Write Definition 2.9

For a real number $r \geq 0$, a set $B_r^n(\mathbf{c})$ of points in n dimensions is the **ball in n dimensions of radius r centered at the point c** *if and only if*

$$B_r^n(\mathbf{c}) = \{n\text{-vectors } \mathbf{x} : ||\mathbf{x} - \mathbf{c}||^2 \leq r^2\}.$$

Note that the notation $B_r^n(\mathbf{c})$ in the foregoing definition indicates the dependence on the radius r, the dimension n, and the center **c**.

A summary of these sequential definitions for the n-dimensional ball of radius r centered at a point is given in Table 2.1. Observe how each successive definition is a generalization of the previous one, which then becomes a special case.

Defining a Bounded Set

The objective of this next example is to define the concept of a *bounded set* in n dimensions, which is a set that is "not too large." However, because you cannot visualize a set in n-dimensions, the idea is to think in two dimensions (where you *can* visualize this concept) and then write appropriate mathematics that applies to n dimensions. So now, follow the four steps given earlier in this section.

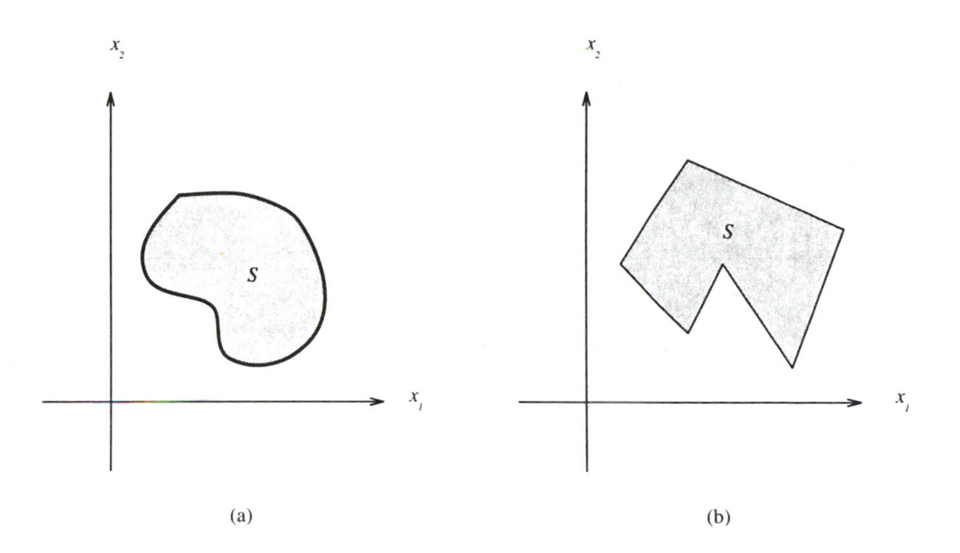

Figure 2.22: Examples of Sets that Are Bounded

Step 1. Identify a common property shared by all the desirable objects. Look at the desirable sets in Figure 2.22, as opposed to the undesirable sets in Figure 2.23. The common property is that the desirable sets are "not too large" whereas the undesirable sets "go off to infinity."

Step 2. Choose a name for those objects having the desirable property. In this case, that name is a *bounded set*.

Step 3. Translate your visual image to symbolic form. To translate the visual image of a bounded set in Figure 2.22 to symbolic form, first try to create a *verbal* statement of the property shared by all bounded sets. You might arrive at the following statement:

The points inside a bounded set are "not going off to infinity."

The question, now, is how to translate the concept of "not going off to infinity" to symbolic form.

Think of each point in the set as an n-vector. One way to specify that these n-vectors are "not going off to infinity" is to say that their *lengths* are less than or equal to some number u. Thus, using the symbol $\mathbf{x}$ for an n-vector in the set (note that $\mathbf{x} = (x_1, \ldots, x_n)$ is an n-vector and not just a point in the plane), you can write that

$$||\mathbf{x}|| \leq u \text{ (recall that } ||\mathbf{x}|| = \sqrt{x_1^2 + \cdots + x_n^2}\text{).}$$

Your first attempt might therefore be the following definition:

A set S is *bounded* if and only if $||\mathbf{x}|| \leq u$.

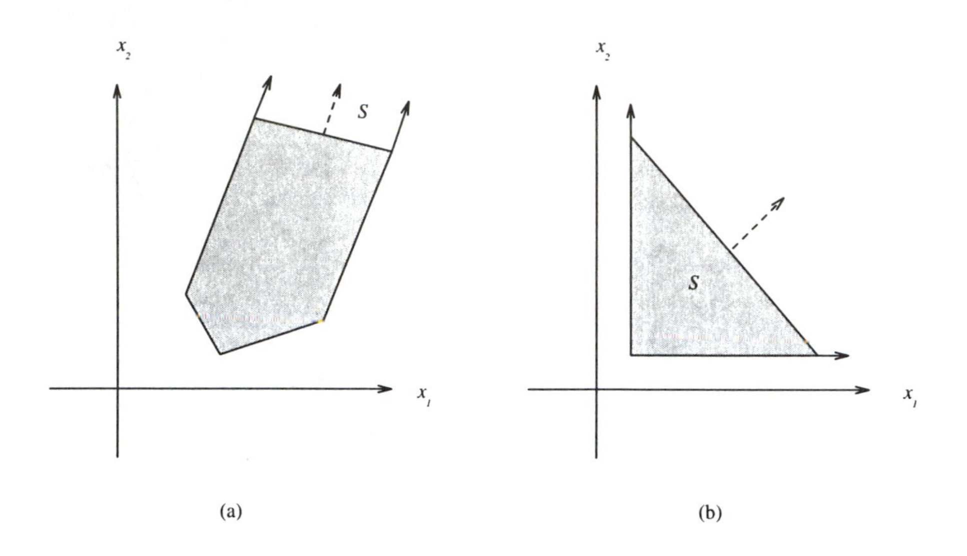

Figure 2.23: Examples of Sets that Are Not Bounded

Checking for syntax errors, you can see that the symbol S is understood – S is the set for which boundedness is being defined – but, what is **x**? Is **x** just *one* point in the set? Careful thought should lead you to recognize that the statement $||\mathbf{x}|| \leq u$ needs to be true for *every* n-vector **x** in S. You should therefore use the quantifier *for all* to write the following attempt at the definition:

> A set S is *bounded* if and only if for all elements $\mathbf{x} \in S$, $||\mathbf{x}|| \leq u$.

Once again checking for syntax errors, observe that the symbol u is undefined. Of course it represents a positive real number, but is it *for some* u or is it *for all* u? In this case, it is *for some* u, thus suggesting the use of the quantifier *there is*. You might therefore create the following attempt at the definition:

> A set S is *bounded* if and only if for all elements $\mathbf{x} \in S$, there is a real number $u > 0$ such that $||\mathbf{x}|| \leq u$.

At this point there are no more syntax errors, so proceed to Step 4.

Step 4. Verify that the resulting definition works. Test this final definition against the examples of bounded sets in Figure 2.22. You can check that for both sets, it is *true* that

> For all elements $\mathbf{x} \in S$, there is a real number $u > 0$ such that $||\mathbf{x}|| \leq u$.

However, is it *also* the case that every set satisfying this property is what you think of as a bounded set? No, for consider the sets in Figure 2.23, which are *not* what you think of as bounded sets. Those sets unfortunately satisfy the property that for every element $\mathbf{x} \in S$, there is a real number $u > 0$ such that $||\mathbf{x}|| \leq u$ (for example, $u = ||\mathbf{x}||+1$). The fact that undesirable objects – the sets in Figure 2.23 – satisfy the stated property indicates that there is an error in the definition. Can you identify and correct the mistake?

The answer is that the quantifiers *for all* and *there is* in the last definition are in the wrong *order*. By reversing that order, you obtain the following correct definition (which you should verify with the examples in Figure 2.22 and Figure 2.23).

Definition 2.10 *A set S of points in n dimensions is* **bounded** *if and only if there is a real number $u > 0$ such that for all elements $\mathbf{x} \in S$, $||\mathbf{x}|| \leq u$.*

The final definition you end up with need not be unique – there may be different, but *equivalent*, ways to define the same concept. It all depends on how you translate the visual image. For example, another way to express the concept of a bounded set is to observe that a bounded set – such as the ones in Figure 2.22 – can be made to lie inside a ball of some large-enough radius centered at the origin, whereas other sets – such as the ones in Figure 2.23 – cannot. By using Definition 2.9, an appropriate definition for a bounded set, based on this observation, is as follows.

Alternative Definition of a Bounded Set

A set S of points in n dimensions is **bounded** if and only if there is a real number $r > 0$ such that $S \subseteq B_r^n(\mathbf{0})$, where $\mathbf{0} = (0, \ldots, 0)$ and $B_r^n(\mathbf{0}) = \{n\text{-vectors } \mathbf{x} : ||\mathbf{x}|| \leq r^2\}$.

Here you see the use of a previous definition – that of $B_r^n(\mathbf{0})$ – to define a *new* concept – that of a bounded set. However, keep in mind that there can be only one definition for each concept. Thus, you must select one of these as the definition of a bounded set – whichever one suits your purpose. Suppose you choose the one in Definition 2.10. Then what happens to the alternative definition? Realizing that the alternative is, for all practical purposes, an equally valid description of a bounded set, you should use a *proof* to establish the equivalence of these two ideas. That is, you should create the following two statements associated with the definition and its alternative, respectively:

A: There is a real number $u > 0$ such that for all elements $\mathbf{x} \in S$, $||\mathbf{x}|| \leq u$.

B: There is a real number $r > 0$ such that $S \subseteq B_r^n(\mathbf{0})$.

Then you should prove that **A** is *true* if and only if **B** is *true*. In this way, you will know that the definition and its alternative are *equivalent*.

Summary of Creating Definitions

A summary of the process of creating definitions follows.

How to Create Definitions

Step 1. Identify a common property shared by all desirable objects you are considering.

Step 2. Choose a representative name for those objects having the desirable property.

Step 3. Translate your visual image of this property to a corresponding symbolic form that then becomes the definition.

Step 4. Verify that the resulting definition works both ways: (1) all desirable objects should satisfy the stated property and (2) any object satisfying the stated property should be a desirable object.

Remember to check for syntax errors that arise when a symbol or operation is not properly defined or understood. Be sure to test your definition thoroughly with many different examples so that you can discover and correct any mistakes. Only then can you have confidence that the definition correctly separates objects that have the property from objects that do not.

In this section, you have seen how to create and work with definitions. In the next section, you will see how these ideas are used together with the techinques of unification, generalization, and abstraction from Chapter 1.

2.3 Developing Axiomatic Systems

In Chapter 1, you saw that unification, generalization, and abstraction enable you to understand many different concepts while having to study and work with only one encompassing problem. To study that one problem involves not only identifying similarities and differences, translating visual images, and working with definitions, but also using the techniques you will learn in this section.

2.3.1 What is an Axiomatic System?

Consider the operations indicated by the italicized words in each of the following mathematical contexts. What similarities and differences do you see?

1. $-n$, meaning the *negative* of the integer n.
2. A^c, meaning the *complement* of the set A (see Example 2.8 in Section 2.1.1).
3. r^{-1}, meaning the *reciprocal* of the real number $r \neq 0$.

These operations are different in that they are performed on different items – in (1), the *negative* is applied to an *integer*; in (2), the *complement* is performed on a *set*; and in (3), the *reciprocal* is applied to a *real number*. Despite these differences, there are some similarities. Within each context, the operation is applied to *one* item and the result is an item *of the same kind*. Also, in each case, the operation produces, in some sense, the "opposite" of the item. These similarities merit further investigation.

Creating an Abstract System for a Unary Operator

Perhaps the best approach for studying these similarities is to use abstraction, as described in Section 1.5. In that regard, think about performing an operation on an *object* rather than on a specific item.

Now is the time to introduce some appropriate notation. Let x be a general object of some kind (an integer, a set, a real number, or something else). Because the symbol x represents an object, it is natural to consider x as belonging to some set – say, S – consisting of all possible objects of the same kind on which you can perform the operation. In general, because abstraction involves objects, a set is used to represent all possible objects under consideration.

A symbol – say, ν – is also needed to indicate that you are performing an operation on the object $x \in S$. The notation

$$\nu(x) \tag{2.5}$$

is understood to mean the *result* of performing the operation ν on x and is read as "new of x." The symbol ν is called a **unary operator**, meaning that ν is applied to *one* object, in contrast to a *binary operator*, which is applied to *two* objects.

The abstract operator ν includes each of the three special cases presented at the beginning of this section, as shown by the following substitutions:

1. By replacing ν in (2.5) with the minus sign ($-$) and x with the integer n, the abstract operation becomes

   $$\nu(x) = -(n) = -n \text{ (the negative of the integer } n).$$

2. By rewriting $\nu(x)$ in (2.5) as $(x)^c$ and then replacing x with the set A, the abstract operation becomes

   $$\nu(x) = (A)^c = A^c \text{ (the complement of the set } A).$$

3. By rewriting $\nu(x)$ in (2.5) as $(x)^{-1}$ and then replacing x with the real number $r \neq 0$, the abstract operation becomes

$$\nu(x) = (r)^{-1} = r^{-1} \text{ (the reciprocal of the real number } r \neq 0).$$

The advantage of abstraction, as you will see later in this section, is the following:

The Advantage of Abstraction

Whatever knowledge you obtain for the abstract system, you also obtain for each of the special cases, *with no additional work other than an appropriate substitution*.

So now you have the set S and the ability to perform an operation on an object $x \in S$ with the unary operator ν. The pair (S, ν) is one example of an *abstract system*, meaning that you have some objects (those in the set S, in this case) and some way to work with them (by performing the operation ν, in this case).

Developing Axioms for a Unary Operator

Remember that the purpose of this abstract system (S, ν) is to study unary operators. So what you do next depends on what, exactly, you want to study. In this case, recall that you are trying to understand the similarities between the use of the unary operators in each of the three special cases given at the beginning of this section. To this end you should ask, What properties would a general unary operator ν need to have to be considered in the same class (that is, more or less the same) as those in the special cases?

Answering this question is not easy. How you do so depends on the specific examples you are working with, the properties you feel are important, the kind of results you eventually want to obtain about the abstract system, and more. In this case, the objective is to isolate and capture the concept of a unary operator that generates the "opposite" object. After working some time with the specific examples as a guide, you might come to the conclusion that the following are the most important properties for the unary operator ν to be considered similar to the unary operators in the special cases (throughout, let $x \in S$):

1. The result of performing the operation on x should be an object of the *same kind* as x.

2. If you perform the operation on x and then perform the operation on the resulting object, you obtain the original x.

You should now translate each of these properties to symbolic form using the language of mathematics and the symbols x, S, and ν. By replacing the words "performing the operation on" with the unary operator ν in the two foregoing properties, you obtain the following corresponding definitions (pay careful attention to the use of the quantifier *for all*).

Definition 2.11 *A unary operator ν is* **closed on a set** *S if and only if for all $x \in S$, $\nu(x) \in S$. (In this case, ν is also called a* **closed unary operator on** *S.)*

Definition 2.12 *A closed unary operator ν on a set S is an* **invertible operator** *if and only if for all $x \in S$, $\nu(\nu(x)) = x$.*

In the context of the abstract system (S, ν), Definition 2.11 and Definition 2.12 are each called an **axiom**, meaning a property that you *assume* holds true. An **axiomatic system** is an abstract system that is assumed to satisfy a given list of axioms. You have seen axioms in geometry – for example, the axiom that the shortest distance between two points is a straight line. Observe that an axiom by itself is neither *true* nor *false*. Rather, an axiom is a statement you are *assuming* is *true*. On a historical note, the *parallel postulate* is an axiom in geometry stating that, through a point P *not* on a line L in the plane, there is one and only one line parallel to L. In Euclidean geometry, this axiom is *assumed* to be *true*. However, by assuming instead that there are an *infinite* number of such lines through P parallel to L, the mathematician Lobatchewsky developed an equally valid, but different axiomatic system for geometry.

Returning to the current example of a unary operator, by putting together the pieces, the following constitutes an axiomatic system for the unary operator ν on the set S, in which the two axioms are the two properties of ν being closed and invertible.

Definition 2.13 *The unary operator ν on a set S is an* **inverse operator** *if and only if the unary operator ν on S is closed and invertible.*

Recall that the axioms of an inverse operator are designed to capture the properties needed for ν to be considered like those in the special cases. In that regard, you can verify that each of the unary operators in the special cases are closed and invertible.

Another property of an axiomatic system is that the system should include the *fewest* number of axioms needed to obtain the desired result. For example, another property of an inverse operator is that for each element $y \in S$, you can find an element $x \in S$ for which $\nu(x) = y$. However, this property is *not* included as an axiom because you can derive this property using the already-chosen axioms of ν being closed and invertible, as shown in the following proposition.

Proposition 2.1 *Suppose that ν is an inverse operator on a set S. For every element $y \in S$, there is an element $x \in S$ such that $\nu(x) = y$.*

Proof. Let $y \in S$. It is necessary to find an element $x \in S$ for which $\nu(x) = y$. Because $y \in S$, it is possible to apply ν to y, so construct

$$x = \nu(y).$$

This value of x is an element of S because ν is closed on S. Furthermore, by applying ν to both sides of $x = \nu(y)$ and using the fact that ν is invertible, you have that

$$\nu(x) = \nu(\nu(y)) = y,$$

thus completing the proof. QED

Continuing the discussion of including only those axioms that are really necessary, it takes some work to show that neither of the two axioms in Definition 2.13 can be eliminated. In other words, neither of those axioms can be proved from the remaining one. Thus, both axioms are needed.

2.3.2 Working with an Axiomatic System

As mentioned previously, the primary advantage of an axiomatic system is that when you obtain knowledge about that system, you obtain the same knowledge for each special case by an appropriate substitution. You can also apply that same knowledge to a *new* special case, with no additional work. You will see these features of an axiomatic system in this section.

Creating a Visual Image

As you know, whenever you encounter a new mathematical concept (such as a unary operator), you should create an associated visual image to help you picture and work with that concept. One way to do so for a unary operator on a set S is to use a *directed graph* (see Section 1.5.1), in which there is one vertex for each element of S whose label is that of the corresponding element. An edge from a vertex x pointing to a vertex y means that $\nu(x) = y$. For example, the graphs in Figure 2.24 and Figure 2.25 correspond to two different unary operators on the set $S = \{t, u, v, w\}$.

Having learned the concept of an inverse operator, you should translate the two axioms of ν being closed and invertible to a picture involving a directed graph.

Translating the Property of Being Closed. The property of being closed in Definition 2.11 states that for all $x \in S$, $\nu(x) \in S$. This property is reflected in the graph by the fact that an edge from each vertex ends at some vertex also in the graph. This is the case for both the graphs in Figure 2.24 and Figure 2.25.

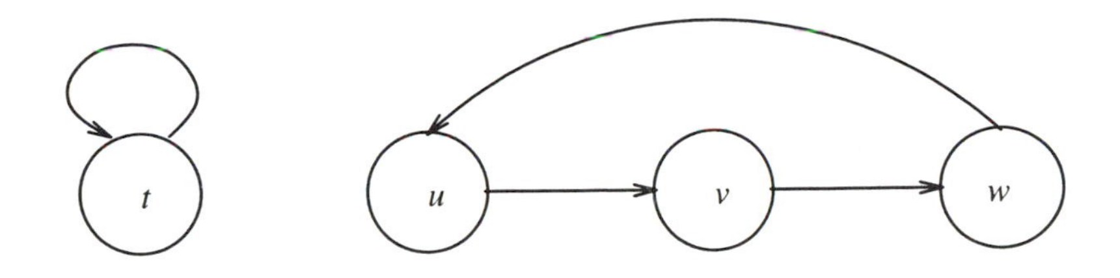

Figure 2.24: A Visual Representation of a Unary Operator that Is Not an Inverse Operator

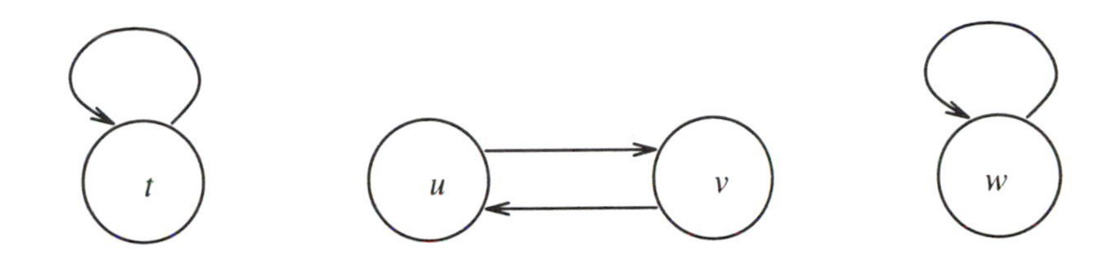

Figure 2.25: A Visual Representation of a Unary Operator that Is an Inverse Operator

Translating the Property of Being Invertible. The property of being invertible in Definition 2.12 is that $\nu(\nu(x)) = x$. Consequently, if there is an edge from a vertex x to a vertex y ($\nu(x) = y$), then there should be an edge from the vertex y back to x ($\nu(y) = x$). This is the case for the graph in Figure 2.25 but not for the graph in Figure 2.24.

Because the graph in Figure 2.25 reflects a unary operator that is both closed and invertible, the associated unary operator is an inverse operator. The unary operator associated with the graph in Figure 2.24 is *not* an inverse operator because the property of being invertible fails to hold.

Obtaining New Results with an Axiomatic System

Now that you have a visual representation of a unary operator, it is time to see what properties you can derive about this abstract concept and its associated axiomatic system. For example, you have already seen the result in Proposition 2.1. As another example, you can prove that if x and y are two different elements in S, then $\nu(x)$ and $\nu(y)$ are different, as shown in the following proposition.

Proposition 2.2 *Suppose that ν is an inverse operator on a set S and let $x, y \in S$. If $x \neq y$, then $\nu(x) \neq \nu(y)$.*

Proof. By the contrapositive method, assume that

$$\nu(x) = \nu(y). \tag{2.6}$$

Because ν is closed on S, both $\nu(x)$ and $\nu(y)$ are in the set S, so you can apply the operator ν to both sides of (2.6), resulting in the following:

$$\nu(\nu(x)) = \nu(\nu(y)). \tag{2.7}$$

Now, because ν is invertible, $\nu(\nu(x)) = x$ and $\nu(\nu(y)) = y$ so, from (2.7), it follows that

$$x = y.$$

Thus, the proposition is *true*. QED

By using substitution, you can apply Proposition 2.2 to the three special cases given at the beginning of this section. Specifically, for the negative of an integer, Proposition 2.2 results in the following fact:

> If m and n are two different integers, then $-m$ and $-n$ are different integers.

Similarly, for the complement of a set, Proposition 2.2 states that

> If A and B are two different sets, then their complements are different sets.

Finally, for the reciprocal of a real number, from Proposition 2.2 you have that

> If r and s are two different real numbers, neither of which are 0, then their reciprocals are different real numbers.

You can also apply Proposition 2.2 to any *new* unary operator that is closed and invertible. To illustrate, consider a mathematical statement p (see Section 1.6.1). You know what it means to say that p is *not true*, henceforth written as

$$\neg p.$$

For example, if

> p is the statement that the real number $r > 0$,

then

> $\neg p$ is the statement that $r \leq 0$.

Here you see that $\neg$ is a unary operator on the set S of all mathematical statements.

You can also verify that $\neg$ is an inverse operator. This operator is closed on S because, if p is a mathematical statement, then so is $\neg p$. Also, this operator is invertible because $\neg(\neg p)$ is equivalent to the original statement p. You can now apply Proposition 2.2 to this unary operator to state that,

If p and q are two different statements, then $\neg p$ and $\neg q$ are different statements.

That is, if p and q are two different statements, then so are their negations.

In this section, you have seen how to work with an axiomatic system to derive results that are applicable not only to the existing special cases but also to any new special case. In the next section, you will see several other examples of axiomatic systems.

2.3.3 Other Examples of Axiomatic Systems

In this section, two additional examples of developing axiomatic systems are presented. As before, this is done by using special cases to create an axiomatic system that reflects certain common properties of those special cases. Results are obtained for the axiomatic system that are then applied not only to the existing special cases but also to a new one.

An Axiomatic System for the Length of a Vector

In Section 1.4.3, you learned that one way to measure the length of an n-vector $\mathbf{x} = (x_1, \ldots, x_n)$ is by the formula

$$||\mathbf{x}|| = \sqrt{x_1^2 + \cdots + x_n^2}. \tag{2.8}$$

However, this is not the *only* way in which you might do so. For example, another possibility is the following:

$$||\mathbf{x}|| = |x_1| + \cdots + |x_n|. \tag{2.9}$$

By identifying similarities in (2.8) and (2.9), you can use unification to develop a *single* formula that contains each of these two measures as a special case. For example, you can rewrite (2.8) as

$$||\mathbf{x}|| = (|x_1|^2 + \cdots + |x_n|^2)^{\frac{1}{2}} \tag{2.10}$$

and (2.9) as

$$||\mathbf{x}|| = (|x_1|^1 + \cdots + |x_n|^1)^{\frac{1}{1}}. \tag{2.11}$$

From (2.10) and (2.11) it is easier to arrive at the following unified formula involving the positive integer p:

$$||\mathbf{x}|| = (|x_1|^p + \cdots + |x_n|^p)^{\frac{1}{p}}. \tag{2.12}$$

When you replace p with 1 in (2.12), you obtain the special case in (2.9), and when you replace p with 2 in (2.12), you obtain the special case in (2.8).

Now consider generalizing the concept to a more general method for computing the length of an n-vector $\mathbf{x}$. (It is also possible to apply abstraction to develop the concept of the length of an *object*, however, to keep the discussion more tangible, only the length of an n-vector is considered here.)

Creating an Abstract System. Now is the time to introduce some appropriate notation. Let $\mathbb{R}^n$ be the set of all n-vectors – the set of all objects under consideration – and let $\mathbf{x} \in \mathbb{R}^n$. A symbol is also needed to indicate that you are computing the length of $\mathbf{x}$. So, let

$$\text{length}(\mathbf{x}) = \text{the length of the } n\text{-vector } \mathbf{x}.$$

The general expression length($\mathbf{x}$) includes the special cases in (2.8), (2.9), and even (2.12), as well as other methods for computing the length of $\mathbf{x}$.

You now have the set $\mathbb{R}^n$ and the ability to compute the length of any n-vector $\mathbf{x} \in \mathbb{R}^n$ by length($\mathbf{x}$). The pair ($\mathbb{R}^n$, length) constitutes an abstract system.

Developing Axioms. What you do next depends on what, exactly, you want to study with this abstract system. In this case, recall that you are trying to understand the properties pertaining to the length of an n-vector. Thus you should ask, What properties would a general function such as length($\mathbf{x}$) need to have to be considered in the same class (that is, more or less the same) as the length function in (2.8), for example?

Answering this question is not easy. How you do so depends on the specific examples you are working with, the properties you feel are important, the kind of results you eventually want to obtain about the abstract system, and more. It takes some time working with the formulas in (2.8), (2.9), and (2.12) to come to the conclusion that the following are the most important properties associated with the length of an n-vector (throughout, let $\mathbf{x}, \mathbf{y} \in \mathbb{R}^n$):

1. The length of an n-vector $\mathbf{x}$ should always be nonnegative.

2. The length of the n-vector $\mathbf{x}$ is 0 if and only if the vector $\mathbf{x}$ is the origin.

3. The length of an n-vector $\mathbf{x}$ should be the same as that of $-\mathbf{x}$ (see Example 2.10 in Section 2.1.1).

4. The length of the sum of two n-vectors should not exceed the sum of their individual lengths.

Now translate each of these properties to symbolic form using the language of mathematics and the symbols $\mathbf{x}$, $\mathbf{y}$, $\mathbb{R}^n$, and length. You might also find it helpful to use the notation that $\mathbf{0} = (0, \ldots, 0)$ and $\mathbf{x} = \mathbf{y}$, which means that for all $i = 1, \ldots, n$, $x_i = y_i$. You will then obtain the following axiomatic system.

Definition 2.14 *The function denoted by* length *is called a* **norm on $\mathbb{R}^n$** *if and only if*

1. *For all* $\mathbf{x} \in \mathbb{R}^n$, $\text{length}(\mathbf{x}) \geq 0$.
2. *For all* $\mathbf{x} \in \mathbb{R}^n$, $\text{length}(\mathbf{x}) = 0$ *if and only if* $\mathbf{x} = \mathbf{0}$.
3. *For all* $\mathbf{x} \in \mathbb{R}^n$, $\text{length}(\mathbf{x}) = \text{length}(-\mathbf{x})$.
4. *For all* $\mathbf{x}, \mathbf{y} \in \mathbb{R}^n$, $\text{length}(\mathbf{x} + \mathbf{y}) \leq \text{length}(\mathbf{x}) + \text{length}(\mathbf{y})$.

These axioms of a norm are designed to capture the properties of the length of an n-vector. In that regard, you can verify that each of the special cases satisfies all four of these axioms. Furthermore, recall that an axiomatic system should include the *fewest* number of axioms needed to obtain the desired result. For example, another property of a norm is that

$$\text{For all } \mathbf{x}, \mathbf{y}, \mathbf{z} \in \mathbb{R}^n, \ \text{length}(\mathbf{x} - \mathbf{z}) \leq \text{length}(\mathbf{x} - \mathbf{y}) + \text{length}(\mathbf{y} - \mathbf{z}).$$

However, this property is *not* included as an axiom because you can derive this property by using Axiom (4) in Definition 2.14, as shown in the following proposition.

Proposition 2.3 *Suppose that* length *is a* norm *on* $\mathbb{R}^n$. *If* $\mathbf{x}, \mathbf{y}, \mathbf{z} \in \mathbb{R}^n$, *then* $\text{length}(\mathbf{x} - \mathbf{z}) \leq \text{length}(\mathbf{x} - \mathbf{y}) + \text{length}(\mathbf{y} - \mathbf{z})$.

Proof. Let $\mathbf{x}, \mathbf{y}, \mathbf{z} \in \mathbb{R}^n$. Then,

$$\text{length}(\mathbf{x} - \mathbf{z}) = \text{length}(\mathbf{x} - \mathbf{y} + \mathbf{y} - \mathbf{z}).$$

Using $\mathbf{x} - \mathbf{y}$ in place of $\mathbf{x}$ in Axiom (4) of Definition 2.14 and $\mathbf{y} - \mathbf{z}$ in place of $\mathbf{y}$, it follows by specialization that

$$\begin{aligned} \text{length}(\mathbf{x} - \mathbf{z}) &= \text{length}(\mathbf{x} - \mathbf{y} + \mathbf{y} - \mathbf{z}) \\ &\leq \text{length}(\mathbf{x} - \mathbf{y}) + \text{length}(\mathbf{y} - \mathbf{z}), \end{aligned}$$

thus completing the proof. QED

It is substantially more difficult to show that none of the four axioms in Definition 2.14 can be proved from the remaining three, but this is in fact the case. Thus, all four axioms are necessary.

Working with an Axiomatic System. You will now see how to work with the axiomatic system of a norm to obtain results that are applicable to all the special cases.

Creating a Visual Image. First create a visual image associated with the norm to help you picture and work with that concept. You can picture length(**x**) as the usual distance from **x** to the origin, that is, by the formula

$$\text{length}(\mathbf{x}) = \sqrt{x_1^2 + \cdots + x_n^2}.$$

This specific formula for computing the length of an n-vector satisfies the four axioms of a norm, but so do many other *new* formulas for the norm, as you can verify for the following formula:

$$\text{length}(\mathbf{x}) = \max\{|x_i| : i = 1, \ldots, n\}.$$

Obtaining New Results with an Axiomatic System. Now that you have a visual representation of a norm, it is time to see an example of what can be done with this axiomatic system. With regard to n-vectors, you are often interested not only in their length, but also in the *distance* from an n-vector **x** to an n-vector **y**. How, then, do you measure this distance? One answer is that you can use any length function that satisfies the property of a norm to compute the distance between **x** and **y**, as follows:

$$\text{Distance from } \mathbf{x} \text{ to } \mathbf{y} = \text{length}(\mathbf{y} - \mathbf{x}).$$

But, how do you know that this formula satifies all the desirable properties that you normally associate with measuring distance?

You can use a proof to establish this fact. However, you must first identify exactly what the desirable properties of a measure of distance are. That is, you must first develop an axiomatic system of the concept of the "distance between two n-vectors." Again, after some thought, you might identify the following important properties for a function d that associates to two n-vectors **x** and **y**, a number that can be considered as the distance from **x** to **y**.

Definition 2.15 *A function d that associates to each ordered pair of n-vectors* **x** *and* **y** *the number $d(\mathbf{x}, \mathbf{y})$ is a* **distance function** *if and only if*

1. *For all* $\mathbf{x}, \mathbf{y} \in \mathbb{R}^n$, $d(\mathbf{x}, \mathbf{y}) \geq 0$.
2. *For all* $\mathbf{x}, \mathbf{y} \in \mathbb{R}^n$, $d(\mathbf{x}, \mathbf{y}) = 0$ *if and only if* $\mathbf{x} = \mathbf{y}$.
3. *For all* $\mathbf{x}, \mathbf{y} \in \mathbb{R}^n$, $d(\mathbf{x}, \mathbf{y}) = d(\mathbf{y}, \mathbf{x})$.
4. *For all* $\mathbf{x}, \mathbf{y}, \mathbf{z} \in \mathbb{R}^n$, $d(\mathbf{x}, \mathbf{z}) \leq d(\mathbf{x}, \mathbf{y}) + d(\mathbf{y}, \mathbf{z})$.

You can now prove that, for any function *length* that satisfies the properties of a norm, the function d defined by

$$d(\mathbf{x}, \mathbf{y}) = \text{length}(\mathbf{y} - \mathbf{x})$$

satisfies the properties of a distance function, as shown in the following.

Proposition 2.4 *If* length *is a norm on* $\mathbb{R}^n$*, then the function* d *defined by* $d(\mathbf{x}, \mathbf{y}) = \text{length}(\mathbf{y} - \mathbf{x})$ *for the* n*-vectors* $\mathbf{x}, \mathbf{y} \in \mathbb{R}^n$ *is a distance function.*

Proof. According to Definition 2.15 for a distance function, four properties must hold. To see that they do, let $\mathbf{x}, \mathbf{y}, \mathbf{z} \in \mathbb{R}^n$. The fact that $d(\mathbf{x}, \mathbf{y}) = \text{length}(\mathbf{y} - \mathbf{x}) \geq 0$ follows directly from Axiom (1) of a norm in Definition 2.14.

The fact that $d(\mathbf{x}, \mathbf{y}) = 0$ if and only if $\mathbf{x} = \mathbf{y}$ follows from Axiom (2) of a norm because

$$\begin{aligned} d(\mathbf{x}, \mathbf{y}) = 0 & \text{ if and only if } \text{length}(\mathbf{y} - \mathbf{x}) = 0 \\ & \text{ if and only if } \mathbf{y} - \mathbf{x} = \mathbf{0} \\ & \text{ if and only if } \mathbf{y} = \mathbf{x}. \end{aligned}$$

To see that $d(\mathbf{x}, \mathbf{y}) = d(\mathbf{y}, \mathbf{x})$, use Axiom (3) of a norm to conclude that

$$d(\mathbf{x}, \mathbf{y}) = \text{length}(\mathbf{y} - \mathbf{x}) = \text{length}(\mathbf{x} - \mathbf{y}) = d(\mathbf{y}, \mathbf{x}).$$

Finally, the fact that $d(\mathbf{x}, \mathbf{z}) \leq d(\mathbf{x}, \mathbf{y}) + d(\mathbf{y}, \mathbf{z})$ is precisely the result obtained in Proposition 2.3. Thus, d is a distance function. QED

An Axiomatic System for Equivalence Relations

As a final example of an axiomatic system, consider the use of the word "equal" in each of the following mathematical contexts. What similarities and differences do you see?

1. The two real numbers r and s are *equal* if and only if $r \leq s$ and $s \leq r$.

2. Two n-vectors $\mathbf{u} = (u_1, \ldots, u_n)$ and $\mathbf{v} = (v_1, \ldots, v_n)$ are *equal* if and only if for all $i = 1, \ldots, n$, $u_i = v_i$.

3. Two sets A and B are *equal* if and only if $A \subseteq B$ and $B \subseteq A$.

4. Two real-valued functions f and g of one variable are *equal* if and only if for all real numbers t, $f(t) = g(t)$.

The use of the same word "equal" is somewhat misleading because in each case its meaning is different. In (1), the "equal" is used to compare two *real numbers;* in (2), equality is used to compare two n*-vectors;* in (3), two *sets* are compared; and in (4), the "equal" compares two *functions*. Yet, despite these differences, there are some similarities. Within each context, the word "equals" always compares *two items of the same kind* and the result of the comparison is either *true* or *false*. Furthermore, the use of the word *equal* in each case is meant to express the concept of two items being the same. These similarities merit further investigation.

Creating an Abstract System. Abstraction is used to study these similarities, so you should think about comparing two *objects*, rather than two specific items. The result of such a comparison is always *true* or *false*. You may recognize that this is the concept of a *binary relation* presented in Section 1.5.2.

Now is the time to introduce some appropriate notation. Let x and y be two general objects of the same kind (two real numbers, two n-vectors, two sets, two functions, or two of something else). Because the symbols x and y represent two objects of the same kind, it is natural to consider them as belonging to some set – say, S – consisting of all possible objects that can be compared. So, $x, y \in S$.

A symbol is also needed to indicate that you are comparing the two objects x and y. Henceforth, the symbol R is used to compare two general objects, the result of which is understood to be *true* or *false*. Thus, the notation

$$x \; R \; y$$

should be read as "x is related to y" and means that the result of the comparison is *true*.

The abstract statement $x \; R \; y$ includes each of the four special cases presented at the beginning of this section. For example, by replacing the object x with the set A, y with the set B, and the symbol R with =, the abstract statement

$x \; R \; y$ (the object x is related to the object y)

becomes

$A = B$ (the set A is equal to the set B).

A similar substitution is used if you want to compare two real numbers, two n-vectors, two functions, or two of something else.

So now you have the abstract system (S, R) consisting of the set S and the ability to compare any two objects $x, y \in S$ with the binary relation R.

Developing Axioms. Remember that the purpose of this abstract system is to study binary relations. So, what you do next depends on what, exactly, you want to study. In this case, recall that you are trying to understand the similarities between the use of the binary relation = in each of the four special cases given at the beginning of this section. To this end, you should ask, What properties would a general binary relation R need to have to be considered in the same class (that is, more or less the same) as the equality sign?

After working some time with the binary relation = in the special cases as a guide, you might come to the conclusion that the following are the most important properties for the binary relation R to be considered similar to = (throughout, let $x, y, z \in S$):

1. Every object should always be equal to itself.

2. If x is equal to y, then y should be equal to x.

3. If x is equal to y and if y is equal to z, then x should be equal to z.

Now translate each of these properties to symbolic form using the language of mathematics and the symbols x, y, z, S, and R. By replacing the words "is equal to" with the general binary relation R in the three foregoing properties, you obtain the following corresponding definitions (pay careful attention to the use of the quantifier *for all*).

Definition 2.16 *A binary relation R on a set S is* **reflexive** *if and only if for all $x \in S$, $x \; R \; x$.*

Definition 2.17 *A binary relation R on a set S is* **symmetric** *if and only if for all $x, y \in S$, if $x \; R \; y$, then $y \; R \; x$.*

Definition 2.18 *A binary relation R on a set S is* **transitive** *if and only if for all $x, y, z \in S$, if $x \; R \; y$ and $y \; R \; z$, then $x \; R \; z$.*

In the context of the abstract system (S, R), each definition becomes an *axiom*. Thus, the following constitutes an axiomatic system for the binary relation R on the set S, in which the three axioms are the three properties of R being reflexive, symmetric, and transitive.

Definition 2.19 *A binary relation R on a set S is an* **equivalence relation** *if and only if R is reflexive, symmetric, and transitive.*

These axioms of an equivalence relation are designed to capture the properties needed for a binary relation R to be considered like the usual concept of equality. In that regard, you can verify that each of the binary relations in the special cases satisfies the three axioms of an equivalence relation. Furthermore, recall that an axiomatic system should include the *fewest* number of axioms needed to obtain the desired result. For example, another property of equality is that if $x = y$ and $x = z$, then $y = z$. However, this property is *not* included as an axiom because you can derive this property using the axioms of symmetry and transitivity, as shown in the following proposition.

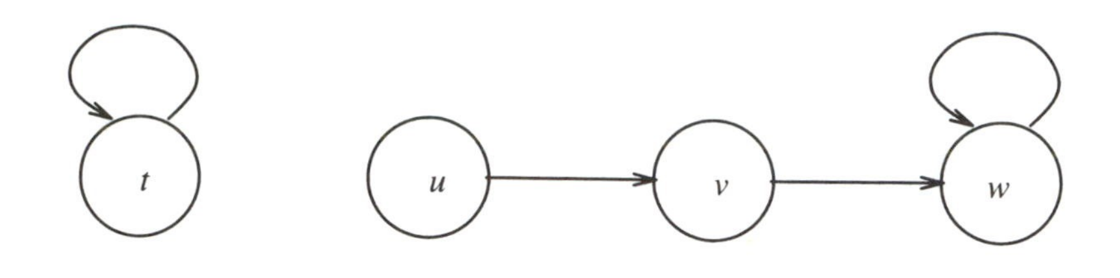

Figure 2.26: A Graphical Representation of a Binary Relation that Is Not an Equivalence Relation

Proposition 2.5 *Suppose that R is an equivalence relation on a set S and let $x, y, z \in S$. If $x \; R \; y$ and $x \; R \; z$, then $y \; R \; z$.*

Proof. It is given in the hypothesis that $x \; R \; y$ so, by the symmetric property of an equivalence relation, $y \; R \; x$. It is also given that $x \; R \; z$, so, by the transitive property, $y \; R \; z$, thus completing the proof. QED

Once again, with some effort it can be shown that none of the three axioms in Definition 2.19 can be eliminated. In other words, no single one of those axioms can be proved from the remaining two. Thus, all three axioms are necessary.

Working with an Axiomatic System. You will now see how results are derived for the foregoing axiomatic system.

Creating a Visual Image. First create a visual image of a binary relation to help you picture and work with that concept. One way to do so for a binary relation R on a set S is to use a *directed graph* (see Section 1.5.1), in which there is one vertex for each element of S whose label is that of the corresponding element. An edge from a vertex x pointing to a vertex y means that $x \; R \; y$. For example, the graphs in Figure 2.26 and Figure 2.27 correspond to two different binary relations on the set $S = \{t, u, v, w\}$.

Having learned the concept of an equivalence relation, you should translate the reflexive, symmetric, and transitive properties to a picture involving a directed graph, as follows:

1. The reflexive property in Definition 2.16 states that for all $x \in S$, $x \; R \; x$. Consequently, there should be a "loop," in the form of an edge, from each vertex to itself. This is the case for the graph in Figure 2.27 but not for the graph in Figure 2.26.
2. The symmetric property in Definition 2.17 states that if $x \; R \; y$, then $y \; R \; x$. Consequently, if there is an edge in the graph from a vertex x to a vertex y ($x \; R \; y$), then there should be an edge from y back to x ($y \; R \; x$). This is the case for the graph in Figure 2.27 but not for the graph in Figure 2.26.

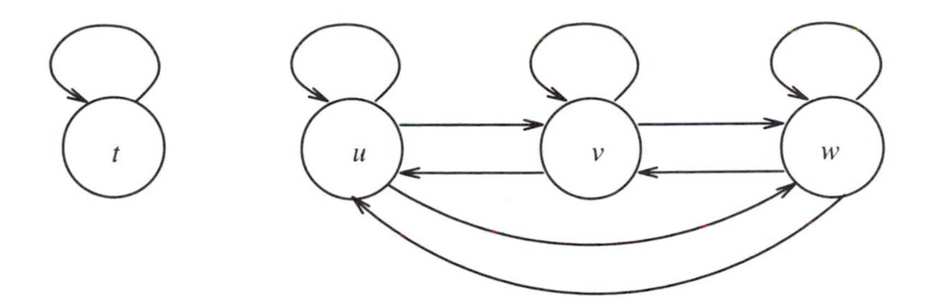

Figure 2.27: A Graphical Representation of a Binary Relation that Is an Equivalence Relation

3. The transitive property in Definition 2.18 states that if $x \; R \; y$ and $y \; R \; z$, then $x \; R \; z$. Consequently, if there is an edge in the graph from a vertex x to a vertex y ($x \; R \; y$) and an edge from the vertex y to a vertex z ($y \; R \; z$), then there should be an edge from vertex x to vertex z ($x \; R \; z$). This is the case for the graph in Figure 2.27 but not for the graph in Figure 2.26.

Putting together the pieces, the graph in Figure 2.27 reflects an equivalence relation and the graph in Figure 2.26 does not.

Obtaining New Results with an Axiomatic System. Now that you have a visual representation of an equivalence relation, you can derive new results pertaining to this axiomatic system. The following material requires a working knowledge of sets, which the reader can find in Section 3.1.

Let $\sim$ be an equivalence relation on a set S and $x \in S$. Recall that an equivalence relation is designed to capture the essential properties of two objects being the same. It is therefore reasonable to create the set of those objects in S that are considered the same, under the equivalence relation $\sim$, as the selected object x. This set of objects is formalized in the following definition.

Definition 2.20 *Let $\sim$ be an equivalence relation on a set S and let $x \in S$. The following set is the* **equivalence class generated by** x*:*

$$[x] = \{y \in S : x \sim y\}.$$

That is, to each element $x \in S$ is associated a subset, $[x]$, of S. The set $[x]$ consists of all those elements of S that are related to x by $\sim$. The collection of all these subsets constitutes a *partition* (see Definition 2.5 in Section 2.2.1) of S, as shown in the following theorem.

Theorem 2.1 *If $\sim$ is an equivalence relation on S, then $\{[x] : x \in S\}$ is a partition of S.*

Proof.

According to Definition 2.19, there are three properties to prove:

(1) For each $x \in S$, $[x] \neq \emptyset$.

(2) For all $x, y \in S$ with $[x] \neq [y]$, $[x] \cap [y] = \emptyset$.

(3) $\bigcup \{[x] : x \in S\} = S$.

Throughout the rest of this proof, let $x \in S$.

By using the property that $\sim$ is reflexive, you can see that $[x] \neq \emptyset$ because $x \in [x]$. Thus, condition (1) holds.

To see (2), let $x, y \in S$ with $[x] \neq [y]$. It must be shown that $[x] \cap [y] = \emptyset$. Following the contrapositive method, suppose that $[x] \cap [y] \neq \emptyset$, so there is an element $z \in [x]$ such that $z \in [y]$. Hence $x \sim z$ and $y \sim z$. It is shown that $[x] = [y]$, or equivalently, that $[x] \subseteq [y]$ and $[y] \subseteq [x]$. To prove that $[x] \subseteq [y]$, let $w \in [x]$, so $x \sim w$. It follows that $w \in [y]$ because

(a)	$x \sim z$	$(z \in [x])$
(b)	$y \sim z$	$(z \in [y])$
(c)	$x \sim w$	$(w \in [x])$
(d)	$z \sim x$	(from (a) and symmetry)
(e)	$y \sim x$	(from (b), (d), and transitivity)
(f)	$y \sim w$	(from (e), (c), and transitivity)
(g)	$w \in [y]$	(definition of $[y]$).

Thus, $[x] \subseteq [y]$. A similar argument shows that $[y] \subseteq [x]$ and hence $[x] = [y]$. Thus, condition (2) holds.

For (3), you can see that $\bigcup \{[x] : x \in S\} \subseteq S$ because $[x] \subseteq S$ and the union of subsets of S is again a subset of S. To see that $S \subseteq \bigcup \{[x] : x \in S\}$, let $y \in S$, for which it must be shown that $y \in \bigcup \{[x] : x \in S\}$, that is, that y belongs to one of these sets. But because $\sim$ is reflexive, $y \in [y]$ and so in fact $y \in \bigcup \{[x] : x \in S\}$. Thus $S = \bigcup \{[x] : x \in S\}$.

So, (1), (2), and (3) hold, and indeed $\{[x] : x \in S\}$ forms a partition of S, as desired.

QED

Theorem 2.1 states that an equivalence relation provides a partition of the set S into its equivalence classes. You can apply this result to any special case, including new ones you have not yet encountered. For example, consider the binary relation of *congruent triangles*, that is, two triangles are **congruent** if and only if each of the three sides of one triangle is equal in length to a corresponding side of the other triangle. It is not hard to show that *congruence* is an equivalence relation on the set of all triangles. After doing so, you can apply Theorem 2.1 to conclude that this equivalence relation of *congruence* partitions the set of all triangles into equivalence classes. To understand this partition, consider any triangle, say, ABC. The equivalence class generated by this traingle is the set of all triangles that are *congruent* to triangle ABC. Now consider any other triangle, say, RST, that is *not* congruent to ABC. This new triangle also generates an equivalence class – the set of all triangles that are *congruent* to triangle RST. According to Theorem 2.1, the collection of *all* such equivalence classes forms a partition of the set of all triangles. Another example of the application of Theorem 2.1 to a new equivalence relation is presented in Chapter 5.

In this section, you have seen how to work with an axiomatic system to derive results that are applicable not only to the existing special cases but also to any new special case. A section and chapter summary follow.

Summary of Developing Axiomatic Systems

In this section, you have learned how to create and work with axiomatic systems. A summary of this process follows.

How to Develop an Axiomatic System

Step 1. Create an abstract system by performing abstraction on a mathematical concept you wish to study – say, C. This is done by replacing the specific items in each special case with general objects, each of which belongs to some set – say, S.

Step 2. Identify the axioms you want the general concept C to satisfy so that C reflects the common properties of the special cases.

Step 3. Put the properties identified in Step 2 together to form an axiomatic system.

Step 4. By an appropriate substitution, verify that each of the special cases satisfies the axioms listed in Step 3.

Step 5. Derive results pertaining to the axiomatic system that you can then apply to any special case, including new ones not identified in Step 1.

Chapter Summary

In Section 2.1, you learned how to create visual images of mathematical concepts and how to translate visual images first to verbal form and then to symbolic form expressed in the language of mathematics. This process is critical for your ability to work with mathematical ideas.

In Section 2.2, you saw how to create definitions by comparing and contrasting objects that have desirable properties against objects that do not. Through the process of identifying similarities and differences, you learned to isolate and to state the desirable properties in symbolic form so that the resulting definition work both ways: (1) all desirable objects should satisfy the property in the definition and (2) any object satisfying the property should be a desirable object.

In Section 2.3, you learned how to develop and work with an axiomatic system. To do so, you first apply abstraction to a number of special cases to create an abstract system. Then, by identifying the similarities inherent in the special cases, you isolate and define appropriate axioms in the form of properties you want the abstract system to satisfy. You can then study the resulting axiomatic system and apply any knowledge thus obtained not only to each of the special cases, but also to any new special case you have yet to encounter, with no additional work other than an appropriate substitution.

Now it is time to see how the ideas in Chapters 1 and 2 are applied in specific areas of mathematics. In the remaining chapters of this book, selected topics in discrete mathematics, linear algebra, abstract algebra, and real analysis are presented.

Exercises

Exercise 2.1 Create a picture for each of the following mathematical concepts.

(a) Suppose that $\mathbf{d} = (d_1, d_2)$ is a vector starting at the origin of the plane and t is a real number whose value can vary. Draw the vectors $t\mathbf{d} = (td_1, td_2)$ for all possible values of t (positive and negative).

(b) Suppose that A and B are overlapping sets in the plane. Draw a picture to indicate the set of points that are in A but not in B together with the set of points that are in B but not in A.

(c) Suppose that f is a real-valued function of one variable. Draw an example of the graph of a function f having several points x^* for which $f(x^*) = x^*$. Indicate where the points x^* are in your picture.

(d) Among the first six positive integers, some integers divide others evenly. For example, 3 divides 6 evenly whereas 3 does *not* divide 4 evenly. Draw a graph using vertices and edges (see Example 2.4 in Section 2.1.1) to illustrate which of these integers divide which other integers evenly. (Hint: Use a *directed* graph. What do the vertices and edges mean in the context of this problem?)

Exercise 2.2 Create a picture for each of the following mathematical concepts.

(a) Given two sets A and B of points in the plane, a *separating line* is a line for which all of A is on one side and all of B is on the other side. Draw an example of two sets in the plane together with a separating line. Also draw an example of two sets in the plane for which there is no separating line.

(b) Suppose that $\mathbf{x} = (x_1, \ldots, x_n)$ is an n-vector starting at the origin and imagine a plane through the origin that is perpendicular to $\mathbf{x}$. Now consider another n-vector $\mathbf{y} = (y_1, \ldots, y_n)$ that also starts at the origin. The result of multiplying the two n-vectors $\mathbf{x}$ and $\mathbf{y}$ is the *number* $\mathbf{x} \cdot \mathbf{y}$ computed as follows:

$$\mathbf{x} \cdot \mathbf{y} = x_1 y_1 + \cdots + x_n y_n .$$

The sign of $\mathbf{x} \cdot \mathbf{y}$ has the following geometric meaning:

$$\begin{cases} \mathbf{y} \text{ lies on the same side of the plane as } \mathbf{x}, & \text{if } \mathbf{x} \cdot \mathbf{y} > 0 \\ \mathbf{y} \text{ lies on the opposite side of the plane as } \mathbf{x}, & \text{if } \mathbf{x} \cdot \mathbf{y} < 0 \\ \mathbf{y} \text{ lies on the plane}, & \text{if } \mathbf{x} \cdot \mathbf{y} = 0. \end{cases}$$

For a given vector $\mathbf{x}$ in two dimensions, draw three different $\mathbf{y}$ vectors that illustrate the three possible values for the sign of $\mathbf{x} \cdot \mathbf{y}$. What is the geometric relation of $\mathbf{x}$ to $\mathbf{y}$ when $\mathbf{x} \cdot \mathbf{y} = 0$? (Hint: In two dimensions, a plane becomes a line.)

(c) Suppose that f is a real-valued function of one variable and that $x < z$ are two points on the x-axis. Draw an example of a function f having the property that the graph of f at all points between x and z lies below the line segment connecting the points $(x, f(x))$ and $(z, f(z))$ on the graph of the function.

(d) A real-valued function f of one variable associates to each real number x, a real number, $f(x)$. One way to picture this function is by drawing its graph. As a generalization, consider a function g that associates to each real number x, a *set* of real numbers, denoted by $g(x)$. Draw a picture – similar to a graph – to illustrate this concept of a *point-to-set map*. (Hint: To simplify, suppose that for each value of x, $g(x)$ is an *interval*.)

Exercise 2.3 Convert each of the following desired actions to symbolic form. Introduce appropriate notation and symbols as necessary.

(a) You want to shift the entire graph of the real-valued function f up by some fixed positive amount.

(b) Suppose that $\mathbf{x} = (x_1, x_2)$ and $\mathbf{y} = (y_1, y_2)$ are two points in the plane. You want to find the vector that points from $\mathbf{x}$ to $\mathbf{y}$. Then generalize your result to the case when $\mathbf{x}$ and $\mathbf{y}$ are n-vectors.

(c) Suppose that S is a nonempty set of real numbers and that u is a given real number. You want to express the fact that the whole set S is strictly to the left of u.

Exercise 2.4 Convert each of the following desired actions to symbolic form. Introduce appropriate notation and symbols as necessary.

(a) You want conditions to insure that the parabola $ax^2 + bx + c$ crosses the x-axis in two different places.

(b) Given two sets A and B, you want to create a new set by removing from A those elements that are also in B.

(c) Suppose that S is a nonempty set of real numbers. You want to express the fact that, of all the real numbers z that lie to the right of the entire set S, u is the *smallest* such real number.

Exercise 2.5 Convert each of the following desired actions to symbolic form. Introduce appropriate notation and symbols as necessary.

(a) Suppose that $\mathbf{x}$ is an n-vector, each of whose components is 0 or 1. You want to create a new n-vector $\mathbf{y}$ in which the component y_i is defined as follows:

$$y_i = \begin{cases} 1, & \text{if } x_i = 0 \\ 0, & \text{if } x_i = 1. \end{cases}$$

Express y_i in terms of a *single* computation involving x_i.

(b) Suppose that $\mathbf{x}$ is a given point in the plane and $\mathbf{d}$ is a direction of movement in the form of a vector that points out of $\mathbf{x}$. You want to move from $\mathbf{x}$ in the direction $\mathbf{d}$ by an amount of t multiples of the vector $\mathbf{d}$ to get to the new point $\mathbf{x}'$. (Here, t is a real number.)

(c) Suppose that A and B are sets of points in 3 dimensions and that P is a plane in 3 dimensions. You want to express the fact that all of A is on one side of P and all of B is on the other side. (Hint: First write an equation for the plane P.)

Exercise 2.6 Convert each of the following desired actions to symbolic form. Introduce appropriate notation and symbols as necessary.

(a) Suppose that s_1, s_2, and s_3 are variables whose values are 1, if a particular stock is to be purchased and 0, otherwise. You want to use these variables to write a *single* mathematical expression to insure that at least *one* of the three stocks is purchased.

(b) Suppose that L is a line in the plane that goes through the origin and that the vector from the origin to a point $(\bar{x}, \bar{y})$ is perpendicular to L. You want to write the fact that the vector from the origin to any point (x, y) on the line L is perpendicular to the vector $(\bar{x}, \bar{y})$, without using the word *perpendicular*. (Hint: Use Exercise 2.2(b).)

(c) Generalize the concept in Exercise 2.5(c) to sets A and B and a plane P in n-dimensions.

Exercise 2.7 For each of the following definitions in the Glossary at the end of this book, identify the *objects* being defined, the *name* given to those objects having the property, and the *property* itself.

(a) The definition of *divides*.

(b) The definition of an *infimum of a set of real numbers*.

(c) The definition of *linearly independent vectors*.

Exercise 2.8 For each of the following definitions in the Glossary at the end of this book, identify the *objects* being defined, the *name* given to those objects having the property, and the *property* itself.

(a) The definition of a *greatest common divisor*.

(b) The definition of an *injective function*.

(c) The definition of *convergence of a sequence of real numbers to a real number*.

Exercise 2.9 For each of the definitions in Exercise 2.7(a) and (b), give two examples: one of an object that satisfies the property and one of an object that does not satisfy the property.

Exercise 2.10 For each of the definitions in Exercise 2.8(a) and (b), give two examples: one of an object that satisfies the property and one of an object that does not satisfy the property.

Exercise 2.11 Convert the following symbolic statements to visual form:

(a) For all real numbers x, $f(x) \leq g(x)$ (where f and g are real-valued functions of one variable).

(b) For all real numbers x, $f(x) \geq mx + b$ (where f is a real-valued function of one variable and m and b are given real numbers).

Exercise 2.12 Convert the following symbolic statements to visual form:

(a) For all real numbers y, there is a real number x such that $f(x) > y$ (where f is a real-valued functions of one variable).

(b) For all real numbers x and $\bar{x}$, $f(x) \geq f(\bar{x}) + f'(\bar{x})(x - \bar{x})$ (where f is a real-valued function of one variable and $f'(\bar{x})$ is the derivative of f at $\bar{x}$).

Exercise 2.13 Identify all syntax errors, if any, that arise in the following definitions.

(a) A real-valued function f of one variable is *increasing* if and only if for all real numbers y with $y > x$, $f(y) > f(x)$.

(b) A set S of real numbers has an *upper bound* if and only if there is a real number u such that for all elements $x \in S$, $x \leq u$.

(c) Suppose that $\mathbf{x}'$ is a point in the plane and L is a line in the plane. The point $\mathbf{x}$ on the line L is the *projection of* $\mathbf{x}'$ *onto* L if and only if for all points $\mathbf{y}$ on the line, $|\mathbf{x} - \mathbf{x}'| \leq |\mathbf{y} - \mathbf{x}'|$.

Exercise 2.14 Identify all syntax errors, if any, that arise in the following definitions.

(a) The n-vector $\mathbf{x}$ is $\leq$ the n-vector $\mathbf{y}$ if and only if $x_i \leq y_i$.

(b) A set S of points in n dimensions is a *cone* if and only if for all elements $\mathbf{t}, \mathbf{x} \in S$, $\mathbf{t} \cdot \mathbf{x} \in S$, where $\mathbf{t} \cdot \mathbf{x} = t_1x_1 + \cdots + t_nx_n$.

(c) The real-valued function f of one variable is *convex* if and only if for all real numbers x and y, and for all real numbers t with $0 \leq t \leq 1$, $f(tx + (1-t)y) \leq tf(x) + (1-t)f(y)$.

Exercise 2.15 Suppose that S is a set of points in the plane. To capture the concept of

A: S contains the line segment connecting any two points in S,

the following definition was made (in which the *property* is the statement labeled **B**):

A set S of points in the plane is *convex* if and only if **B:** for all elements $\mathbf{x}, \mathbf{z} \in S$, there is a point $\mathbf{y}$ on the line segment between $\mathbf{x}$ and $\mathbf{z}$ such that $\mathbf{y} \in S$.

(a) Is it *true* that if a set S satisfies the property in **A**, then S also satisfies the property in **B**? If not, draw an example of a set S that satisfies **A** but not **B**.

(b) Is it *true* that if a set S satisfies the property in **B**, then S also satisfies the property in **A**? If not, draw an example of a set S that satisfies **B** but not **A**.

(c) On the basis of your results in parts (a) and (b), rewrite statement **B** to create a correct definition that captures the concept in **A**.

Exercise 2.16 Suppose that S is a nonempty set of real numbers. To capture the concept of

A: The largest element in the set S,

the following definition was made (in which the property is the statement labeled **B**):

The real number u is the *maximum element* of the set S of real numbers if and only if **B:** for all elements $x \in S$, $x \leq u$.

(a) Is it *true* that if u satisfies the property in **A**, then u also satisfies the property in **B**? If not, give an example of a set S and a real number u that satisfies **A** but not **B**.

(b) Is it *true* that if u satisfies the property in **B**, then u also satisfies the property in **A**? If not, give an example of a set S and a real number u that satisfies **B** but not **A**.

(c) On the basis of your results in parts (a) and (b), rewrite statement **B** to create a correct definition that captures the concept in **A**.

Exercise 2.17 The objective of this exercise is to develop a definition that captures the concept of a "U-shaped" parabola, such as the following:

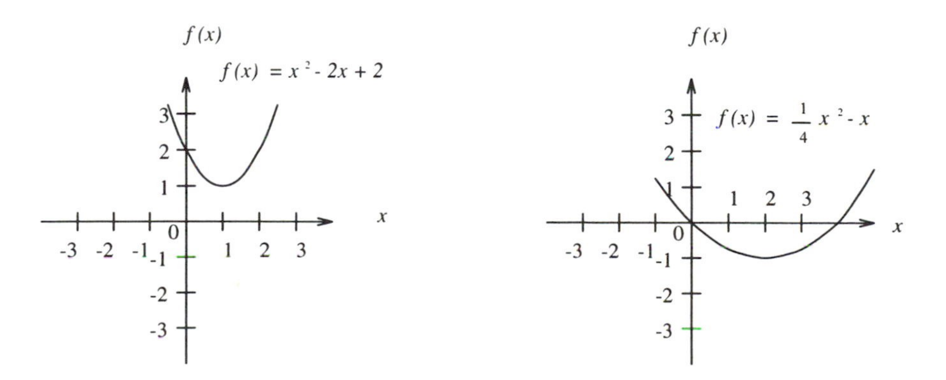

In contrast, the following parabolas are *not* U-shaped:

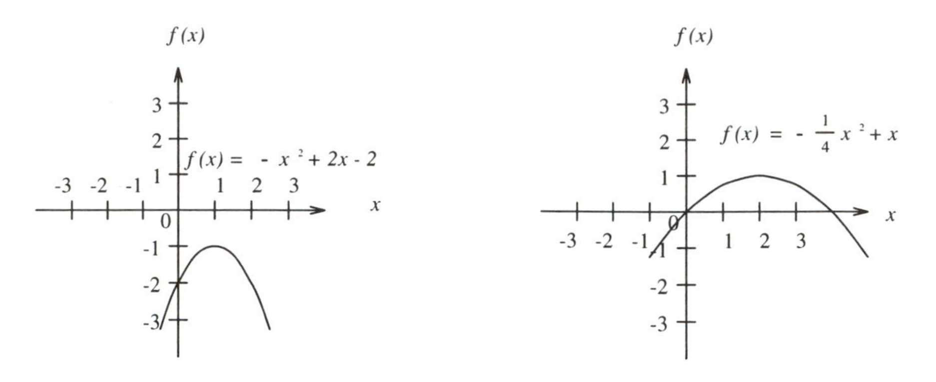

On the basis of the similarities and differences in these examples (and others, if necessary) complete the following definition (in which a, b, and c are real numbers):

The parabola $ax^2 + bx + c$ is *U-shaped* if and only if ....

Verify your definition against the examples above to make sure that all of those in the first group satisfy your property and none of those in the second group satisfy your property.

Exercise 2.18 The objective of this exercise is to develop a definition that captures the concept of a real-valued function "covering the whole y-axis," such as the following functions:

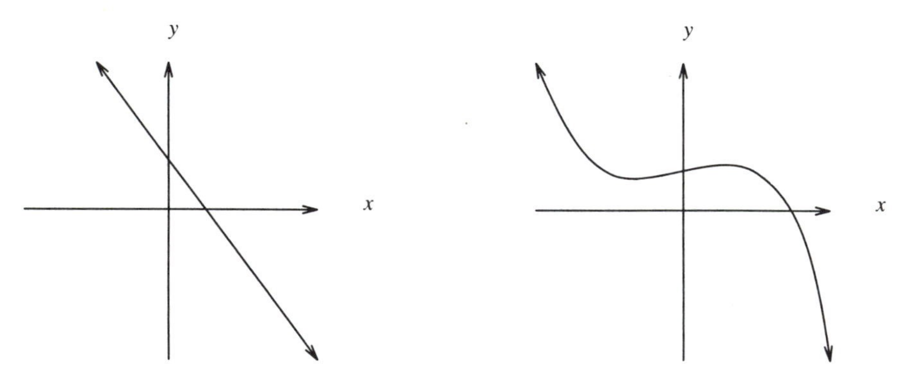

In contrast, the following functions do *not* cover the whole y-axis:

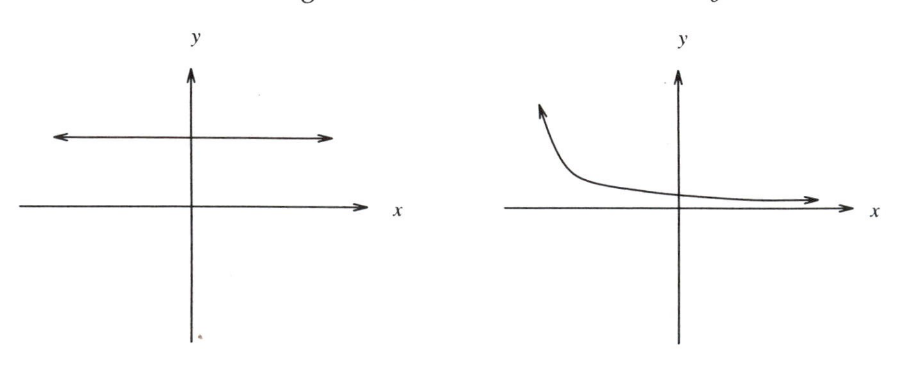

On the basis of the similarities and differences in these examples (and others, if necessary) complete the following definition:

> The function f *covers the whole y-axis* if and only if ....

Verify your definition against the examples above to make sure that all of those in the first group satisfy your property and none of those in the second group satisfy your property. (Hint: Use appropriate quantifiers.)

Exercise 2.19 The objective of this exercise is to develop a definition that captures the concept of the "boundary of a set S of points in n dimensions," such as the boundaries of the sets in 2 dimensions indicated by the bold lines surrounding the shaded regions in the following figures:

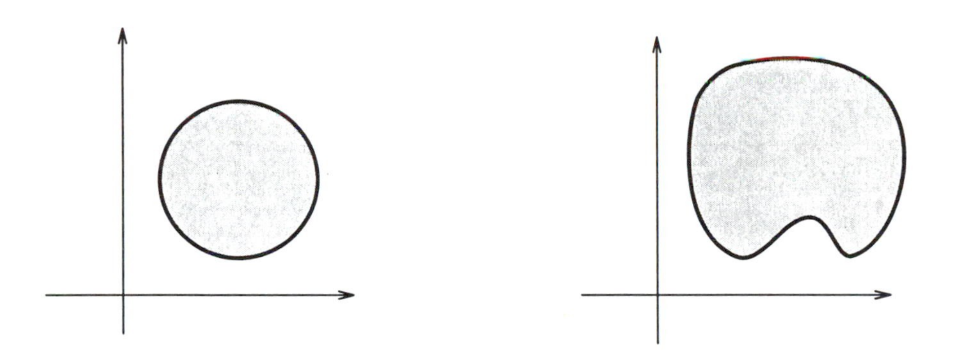

In contrast, all *other* points in these figures (both in the set and not in the set) are *not* boundary points.

On the basis of the similarities between boundary points in these examples (and others, if necessary) complete the following definition:

> An n-vector $\mathbf{x}$ is in the *boundary of a set S of points in n dimensions* if and only if ....

Verify your definition against the examples above to make sure that all of the points on the bolded lines in the figures satisfy your property and that none of the other points in those figures do. (Hint: Use appropriate quantifiers together with the concept of balls of radius r centered at $\mathbf{x}$, as described in Definition 2.9.)

Exercise 2.20 The objective of this exercise is to develop a definition that captures the concept of a real-valued function "never changing directions," such as the following functions:

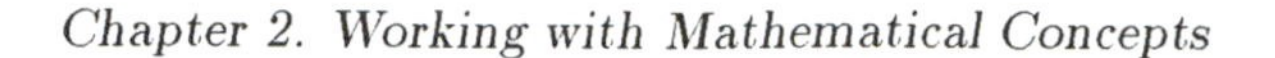

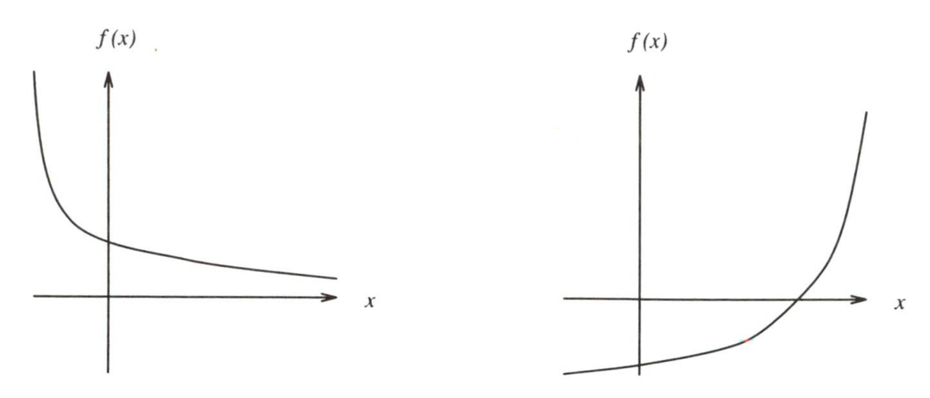

In contrast, the following functions *do* change directions:

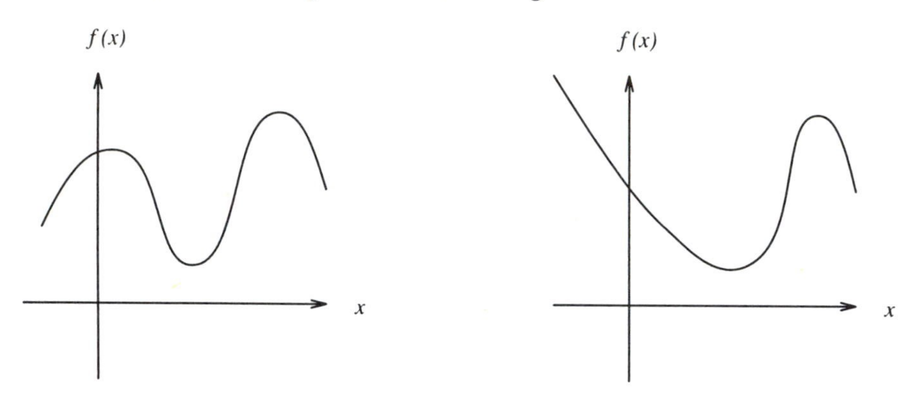

On the basis of the similarities and differences in these examples (and others, if necessary) complete the following definition:

> The function f *never changes directions* if and only if ....

Verify your definition against the examples above to make sure that all of those in the first group satisfy your property and none of those in the second group satisfy your property.

Exercise 2.21 Suppose that p and q are two mathematical statements. As you know, $p \Leftrightarrow q$ is *true* if and only if $p \Rightarrow q$ and $q \Rightarrow p$ are both *true*. Is the binary relation $\Leftrightarrow$ an equivalence relation? If so, prove it. If not, use specific examples of statements p and q to show which of the reflexive, symmetric, and transitive properties do not hold.

Exercise 2.22 Suppose that $\mathbf{x} = (x_1, \ldots, x_n)$ and $\mathbf{y} = (y_1, \ldots, y_n)$ are two n-vectors. Consider comparing these two vectors by saying that $\mathbf{x} \leq \mathbf{y}$ if and only if there is an integer i with $1 \leq i \leq n$ such that $x_i \leq y_i$.

(a) Is this binary relation reflexive? If so, prove it. If not, provide a specific numerical example of $\mathbf{x}$ for which $\mathbf{x} \not\leq \mathbf{x}$.

(b) Is this binary relation symmetric? If so, prove it. If not, provide a specific numerical example of $\mathbf{x}$ and $\mathbf{y}$ for which $\mathbf{x} \leq \mathbf{y}$ but $\mathbf{y} \not\leq \mathbf{x}$.

(c) Is this binary relation transitive? If so, prove it. If not, provide a specific numerical example of **x**, **y**, and **z** for which $\mathbf{x} \le \mathbf{y}$ and $\mathbf{y} \le \mathbf{z}$ but $\mathbf{x} \not\le \mathbf{z}$.

Exercise 2.23 Suppose you have developed an axiomatic system consisting of a set S, a binary operation $\odot$ on S, and three axioms, say, Axiom 1, Axiom 2, and Axiom 3. Now consider a *new* axiomatic system consisting of $(S, \odot)$ and Axioms 1 and 2 only. Is this new axiomatic system a special case or a generalization of the original axiomatic system? Explain.

Exercise 2.24 Suppose you have developed an axiomatic system consisting of a set S, a binary operation $\odot$ on S, and three axioms, say, Axiom 1, Axiom 2, and Axiom 3. Now consider a *new* axiomatic system consisting of $(S, \odot)$ and Axioms 1, 2, 3, and a new Axiom 4. Is this new axiomatic system a special case or a generalization of the original axiomatic system? Explain.

Exercise 2.25 Create an axiomatic system for the concept of *strict inequality* on the basis of your knowledge of the following special cases:

(i) $x < y$, meaning that the real number x is strictly less than the real number y.

(ii) $A \subset B$, meaning that the set A is a strict subset of the set B (that is, A is a subset of B but A is not equal to B).

Specifically, let S be a set of objects and $\prec$ be a binary relation on S. What properties must $\prec$ satisfy to be considered similar to the binary relations of $<$ and $\subset$ in the foregoing special cases?

Exercise 2.26 Create an axiomatic system for the binary operator of *intersection* that captures the property that, whenever you intersect two sets, you get another set that is a *subset* of each of the original sets. Specifically, think about a set S of objects instead of sets. Replace the concept of *intersection* with a more general binary operator, say, $\odot$, and the binary relation of *subset* with a more general binary relation, say, $\preceq$. What properties do $\odot$ and $\preceq$ need to have on the elements of S to be considered in the same way as *intersection* and *subset* on sets?

Chapter 3

SELECTED TOPICS IN DISCRETE MATHEMATICS

The objective of the remaining chapters of this book is to show you how the ideas from Chapters 1 and 2 are applied in various areas of mathematics. In this chapter, selected topics from **discrete mathematics** – the study of mathematical problems involving a *finite* number of items – illustrate these principles.

3.1 Sets

As you have already seen, a **set** is a collection of objects that are generally, but not necessarily, related to each other. Each object in the set is called an **element**, or **member**, of the set and each such element is said to **be in** or **belong to** the set. A set is a unification in that the various individual elements are put together in a single set. Sets play a central role in many areas of mathematics – including abstraction and developing axiomatic systems – because they allow you to group related items together so that you can work with them and study their common properties. In this section, you will learn more precisely how to do so by developing visual images of sets, translating those images to symbolic form, creating definitions, performing operations on sets, and more. Some of the material is elementary in nature and is included for completeness and also to illustrate the ideas from Chapters 1 and 2.

3.1.1 Sets and Their Representations

The first issue is how to represent a set so that you can both think about and work with that set. Correspondingly, there is a *visual* representation that is helpful for picturing a set and a more formal *symbolic* representation that is used to process and work with sets, both of which are described next.

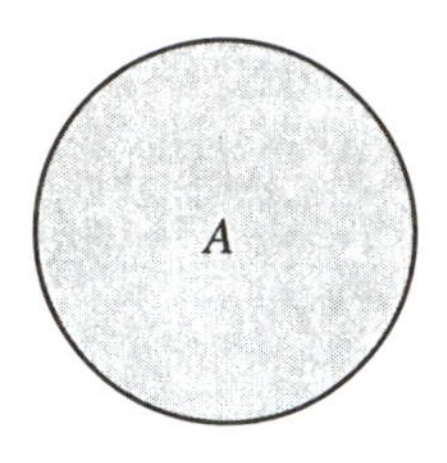

Figure 3.1: A Venn Diagram for Representing a Set

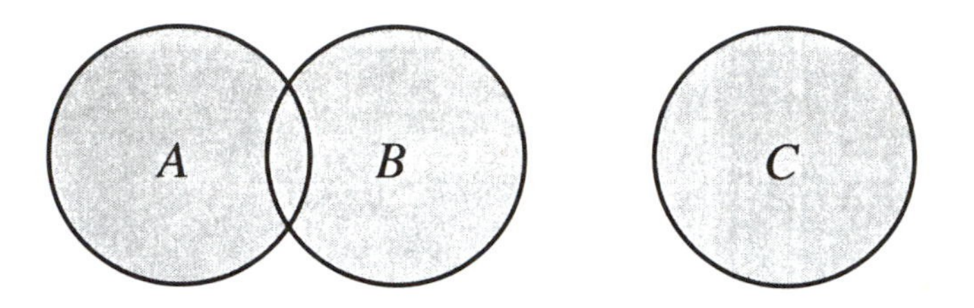

Figure 3.2: A Venn Diagram Illustrating Several Sets

Visual Representation of a Set

As you already saw in Example 2.1 in Section 2.1.1, the most common way to visualize a set is to use a *Venn diagram*, such as the one in Figure 3.1. Everything inside the shaded region is in the set; everything outside the shaded region is *not* in the set. You can easily picture several sets simultaneously in a single Venn diagram, as shown in Figure 3.2. Picturing sets in this way can help you to solve problems, as you will see later in this section. However, to do so, you need to work with a *symbolic* representation of a set, as described next.

Symbolic Representations of a Set

One way to represent a set symbolically is to use a **list notation**, in which you enclose the list of the members of the set – separated from each other by commas – in braces. For example, the set consisting of the three integers 1, 5, and 8 is written as

$$\{1, 5, 8\}.$$

In the event that you need to write this set numerous times, it is convenient to assign a *symbolic* name to the set. For example, you might write

$$A = \{1, 5, 8\}.$$

Then, whenever you refer to the symbol A, it means the set consisting of the three elements 1, 5, and 8.

Some sets have so many elements that you cannot list them all explicitly. For example, consider the set $\mathbb{N}$ of all positive integers. One way to represent such a set is by using three dots, for example:

$$\mathbb{N} = \{1, 2, 3, \ldots\}.$$

Another practical alternative for describing sets with many elements is to use **set-builder notation**, in which you use a verbal and mathematical description of the property the elements of the set must satisfy, called the **defining property**. For example, you can describe the set $\mathbb{N}$ as follows:

$$\mathbb{N} = \{\text{integers } k : k > 0\},$$

in which the colon (:) is read as the words "such that."

As another example of using set-builder notation, you can describe the set of all real numbers between 1 and 3 as follows:

$$A = \{\text{real numbers } x : 1 \leq x \leq 3\}.$$

In this example, the defining property ($1 \leq x \leq 3$) appears to the *right* of the colon but, in some cases, part of that property might also appear to the *left* of the colon. For example,

$$A = \{\text{real numbers } x > 0 : x^2 - x - 2 \leq 0\}.$$

Whatever form the set is written in, the defining property is used to determine which objects *are* and *are not* in the set: *only those objects that satisfy the defining property are members of the set.*

You now have three methods for representing sets: the informal visual image of a Venn diagram and the more formal symbolic list and set-builder notations. When working with sets, you need to convert your visual images to this symbolic form.

Sets with Special Properties

After working with sets for a while, you will discover that there are several special sets that arise frequently. Some of these are described next.

The Universal Set. When you think of a set as a collection of related objects, an important underlying issue is the list of all *possible* items that *could* be included in the set. This encompassing group of items constitutes the **universal set**.

In some cases, the universal set is described *explicitly*. For example, when working with a set of exam scores between 0 and 100, the universal set is

$$U = \{0, 1, \ldots, 100\}.$$

If S is one specific set of exam scores, then you know that each member of S is one of the elements in the universal set, that is, 0 to 100.

In many cases, however, the universal set is understood *implicitly*. For example, if you are working with

$$A = \{\text{real numbers } x > 0 : x^2 - x - 2 \leq 0\},$$

then, implicitly, the universal set is the set of all real numbers. As another example, if you are working with

$$B = \{\text{integers } x : x^2 - x - 2 \leq 0\},$$

then, implicitly, the universal set is the set of all integers.

The Empty Set. When the elements of a set are listed explicitly, you know that there is at least *one* element in the set. However, when set-builder notation is used, the set may have *no* elements. For example, the following set has no elements:

$$\{\text{real numbers } x > 0 : x + 1 \leq 0\}.$$

A set with no elements is called the **empty set** and is denoted by the symbol $\emptyset$. Mathematicians sometimes write $\{\}$ for the empty set.

When working with sets, care is needed to avoid certain operations that cannot be performed when the set is empty. For example, in doing a proof, suppose you make the following statement:

Let x be an element of the set A, ...

In this case, you must be sure that A is not the empty set because, if A *is* empty, then you cannot "let x be an element of A."

Other Special Sets. Five sets of numbers arise frequently in this book and are referred to by the following symbols:

$$\begin{aligned}
\mathbb{N} &= \text{the set of positive integers,}\\
\mathbb{Z} &= \text{the set of all integers,}\\
\mathbb{Q} &= \text{the set of all rational numbers,}\\
\mathbb{R} &= \text{the set of all real numbers,}\\
\mathbb{C} &= \text{the set of all complex numbers.}
\end{aligned}$$

Alternatively, in set-builder notation,

$$\begin{aligned}
\mathbb{N} &= \{1, 2, 3, \ldots\},\\
\mathbb{Z} &= \{\ldots, -2, -1, 0, 1, 2 \ldots\},\\
\mathbb{Q} &= \{\tfrac{p}{q} : p \text{ and } q \text{ are integers with } q \neq 0\},\\
\mathbb{R} &= \{\text{numbers } x : x \text{ is expressible in decimal form}\},\\
\mathbb{C} &= \{a + bi : a \text{ and } b \text{ are real numbers and } i = \sqrt{-1}\}.
\end{aligned}$$

3.1.2 Comparing Sets

Whenever you encounter a new mathematical concept (such as a set), you know that you should develop an assoicated visual image (such as a Venn diagram). However, keep in mind that the reason for creating the concept in the first place is to help you solve certain problems. You will therefore need to work with that concept. To work effectively with sets requires learning about each of the following:

1. How to compare sets to understand their relationship to each other.
2. How to perform unary and binary operations on sets to create new sets.
3. The properties and rules that sets obey.

A comparison involving a set is an operation whose result is either *true* or *false*. In this section, several different types of comparisons pertaining to sets and their elements are presented.

Checking for Membership

As you know, each item belonging to a set is a *member* of that set. Thus, for example, the numbers 2, 4, and 6 are the three members of the set

$$A = \{2, 4, 6\}.$$

However, when set-builder notation is used, you do not "see" all the elements of the set. Therefore, you often want to check whether a particular item is or is not a member of the set. The symbol $\in$ is used to do just that. If x is a particular item and A is a given set, then the statement

$$x \in A$$

means that x *is* an element of A, or equivalently, that the result of checking whether x is an element of A is *true*. Analogously, the statement

$$x \notin A$$

means that x is *not* an element of A, or equivalently, that the result of checking whether x is an element of A is *false*. These ideas are illustrated in the following examples.

Example 3.1 – Checking for Membership in a Set

Suppose that $A = \{1, 2\}$. Then each of the following comparisons is *true*:

$$\begin{aligned} 1 &\in A, \\ 2 &\in A, \\ \{2\} &\notin A. \end{aligned}$$

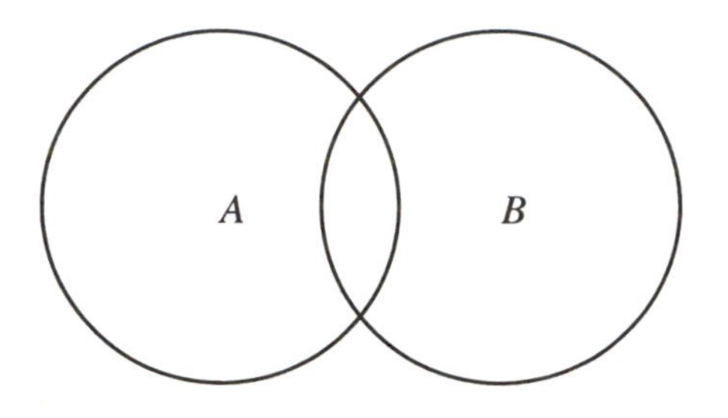

Figure 3.3: A Set A that Is Not Contained in a Set B

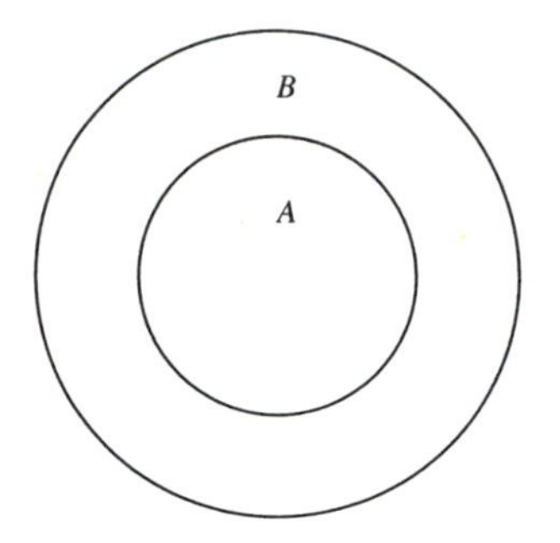

Figure 3.4: A Set A that Is Contained in a Set B

Example 3.2 – Checking for Membership in a Set

Recall that $\mathbb{R}$ is the set of real numbers and suppose $A = \{x \in \mathbb{R} : x^2 - x - 2 \leq 0\}$. Then, by using the defining property of the set, you can determine that each of the following comparisons is *true*:

$$\begin{aligned} 0 &\in A, \\ 1 &\in A, \\ 3 &\notin A, \\ -2 &\notin A. \end{aligned}$$

Checking Whether a Set is a Subset of Another Set

As you just saw, the symbol $\in$ is used to compare a particular item with the elements in a set. Another useful comparison is between two *sets*. In Figure 3.3 and Figure 3.4, you see some of the possible relationships between two sets A and B. In Figure 3.4, the set A is completely "contained in" the set B, which is not the case with the sets in Figure 3.3.

As discussed in Section 2.2.2, after you identify a desirable property

– such as a set A being "contained in" a set B – you should develop a corresponding definition. This is done by translating your visual image to symbolic form using the vocabulary of mathematics. In this case, you can create such a definition by using the concept of *membership* that you just learned. Specifically, the desirable property in Figure 3.4 is that *all* the elements of A are in B. By putting together the pieces and using the quantifier *for all*, you obtain the following definition.

> **Definition 3.1** *A set A is* **contained in** (**included in** or a **subset of**) *a set B, written $A \subseteq B$, if and only if for all $x \in A$, $x \in B$.*

The symbol $\subseteq$ is used to compare two sets. Thus, the statement

$$A \subseteq B$$

means that A *is* a subset of B, or equivalently, that the result of checking whether A is a subset of B is *true*. Analogously, the statement

$$A \not\subseteq B \tag{3.1}$$

means that A is *not* a subset of B, or equivalently, that the result of checking whether A is a subset of B is *false*. By using the rules for negating a statement with quantifiers (see Section 1.6.10), the statement in (3.1) is the same as the following:

There is an element $x \in A$ such that $x \notin B$.

These ideas are illustrated in the following example.

> **Example 3.3 – Checking for a Subset Relation**
>
> Each of the following comparisons is *true* (where A is a set whose elements come from a universal set U):
>
> $\{2\} \subseteq \{0, 1, 2, 3\}$ (because $2 \in \{0, 1, 2, 3\}$),
> $\{2, 1\} \subseteq \{3, 1, 2, 0\}$ (the *order* of elements does not matter),
> $\{1, 4\} \not\subseteq \{0, 1, 2, 3\}$ (because $4 \in \{1, 4\}$ but $4 \notin \{0, 1, 2, 3\}$),
> $A \subseteq U$ (because all elements of A come from U),
> $\emptyset \subseteq A$ (the empty set is a subset of every set).
>
> The last comparison is *true* from Definition 3.1 because every element of $\emptyset$ – of which there are none – belongs to the set A.

Observe that it *is* correct to write

$2 \in \{0, 1, 2, 3, 4\}$ and $\{2\} \subseteq \{0, 1, 2, 3, 4\}$,

but a syntax error results if you write

$2 \subseteq \{0, 1, 2, 3, 4\}$.

This is because you cannot use the symbol $\subseteq$ to compare the *number* 2 and the *set* $\{0, 1, 2, 3, 4\}$, but only to compare two *sets*.

In Example 3.3, you can easily check the subset relationships because the elements of both sets are listed *explicitly*. A more challenging situation arises when the two sets have too many elements to list and are thus described with set-builder notation, as shown in the following example.

Example 3.4 – Using a Proof to Show a Subset Relation

Suppose that

$$\begin{aligned} A &= \{x \in \mathbb{R} : -1 \leq x \leq 2\} \text{ and} \\ B &= \{x \in \mathbb{R} : x^2 - x - 2 \leq 0\}. \end{aligned}$$

In this case, you cannot list all the elements of A and B, so how do you check if $A \subseteq B$?

Solution to Example 3.4

The answer is to use a *proof* to show that $A \subseteq B$, as follows.

Proof. By Definition 3.1, it must be shown that

for all elements $x \in A$, $x \in B$.

The quantifier *for all* suggests choosing an arbitrary element $x \in A$, for which it must be shown that $x \in B$. However, because $x \in A$, by the defining property of A,

$$-1 \leq x \leq 2.$$

Thus, $x - 2 \leq 0$ and $x + 1 \geq 0$, so

$$x^2 - x - 2 = (x - 2)(x + 1) \leq 0.$$

Consequently, x satisfies the defining property of B and so $x \in B$, thus establishing that $A \subseteq B$. QED

The following points are worth noting from Example 3.4:

1. Proofs allow you to establish that certain facts are *true*.

2. The advantage of a symbolic definition is that the you can often identify certain *key words* (such as *for all*, in Definition 3.1), that indicate which proof technique to use. (See Section 1.6 for more details on using proof techniques.)

Comparing Two Sets for Equality

Yet another useful comparison between two sets is to see if they are *equal* (that is, the same). For two sets to be equal, the elements in each set should be the same. On the basis of this concept, you can use the previous notion of a subset in Definition 3.1 to arrive at the following definition of what it means for a set A to be equal to a set B.

Definition 3.2 *Two sets A and B are* **equal**, *written $A = B$, if and only if $A \subseteq B$ and $B \subseteq A$.*

The symbol = is used to compare two sets. Thus, the statement

$$A = B$$

means that A *is* equal to B, or equivalently, that the result of checking whether A is equal to B is *true*. Analogously, the statement

$$A \neq B \tag{3.2}$$

means that A is *not* equal to B, or equivalently, that the result of checking whether A is equal to B is *false*. By using the rules for negating a statement containing the word *and* (see Section 1.6.10), the statement in (3.2) is the same as the following:

$$A \not\subseteq B \text{ or } B \not\subseteq A. \tag{3.3}$$

By applying the rules of negation to Definition 3.1, the statement in (3.3) is the same as the following:

There is an element $x \in A$ such that $x \notin B$ or else
there is an element $x \in B$ such that $x \notin A$.

These ideas are illustrated in the following example.

Example 3.5 – Comparing Two Sets for Equality

The following comparisons are all *true*:

$$\begin{aligned} \{1,3,2\} &= \{3,1,2\} && \text{(the order of elements does not matter)}, \\ \{1,2,3\} &\neq \{1,2\} && \text{(because } 3 \in \{1,2,3\} \text{ but } 3 \notin \{1,2\}), \\ \{2,3\} &\neq \{2,3,4\} && \text{(because } 4 \in \{2,3,4\} \text{ but } 4 \notin \{2,3\}). \end{aligned}$$

As with the concept of subset, when the sets have too many elements to list, you can show that they are equal by means of a *proof*. This approach is illustrated in the following example.

Example 3.6 – Using a Proof to Show that Two Sets are Equal

Suppose that

$$A = \{x \in \mathbb{R} : -1 \le x \le 2\}, \text{ and}$$
$$B = \{x \in \mathbb{R} : x^2 - x - 2 \le 0\}.$$

You cannot list all the elements of these two sets, so how do you check that $A = B$?

Solution to Example 3.6

The answer is to use a proof to show that $A = B$, as follows.

Proof. By Definition 3.2, it must be shown that

$A \subseteq B$ and $B \subseteq A$.

It was already shown that $A \subseteq B$ in Example 3.4, so it remains only to show that $B \subseteq A$, and this is left as an exercise. QED

As a final note regarding the comparison of two sets, it is sometimes necessary to work with a set A that is a *strict* subset of a set B, the precise meaning of which is given in the following definition.

Definition 3.3 *A set A is a* **strict subset** *of a set B, written $A \subset B$, if and only if $A \subseteq B$ and $A \neq B$.*

3.1.3 Unary Operations on Sets

Whenever you encounter new mathematical objects – such as sets – you should learn the different kinds of unary and binary operations that you can perform on those objects to produce new objects of the same kind. In this section, several examples of unary operations on sets are presented.

Complement of a Set

An example of a unary operation on a set A is the *complement* of that set which, as illustrated in Example 2.8 in Section 2.1.2, is the set of points that are in the uiniversal set U but *not* in A. A symbolic definition follows.

Definition 3.4 *Let A be a set of elements from the universal set U. Then the* **complement of** A **in** U, *written A^c, is the following set:*

$$A^c = \{x \in U : x \notin A\}.$$

Example 3.7 – Finding the Complement of a Set

If the universal set is the set $\mathbb{R}$ of real numbers and if

$$A = \{x \in \mathbb{R} : x^2 - x - 2 \leq 0\},$$

then

$$A^c = \{x \in \mathbb{R} : x^2 - x - 2 > 0\}.$$

In some cases, however, the complement of a set A is determined not by the universal set, but rather, with respect to some other set B, in which case, the complement of A has the following meaning.

Definition 3.5 *Let A and B be subsets of elements from the universal set U. Then the* **complement of** A **in** B, *written $B - A$ or $B \backslash A$, is the following set:*

$$B - A = \{x \in B : x \notin A\}.$$

Example 3.8 – Finding the Complement of a Set A in a Set B

Suppose that $A = \{1, 3, 5\}$. Then the following sets are the complements of A in B for various different sets B:

B	A	$B - A$
$\{1, 2, 3, 4, 5\}$	$\{1, 3, 5\}$	$\{2, 4\}$
$\{6, 7\}$	$\{1, 3, 5\}$	$\{6, 7\}$
$\{1, 3\}$	$\{1, 3, 5\}$	$\emptyset$

The Power Set

In the examples so far, each element of a set has been a *number*. However, it is also possible for an element of a set to be another *set*, as shown in the following examples.

Example 3.9 – A Set Containing an Element that Is Also a Set

The set

$$A = \{1, 2, \{1, 2\}\}$$

consists of three elements: the number 1, the number 2, and the *set* $\{1, 2\}$.

Example 3.10 – The Set of All Subsets of a Set

Suppose that $A = \{1, 2\}$. You can create a new set B consisting of all the subsets of A, as follows:

$$B = \{\emptyset,\ \{1\},\ \{2\},\ \{1, 2\}\}.$$

B contains four elements, each of which is a set.

Similarly, if $A = \{1, 2, 3\}$, then the set of all subsets of A is

$$B = \{\emptyset, \{1\}, \{2\}, \{3\}, \{1, 2\}, \{1, 3\}, \{2, 3\}, \{1, 2, 3\}\}.$$

In this case, B contains eight elements, each of which is a set.

By applying generalization to Example 3.10, you can create the concept of "the set of all subsets of a given set A." A formal definition of this concept follows.

Definition 3.6 *The* **power set** *of a given set A, denoted by 2^A, is the following set:*

$$2^A = \{B : B \subseteq A\}.$$

Observe that the sets in Example 3.10 are special cases of Definition 3.6. Note also from these examples that when A has 2 elements, the power set has 4 elements and when A has 3 elements, the power set has 8 elements. In the exercises, you are asked to show that when A has n elements, the power set has 2^n elements (and hence the notation 2^A for the power set of A).

As a final note, observe that the operation of creating the power set is, in fact, a unary operator. That is, the unary operator 2^A creates a new set from A that consists of all the subsets of A.

3.1.4 Binary Operations on Sets

Where unary operations use *one* set to create a new set, binary operations combine *two* sets to create a new set. Several examples of binary operations on sets are presented in this section.

The Union of Two Sets

One way to create a new set from two sets A and B is to combine all of their elements together, as shown by the shaded region in Figure 3.5. Translating the visual image in Figure 3.5 to symbolic form results in the following definition.

Definition 3.7 *The* **union of two sets** *A and B, written $A \cup B$, is the following set:*

$$A \cup B = \{x : x \in A \text{ or } x \in B\}.$$

Example 3.11 – The Union of Two Sets

The following sets are examples of the union of various sets A and B:

A	B	$A \cup B$
$\{1, 2, 3\}$	$\{4\}$	$\{1, 2, 3, 4\}$
$\{1, 2, 3\}$	$\{2, 3, 4\}$	$\{1, 2, 3, 4\}$
$\{1, 2, 3\}$	$\{1, 2\}$	$\{1, 2, 3\}$
$\{x \in \mathbb{R} : 0 \leq x \leq 1\}$	$\{x \in \mathbb{R} : 1 \leq x \leq 3\}$	$\{x \in \mathbb{R} : 0 \leq x \leq 3\}$
$\{x \in \mathbb{R} : x < 0\}$	$\{x \in \mathbb{R} : x \geq 0\}$	$\mathbb{R}$

The Intersection of Two Sets

Another example of a binary operation on two sets A and B is to create a new set consisting of those elements that are in both A and B simultaneously, as shown by the shaded region in Figure 3.6. Translate the visual image in Figure 3.6 to symbolic form to obtain the following definition.

Definition 3.8 *The* **intersection of two sets** *A and B, written $A \cap B$, is the following set:*

$$A \cap B = \{x : x \in A \text{ and } x \in B\}.$$

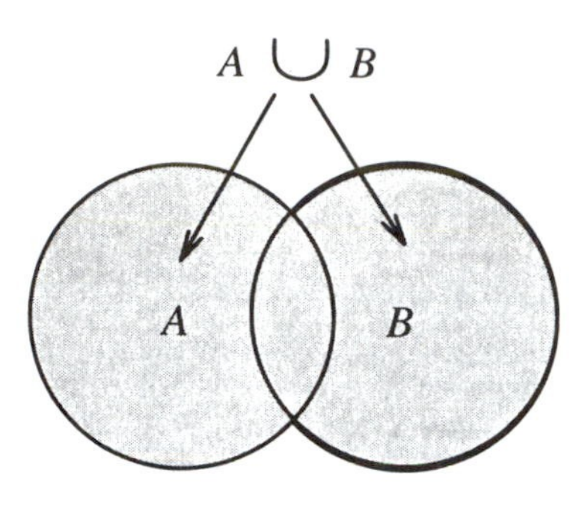

Figure 3.5: The Union of Two Sets

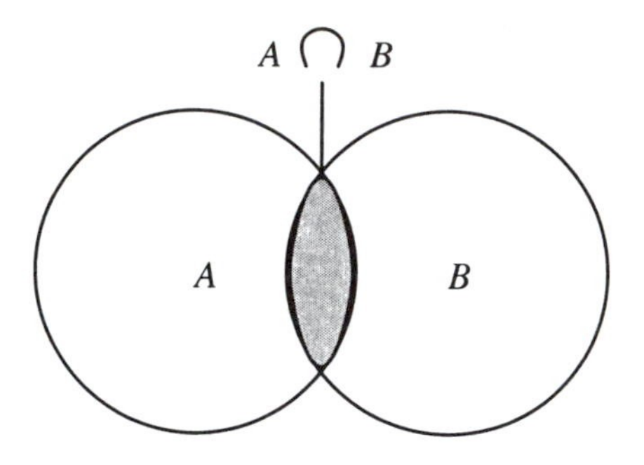

Figure 3.6: The Intersection of Two Sets

Example 3.12 – The Intersection of Two Sets

The following sets are the intersection of the sets A and B in Example 3.11:

A	B	$A \cap B$
$\{1, 2, 3\}$	$\{4\}$	$\emptyset$
$\{1, 2, 3\}$	$\{2, 3, 4\}$	$\{2, 3\}$
$\{1, 2, 3\}$	$\{1, 2\}$	$\{1, 2\}$
$\{x \in \mathbb{R} : 0 \leq x \leq 1\}$	$\{x \in \mathbb{R} : 1 \leq x \leq 3\}$	$\{1\}$
$\{x \in \mathbb{R} : x < 0\}$	$\{x \in \mathbb{R} : x \geq 0\}$	$\emptyset$

The Cartesian Product

Another binary operation on sets arises by applying abstraction. As you learned in Section 1.4.3, one way to generalize the concept of a point (x, y) in the plane is to create an n-vector, $(x_1, \ldots, x_n)$. Another approach, however, is to perform *abstraction* on the point (x, y). Rather than thinking of x and y as real numbers, think of them as *objects* – objects that belong to some set, say, A. Thus, an abstraction of an ordered pair of real numbers is an ordered pair of elements of a *set* A.

A further generalization of an ordered pair of elements arises when you allow the first element of the pair (x, y) to come from one set (say, A) and the second element to come from some *other* set (say, B). Thus, from the original sets A and B, you can use a binary operation to create a new set, denoted by $A \times B$, as follows:

$$A \times B = \{(x, y) : x \in A \text{ and } y \in B\}.$$

Example 3.13 – The Set $A \times B$

The following sets are examples of the set $A \times B$ for various sets A and B:

A	B	$A \times B$
$\{1, 2\}$	$\{3\}$	$\{(1, 3), (2, 3)\}$
$\{3\}$	$\{1, 2\}$	$\{(3, 1), (3, 2)\}$
$\{1, 2\}$	$\{2, 3\}$	$\{(1, 2), (1, 3), (2, 2), (2, 3)\}$

Another generalization of this concept arises when you have n sets, rather than just the two sets A and B. As with n-vectors, a *subscript* notation is more convenient to represent the n sets, say, $A_1, \ldots, A_n$. Now you can create an ordered list of n objects – one from each of the n sets – as stated in the following definition.

Definition 3.9 *Given n sets $A_1, \ldots, A_n$, the* **Cartesian product** *(or* **cross product***), denoted by $A_1 \times \cdots \times A_n$, is the following set:*

$$A_1 \times \cdots \times A_n = \{(x_1, \ldots, x_n) : \textit{for each } i = 1, \ldots, n,\ x_i \in A_i\}.$$

One important special case of the Cartesian product is when each of the sets A_i is the set $\mathbb{R}$ of real numbers. In this case, Definition 3.9 becomes the set of all n-vectors of real numbers because

$$\overbrace{\mathbb{R} \times \cdots \times \mathbb{R}}^{n \text{ times}} = \{(x_1, \ldots, x_n) : \text{for each } i = 1, \ldots, n,\ x_i \in \mathbb{R}\}.$$

This set is so common that, from here on, it is denoted by $\mathbb{R}^n$. Thus, the special case of $\mathbb{R}^1$ is the real line, $\mathbb{R}^2$ is the plane, and $\mathbb{R}^n$ is n-dimensional space.

Any sets A, B, and C satisfy the following properties.

1. $A \cup \emptyset = A$ and $A \cap \emptyset = \emptyset$.
2. $B - A = B \cap A^c$.
3. $A \cup (B \cap C) = (A \cup B) \cap (A \cup C)$.
4. $A \cap (B \cup C) = (A \cap B) \cup (A \cap C)$.
5. $(A \cup B)^c = A^c \cap B^c$.
6. $(A \cap B)^c = A^c \cup B^c$.

Table 3.1: Rules of Operations on Sets.

Properties and Laws of Sets

After working with sets for some time, you will discover certain laws and properties that sets and their operators obey. Some of those rules are summarized in Table 3.1, most of whose proofs are left to the exercises.

In this section you have learned what sets are and how to compare them. You have also seen numerous examples of unary operations (such as complementing a set and finding the power set) and binary operations (such as the union, intersection, and Cartesian product). These operators use one or more sets to create a new set. In the next section, a similar development for functions is presented.

3.2 Functions

You have already seen many examples of *functions*, such as the function $f(x) = 2x + 1$ that associates to each real number x, the real number $2x + 1$. Functions play a central role in many areas of mathematics. For example, the unary operation of complementing a set is actually a function: a function that associates to each set A, another set, namely, the complement of A. In this section, you will learn more precisely what functions are and how to work with them by developing visual images, translating those images to symbolic form, creating definitions, discovering useful properties of functions, and more.

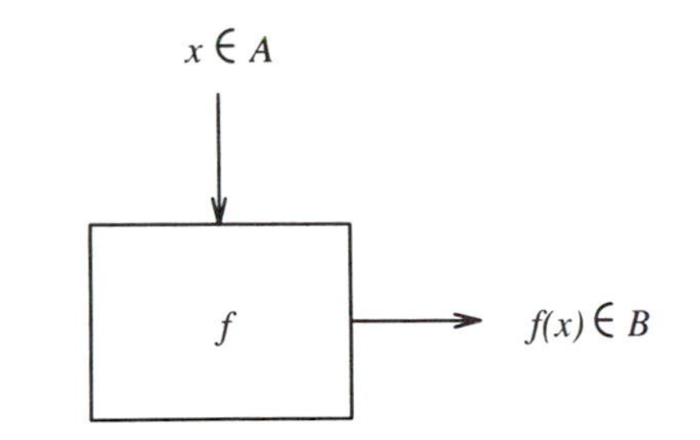

Figure 3.7: A Representation of a Function as a Black Box

3.2.1 Functions and Their Representations

Through the process of abstraction in Section 1.5.2, you have come to think of a function as a rule that associates to each *object*, some other *object*. Because a function works with objects, it is reasonable to group those objects in a set which, as you learned in Section 3.1, is a collection of related objects. One useful way to visualize a function is as a *black box* whose name is that of the function, say, f. You put an element x in a set A in the box and, after performing some computations, you obtain the element $f(x)$ in a set B (see Figure 3.7). In that regard, a function is much like a computer program or algorithm. What goes in the box is often called the **input value** for the function and what comes out is called the **output value**. The operations that go on inside the black box constitute the rule associated with the function.

One of the problems with these informal verbal and visual descriptions is that they do not capture an important property of a function f, namely, that to each element $x \in A$, f must associate a *unique* element $f(x) \in B$. This property *is* included in the following definition.

Definition 3.10 *Given two sets A and B, a* **function** *f is a collection of ordered pairs (x, y) in which $x \in A, y \in B$ and for which the following properties hold:*

1. *For each $x \in A$, there is a $y \in B$ such that $(x, y) \in f$.*
2. *If $(x, y) \in f$ and $(x, z) \in f$, then $y = z$.*

The set A is called the **domain** *of f and the set B is called the* **codomain** *of f. Notationally, the statement*

$$f : A \to B$$

is read as "f is a function from A to B" and means that you can apply the function f to any element $x \in A$ to obtain a unique corresponding element $f(x) \in B$. The associated ordered pair is written $(x, f(x))$.

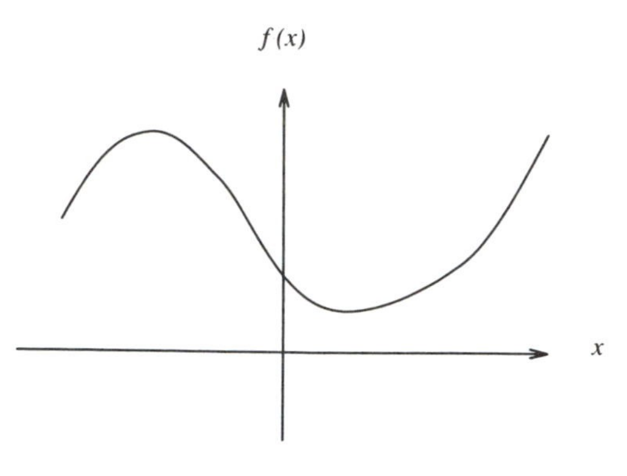

Figure 3.8: Representing a Function by its Graph

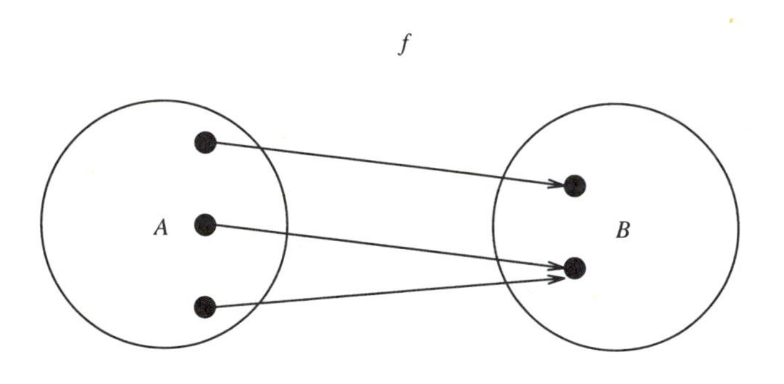

Figure 3.9: Representing a Function as an Association

The next issue is how to represent a function so that you can both think about and work with that function. There are several *visual* representations and also a *symbolic* form that is useful when working with functions, both of which are described next.

Visual Representations of a Function

You can visualize a function informally by the black box in Figure 3.7. But if you want to capture the uniqueness property in Definition 3.10, then you should use either of the visual images presented in Example 2.2 in Section 2.1.1, both of which are repeated here in Figure 3.8 and Figure 3.9. In Figure 3.8, the uniqueness is captured by the fact that for each value of x on the x-axis, there is only *one* point on the graph of f that corresponds to the unique function value of f at x. Likewise, in Figure 3.9, for each $x \in A$, there is one and only one arrow from x pointing to the unique value in B that corresponds to the function value at x.

Symbolic Representations of a Function

Recall that, informally, a function f is a rule that associates to each element $x \in A$, a unique element $f(x) \in B$. The way in which you specify the rule constitutes a symbolic representation of the function, as illustrated in the following examples.

Example 3.14 – A Symbolic Representation of a Function

Suppose that $A = \{1, 2, 3\}$ and $B = \{1, 3, 5, 7\}$. Then A is the domain and B is the codomain of the function $f : A \rightarrow B$ defined by

$$f(1) = 1, \ f(2) = 3, \ f(3) = 5.$$

Here, the rule is a list that tells you the value of the function at each point in the domain. Observe that applying f to all the elements of A does *not* have to result in *every* element of B. In this case, the element 7 in B is not obtained from f.

The set A in Example 3.14 consists of few elements, so you can easily list them all together with their corresponding function values. However, when A has an *infinite* number of elements and is written using set-builder notation (see Section 3.1.1), the function is often described by a closed-form expression, as shown in the next example.

Example 3.15 – Representing a Function in Closed Form

Recall that $\mathbb{R}^1$ is the set of real numbers. The function $f : \mathbb{R}^1 \rightarrow \mathbb{R}^1$ defined by

$$f(x) = x^2$$

associates to each real number $x \in \mathbb{R}^1$, the unique real number $x^2 \in \mathbb{R}^1$. Here, the rule is described by the closed-form expression x^2.

The function f in Example 3.15 associates to each real number x, another real number, $f(x)$. A generalization arises when the input to the function consists of *two* numbers instead of just one. For example, the input to the function

$$g(x, y) = x^2 + 2xy - y^2,$$

is the two numbers x and y and the output is the number $x^2 + 2xy - y^2$.

Another generalization is a function g whose input consists of n numbers. In this case, g associates to each n-vector $\mathbf{x} = (x_1, \ldots, x_n)$, the real number $g(\mathbf{x})$.

A further generalization arises when the function produces an *output* consisting of several numbers. For example, the input to the function

$$h(x, y) = (x + y, \ x - y, \ x^2 + 2xy - y^2)$$

is the two numbers x and y and the output consists of *three* numbers: $x + y$, $x - y$, and $x^2 + 2xy - y^2$. In general, a function h can have an input consisting of n numbers and an output consisting of p numbers. In this case, h associates to each n-vector $\mathbf{x} = (x_1, \ldots, x_n)$, a p-vector $h(\mathbf{x}) = (h_1(\mathbf{x}), \ldots, h_p(\mathbf{x}))$. Each of the outputs, h_i, is a **coordinate function** that associates to each n-vector $\mathbf{x}$, the real number $h_i(\mathbf{x})$. These various types of functions are summarized in the next example.

Example 3.16 – Functions Whose Input and Output Are Vectors

The following functions constitute a sequential generalization of a function that has one input number and one output number:

$$\begin{aligned} f &: \mathbb{R}^1 \to \mathbb{R}^1 \quad \text{(1 input and 1 output)},\\ g &: \mathbb{R}^n \to \mathbb{R}^1 \quad (n \text{ inputs and 1 output}),\\ h &: \mathbb{R}^n \to \mathbb{R}^p \quad (n \text{ inputs and } p \text{ outputs}). \end{aligned}$$

Avoiding Syntax Errors When Generalizing Functions. When generalizing a statement or computation that involves a function $f : \mathbb{R}^1 \to \mathbb{R}^1$, care is needed to avoid syntax errors. For example, suppose that x, y, and t are real numbers with $x \neq y$ and that you want to generalize the statement

$$\left| \frac{f(y) - f(x)}{y - x} \right| \leq t \tag{3.4}$$

to a function $g : \mathbb{R}^n \to \mathbb{R}^1$. By replacing f in (3.4) with g, you obtain the statement

$$\left| \frac{g(y) - g(x)}{y - x} \right| \leq t. \tag{3.5}$$

However, (3.5) has a syntax error because you cannot put the *real numbers* x and y in the function $g : \mathbb{R}^n \to \mathbb{R}^1$, whose input must be an *n-vector*. Even if you think of $\mathbf{x}$ and $\mathbf{y}$ as n-vectors in (3.5), there is still a syntax error. This is because the numerator of the fraction is a real number (since $g(\mathbf{y})$ and $g(\mathbf{x})$ are real numbers), and the denominator is an *n-vector* (since $\mathbf{x}$ and $\mathbf{y}$ are n-vectors), and you cannot divide a number by an n-vector. Can you resolve this problem?

One way to do so is to rewrite (3.4) so that the operation of division does not appear. For example, multiplying both sides of (3.4) by $|y - x|$ results in

$$|f(y) - f(x)| \leq t|y - x|. \tag{3.6}$$

To generalize (3.6), replace $f : \mathbb{R}^1 \to \mathbb{R}^1$ by $g : \mathbb{R}^n \to \mathbb{R}^1$ and think of $\mathbf{x}$ and $\mathbf{y}$ as n-vectors to obtain

$$|g(\mathbf{y}) - g(\mathbf{x})| \leq t|\mathbf{y} - \mathbf{x}|. \tag{3.7}$$

Unfortunately, (3.7) still contains a syntax error – namely, $|\mathbf{y}-\mathbf{x}|$ – because you cannot take the absolute value of an *n-vector*. The final correction is to replace the operation of the absolute value of numbers with the operation of computing the length of an n-vector (see Section 1.4.3), to obtain the following valid generalization of (3.4):

$$|g(\mathbf{y}) - g(\mathbf{x})| \leq t\|\mathbf{y} - \mathbf{x}\|. \tag{3.8}$$

Identifying the Domain and Codomain of a Function. In the examples so far, the domain and codomain of the functions are given. Many times, however, only the *formula* for computing the function is specified. In such cases, you should always identify the domain and codomain *explicitly*, that is:

1. The domain is the set of all *allowable* values you can put in the function.

2. The codomain is *any* set that contains all possible values that the function can produce. There are various different choices for the codomain, as illustrated in Example 3.18.

The process of identifying the domain and codomain of a function is shown in the following examples.

Example 3.17 – Identifying the Domain and Codomain of a Function

Consider the function f defined by

$$f(x) = \log(x).$$

By asking what values you can put in this function, you determine that the domain of f is the set of *positive* real numbers (because you cannot compute the logarithm of 0 or of a negative number). Similarly, by asking what type of value you get as a result of taking the logarithm of a positive number, you should realize that the codomain of this function is $\mathbb{R}$, the set of real numbers. Thus,

$$f : \{x \in \mathbb{R} : x > 0\} \to \mathbb{R}.$$

Example 3.18 – A Function with Various Domains and Codomains

Consider the function f defined by

Continued

$$f(x) = \sqrt{1 - x^2}.$$

Here, there are several possible domains and codomains:

Domain	Codomain	Explanation
$[-1, 1]$	$[0, 1]$	A domain of $[-1, 1]$ means that you can put any real number x between -1 and 1 into f. When you do so for *all* such values of x, the values of $f(x)$ vary between 0 and 1. So, one codomain is $[0, 1]$.
$[-1, 1]$	$\{x \in \mathbb{R} : x \geq 0\}$	The codomain can be *any* set that contains the set of all possible function values – $[0, 1]$, in this case.
$\mathbb{R}$	$\mathbb{C}$	The domain of $\mathbb{R}$ indicates that you can put any real number into f. However, if you do so with a value of $x > 1$ or $x < -1$, the result is a *complex* number. Thus, the codomain must be a set that contains $\mathbb{C}$.

The remaining examples illustrate various functions with domains and codomains other than *numbers*.

Example 3.19 – A Function Whose Input Is a Set and Whose Ouput Is a Number

Let f be a function that associates to each finite set A, the number of elements in A. In this case,

$$f : \{A : A \text{ is a set with a finite number of elements}\} \to \mathbb{N} \cup \{0\}.$$

Example 3.20 – A Function Whose Input Is a Function and Whose Output Is a Function

Let f be a function that associates to each function $g : \mathbb{R} \to \mathbb{R}$ the function $h : \mathbb{R} \to \mathbb{R}$ defined by $h(x) = g(x) + 3$. That is, $f(g) = h$, where for any real number x, $h(x) = g(x) + 3$. (Note that it is *not* correct to write $f(x) = g(x) + 3$ because you cannot put a real number into f.) Here,

$$f : A \to A,$$

where

$$A = \{\text{functions } g : g : \mathbb{R} \to \mathbb{R}\}.$$

Functions with Special Properties

After working with functions for a while, you will discover that there are several special functions that arise frequently. Some of these are described next.

The Identity Function. One type of special function that comes up often assigns to each element of a set A, that same value. That is, given a set A, the **identity function** is the function $i_A : A \to A$ defined by

$$i_A(x) = x, \text{ for each element } x \in A.$$

The Characteristic Function. Given a set A together with a universal set U, you know that the statement

$$x \in A$$

is *true*, provided that x *is* an element of A and *false*, otherwise. Thinking of *true* as the value 1 and *false* as the value 0, the **characteristic function on the set** A is the function $f_A : U \to \{0,1\}$ defined by

$$f_A(x) = \begin{cases} 1, \text{ if } x \in A \\ 0, \text{ if } x \notin A. \end{cases}$$

3.2.2 Comparisons and Operations Involving Functions

Whenever you encounter a new mathematical concept (such as a function), you know to develop an assoicated image (such as a black box). However, keep in mind that the reason for creating that concept in the first place is to help you solve certain problems. For functions, this means that you need to know how to compare two functions to see if they are the same and how to perform operations on functions to create new functions. Both of these topics are addressed in this section.

Comparing Two Functions for Equality

One of the most important comparison between two functions is to determine when they are *equal*, that is, the same function. For two functions to be the same, their domains and codomains must be equal. Additionally, the values of the functions at each point in their domains should be the same. These observations are summarized in the following definition.

Definition 3.11 *Suppose that $f : A \longrightarrow B$ and $g : C \longrightarrow D$. The functions f and g are* **equal**, *written $f = g$, if and only if*

1. $A = C$.
2. $B = D$.
3. *For all elements $x \in A$, $f(x) = g(x)$.*

The symbol = is used to compare two functions f and g. Thus, the statement

$$f = g$$

means that f *is* equal to g, or equivalently, that the result of checking whether f is equal to g is *true*. Analogously, the statement

$$f \neq g \tag{3.9}$$

means that f is *not* equal to g, or equivalently, that the result of checking whether f is equal to g is *false*. Assuming that the domains and codomains of f and g are the same then, by using the rules for negating the quantifier *for all* (see Section 1.6.10), the statement in (3.9) is the same as the following:

There is an element $x \in A$ such that $f(x) \neq g(x)$.

Binary Operations on Functions

A binary operation combines *two* functions to create a new function. You will now see several examples of binary operations on functions.

Combining Functions with Arithmetic Operators. When the output values of two functions are each real numbers, you can combine those real numbers using the arithmetic operators $+, -, \cdot$, and $/$. That is, if $f, g : \mathbb{R} \longrightarrow \mathbb{R}$, then you can combine these two function with the foregoing arithmetic operators to produce the following functions:

$$\begin{aligned}
f + g : \mathbb{R} \longrightarrow \mathbb{R} \text{ defined by } (f + g)(x) &= f(x) + g(x), \\
f - g : \mathbb{R} \longrightarrow \mathbb{R} \text{ defined by } (f - g)(x) &= f(x) - g(x), \\
f \cdot g : \mathbb{R} \longrightarrow \mathbb{R} \text{ defined by } (f \cdot g)(x) &= f(x) \cdot g(x), \\
f/g : \mathbb{R} \longrightarrow \mathbb{R} \text{ defined by } (f/g)(x) &= f(x)/g(x).
\end{aligned}$$

Observe that this last function f/g requires that $g(x)$ never be 0.

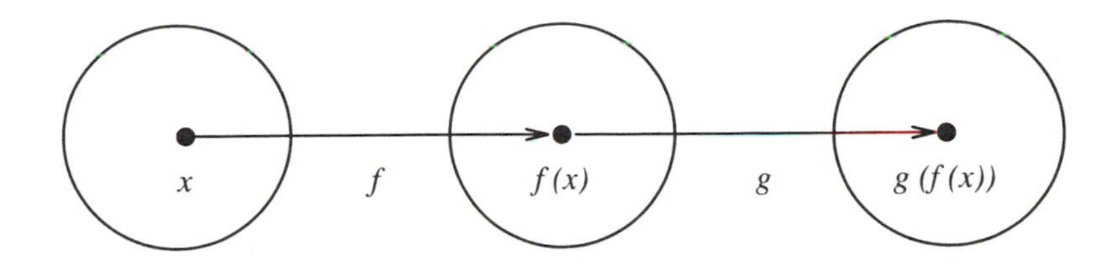

Figure 3.10: The Composition of Two Functions

Example 3.21 – Combining Functions with the Arithmetic Operators

If $f, g : \mathbb{R} \to \mathbb{R}$ are defined by

$$f(x) = 1 - 2x^2 \text{ and } g(x) = 1 + x^2,$$

then

$$(f + g)(x) = f(x) + g(x) = (1 - 2x^2) + (1 + x^2) = 2 - x^2,$$

$$(f - g)(x) = f(x) - g(x) = (1 - 2x^2) - (1 + x^2) = -3x^2,$$

$$(f \cdot g)(x) = f(x) \cdot g(x) = (1 - 2x^2) \cdot (1 + x^2) = 1 - x^2 - 2x^4,$$

$$(f/g)(x) = \frac{f(x)}{g(x)} = \frac{1 - 2x^2}{1 + x^2}.$$

Function Composition. Suppose that $f : A \to B$ and $g : D \to C$. One way to create a new function from f and g is to apply them *sequentially*, as shown in Figure 3.10. That is, first apply f to a point x and then apply g to the resulting point, $f(x)$. The objective now is to translate this image and verbal description to symbolic form by creating an appropriate definition.

Observe in Figure 3.10 that if $x \in A$, a syntax error may result when you write

$$g(f(x)). \tag{3.10}$$

This is because you may not be able to evaluate g at the point $f(x)$. To see why, recall that $f : A \to B$ and so $f(x) \in B$. But $g : D \to C$ and thus, to be able to evaluate g at the point $f(x)$, it must be that $f(x) \in D$ (which may or may not be the case). Because you know that $f(x) \in B$, one way to insure that $f(x) \in D$ is to require that $B = D$, as is assumed from here on. As a result, (3.10) is valid and gives rise to the binary operation on two functions described in the following definition.

Definition 3.12 *Suppose that $f : A \to B$ and $g : B \to C$. Then the* **composition** *of f and g (denoted by $g \circ f$) is the function $g \circ f : A \to C$ defined by*

$$(g \circ f)(x) = g(f(x)).$$

Example 3.22 – The Composition of Two Functions

If $f : \mathbb{R} \to \mathbb{R}$ is defined by $f(x) = 2x + 3$ and $g : \mathbb{R} \to \mathbb{R}$ is defined by $g(x) = x^2$, then $g \circ f : \mathbb{R} \to \mathbb{R}$ is defined by

$$\begin{aligned}(g \circ f)(x) &= g(f(x)) \\ &= g(2x+3) \\ &= (2x+3)^2 \\ &= 4x^2 + 12x + 9.\end{aligned}$$

Thus, for instance, for $x = -5$, you have that

$$f(-5) = 2(-5) + 3 = -7$$

and

$$g(-7) = (-7)^2 = 49,$$

so

$$(g \circ f)(-5) = g(f(-5)) = g(-7) = 49.$$

Example 3.23 – The Composition of Two Functions

Suppose that $f : \{1,2,3\} \to \{1,2\}$ is defined by

$$f(1) = 2, \; f(2) = 1, \; f(3) = 1,$$

and $g : \{1,2\} \to \{4,5\}$ is defined by

$$g(1) = 5, \; g(2) = 4.$$

Then the function $g \circ f : \{1,2,3\} \to \{4,5\}$ is defined by

$$\begin{aligned}(g \circ f)(1) &= g(f(1)) = g(2) = 4, \\ (g \circ f)(2) &= g(f(2)) = g(1) = 5, \\ (g \circ f)(3) &= g(f(3)) = g(1) = 5.\end{aligned}$$

3.2.3 Properties of Functions

In Chapter 1, you saw the problem of solving a linear equation, which was then generalized sequentially to solving a quadratic equation, a polynomial equation, and then a general nonlinear equation. By using a function, you can state this last problem as follows:

> Given a function $f : \mathbb{R}^1 \to \mathbb{R}^1$, find $x \in \mathbb{R}^1$ so that $f(x) = 0$.

Applying generalization by replacing the number 0 with some arbitrary number y, this problem becomes:

> Given a function $f : \mathbb{R}^1 \to \mathbb{R}^1$ and a point $y \in \mathbb{R}^1$, find a point $x \in \mathbb{R}^1$ so that $f(x) = y$.

A further generalization arises when you allow the function to have an input and output consisting of an n-vector. To create this generalization, change the function in the last problem from $f : \mathbb{R}^1 \to \mathbb{R}^1$ to a function $f : \mathbb{R}^n \to \mathbb{R}^n$. You then obtain the following:

> Given a function $f : \mathbb{R}^n \to \mathbb{R}^n$ and a point $y \in \mathbb{R}^1$, find a point $x \in \mathbb{R}^1$ so that $f(x) = y$.

This last problem, however, has a syntax error: if $x \in \mathbb{R}^1$, then you cannot evaluate $f(x)$ because the input to $f : \mathbb{R}^n \to \mathbb{R}^n$ must be an *n-vector*, not a number. One solution is to change x from a number to an n-vector, resulting in the following problem:

> Given a function $f : \mathbb{R}^n \to \mathbb{R}^n$ and a point $y \in \mathbb{R}^1$, find a point $\mathbf{x} \in \mathbb{R}^n$ so that $f(\mathbf{x}) = y$.

Even here there is a syntax error because the value of $f(\mathbf{x})$ is an *n-vector* and y is a *number* – and you cannot compare an n-vector and a number for equality. One way to correct this syntax error is to change y from a number to an n-vector, resulting in the following (correct) problem:

> Given a function $f : \mathbb{R}^n \to \mathbb{R}^n$ and a point $\mathbf{y} \in \mathbb{R}^n$, find a point $\mathbf{x} \in \mathbb{R}^n$ so that $f(\mathbf{x}) = \mathbf{y}$.

However, in Section 3.2.1 you learned the abstract concept of a function that associates to each *object* another *object*, so you can generalize this last problem to allow for a function $f : A \to B$. The resulting problem arises in many areas of science, engineering, economics, statistics, and related fields and is stated in the following example.

Example 3.24 – The Problem of Solving the Equation $f(x) = y$

Given two sets A and B, a function $f : A \to B$, and any point $y \in B$, find a point $x \in A$ for which $f(x) = y$. That is, given any value for $y \in B$, find a value for $x \in A$ that satisfies the following equation:

$$f(x) = y. \tag{3.11}$$

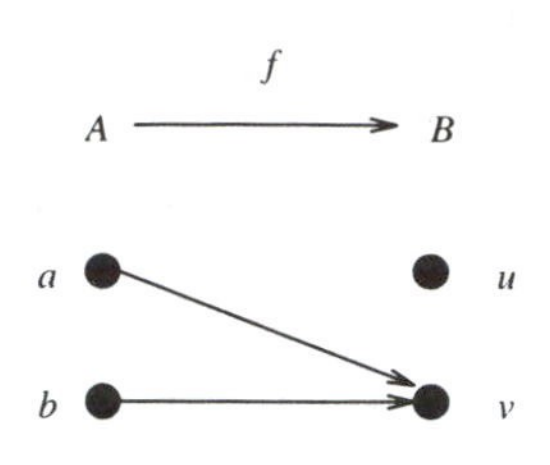

Figure 3.11: Reasons Why Solving the Equation $f(x) = y$ for $x \in A$ Can Be Challenging

As you learned in Chapter 1, when you create a more general problem – such as the one in Example 3.24 – you also need to create a new solution procedure, which is discussed next. As shown in Figure 3.11, there are two reasons why finding a solution to equation (3.11) can be challenging:

1. If you consider the point $u \in B$ in Figure 3.11, then you can see that there is *no* value of $x \in A$ for which $f(x) = u$.

2. If you consider the point $v \in B$ in Figure 3.11, then there are *two* values of $x \in A$ that satisfy $f(x) = v$, namely, $x = a$ and $x = b$. Which point should you choose as the solution?

The objective of the remainder of this section is to develop *conditions* on the function f, together with the sets A and B, so that for each value of $y \in B$, equation (3.11) in Example 3.24 has one and only one solution.

A Surjective Function

As you can see in Figure 3.11, when you evaluate the function f at all points $x \in A$, you do not "cover" every point in B. In particular, you do not cover $u \in B$. In contrast, a function for which you *do* cover all of B is illustrated in Figure 3.12. How do you translate the visual image in Figure 3.12 – and the corresponding verbal description of "covering every point in B" – to a symbolic definition? Two different, but equivalent, ways for doing so are described next.

Using Quantifiers. One approach arises from the observation that, for the function in Figure 3.12, *every* point in the set B is covered by the function. Recognizing the word *every* should encourage you to use the quantifier *for all* to describe the desirable property, as follows:

For all elements $y \in B$, y is *covered* by f.

All that remains is to translate the concept that "y is covered by f" to symbolic form. One way to do so is to use the quantifier *there is* to restate the fact that "y is covered by f," as follows:

There is an element $x \in A$ such that $f(x) = y$.

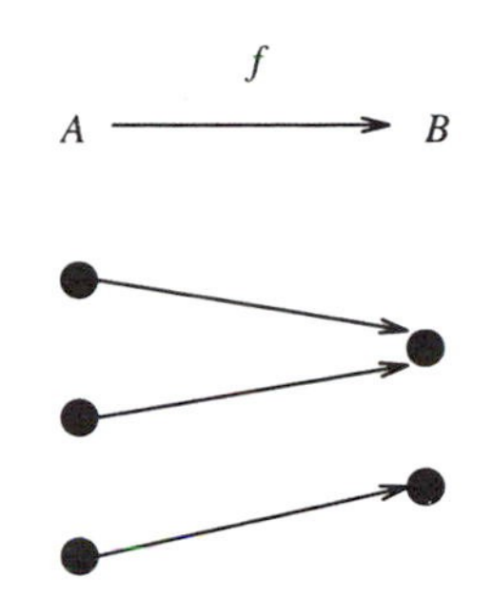

Figure 3.12: A Function that "Covers Every Point in B"

Putting together the pieces results in the following definition.

Definition 3.13 *A function $f : A \to B$ is* **surjective** *(or* **onto***) if and only if for all elements $y \in B$, there is an element $x \in A$ such that $f(x) = y$.*

Using the Range of the Function. An alternative approach to creating the definition of a surjective function is to work with the concept of "the set of all points obtained by evaluating the function at each point in the domain." Translating this verbal description to symbolic form results in the following definition.

Definition 3.14 *Suppose that $f : A \to B$. The* **range of** f*, denoted by* range f*, is the following set:*

$$\begin{aligned} \text{range } f &= \{y \in B : \textit{there is an } x \in A \textit{ for which } f(x) = y\} \\ &= \{f(x) : x \in A\}. \end{aligned}$$

Example 3.25 – The Range of a Function

If $A = \{1, 2, 3\}$ and $B = \{a, b, c\}$ and $f : A \to B$ is defined by

$$f(1) = a, \ f(2) = a, \ f(3) = b,$$

then range $f = \{a, b\}$.

A generalization of the range of f arises when you consider the set of those points obtained by applying the function to all the elements of a *subset* X of the domain A, as summarized in the following definition.

Definition 3.15 *Suppose that $f : A \to B$ and $X \subseteq A$. The* **image of X under** f, *denoted by $f(X)$, is the following set:*

$$f(X) = \{y \in B : \text{there is an } x \in X \text{ for which } f(x) = y\} = \{f(x) : x \in X\}.$$

Example 3.26 – The Image of a Set Under a Function

For the function f in Example 3.25 in which $A = \{1, 2, 3\}$, $B = \{a, b, c\}$, and $f : A \to B$ is defined by

$$f(1) = a, \ f(2) = a, \ f(3) = b,$$

the following are $f(X)$ for various $X \subseteq A$:

X	$f(X)$
$\{1\}$	$\{a\}$
$\{2\}$	$\{a\}$
$\{3\}$	$\{b\}$
$\{1,2\}$	$\{a\}$
$\{1,3\}$	$\{a,b\}$
$\{2,3\}$	$\{a,b\}$
$\{1,2,3\}$	$\{a,b\}$

Observe in Definition 3.15 and Example 3.26 that for the special case of $X = A$, $f(X) = \text{range } f$.

You can use the concept of the range to describe a surjective function. Looking at Figure 3.12, you can see that a function f is surjective if and only if range $f = B$. However, recall that there can be only one definition for a surjective function (namely, the one in Definition 3.13). So what should you do with this *new* concept that range $f = B$? The answer is to *prove* the equivalence of the new concept with the definition, as shown in the following proposition and proof.

Proposition 3.1 *Suppose that $f : A \to B$. The function f is surjective if and only if* range $f = B$.

Proof. Because of the words *if and only if*, two proofs are required. So suppose first that f is surjective. According to Definition 3.13, this means that

for all elements $y \in B$, there is an element $x \in A$ such that $f(x) = y$.

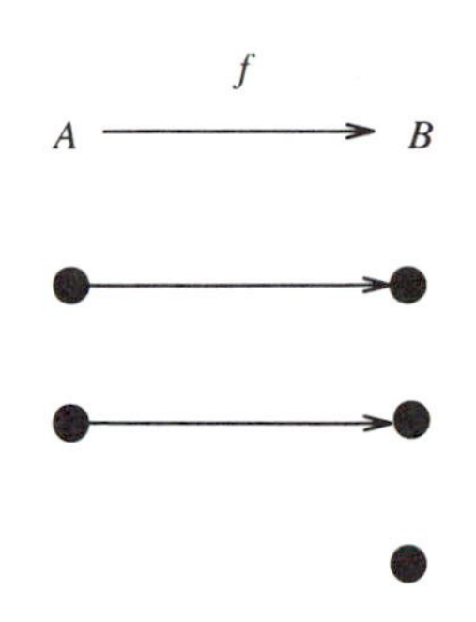

Figure 3.13: A Function for Which No Two Points in the Domain Result in the Same Value

It must be shown that the following two sets are equal:

range $f = B$,

or equivalently, that

range $f \subseteq B$ and $B \subseteq$ range f.

Now range $f = \{f(x) : x \in A\} \subseteq B$ because any value $f(x)$ obtained from the function is necessarily in the codomain B. So you need only show that

$B \subseteq$ range f,

or equivalently, that

for all elements $y \in B$, $y \in$ range f.

Because of the quantifier *for all*, let $y \in B$. By the assumption that f is surjective,

there is an element $x \in A$ such that $f(x) = y$.

But then, by the defining property of range f, this means that $y \in$ range f. It has therefore been shown that if f is surjective, then range $f = B$.

To complete the proof, it remains to show that if range $f = B$, then f is surjective, and this part is left as an exercise. QED

An Injective Function

Again looking at the value of $v \in B$ in Figure 3.11, you can see that another property of a function that is useful for solving equation (3.11) in Example 3.24 is that "no two points in the domain result in the same function value." Such a function is shown in Figure 3.13, which is in contrast to the function in Figure 3.11.

To translate the visual image of the desirable property in Figure 3.13 – and the corresponding verbal description – to a symbolic definition, use the quantifier *there is* to create the following statement:

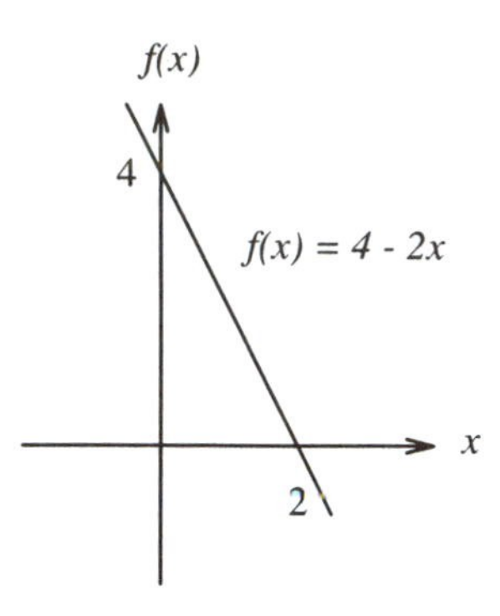

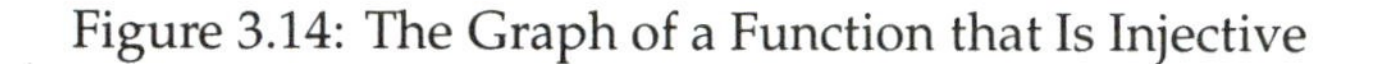

Figure 3.14: The Graph of a Function that Is Injective

> There do not exist elements x and y in A such that $f(x) = f(y)$.

However, if you verify this translation carefully, you will discover a mistake because there *always* exist two elements x and y in A for which $f(x) = f(y)$, namely, $x = y$. To avoid this one case, the statement of the property should be changed to the following:

> There do not exist elements x and y in A with $x \neq y$
> such that $f(x) = f(y)$.

By rephrasing this in a positive way so that the word *not* does not appear explicitly, you arrive at the following definition.

Definition 3.16 *A function $f : A \to B$ is* **injective** *(or* **one-to-one***) if and only if for all elements $x, y \in A$ with $x \neq y$, $f(x) \neq f(y)$.*

For instance, the function $f(x) = 4 - 2x$ is injective because, if $x \neq y$, it follows that $4 - 2x \neq 4 - 2y$, so $f(x) \neq f(y)$ (see Figure 3.14). In contrast, the function $f(x) = x^2$ is *not* injective because, for $x = 2$ and $y = -2$, you have $x \neq y$ and yet $f(x) = f(y) = 4$ (see Figure 3.15).

A visual image of an injective function is one with the property that no matter which value of y you choose on the y-axis, the horizontal line through y intersects the graph of the function in at most one point. This is the case for the graph in Figure 3.14 but not for the graph of Figure 3.15.

A Bijective Function

To solve the problem in Example 3.24, you need a function that is both surjective and injective. Such a function merits its own definition, which follows.

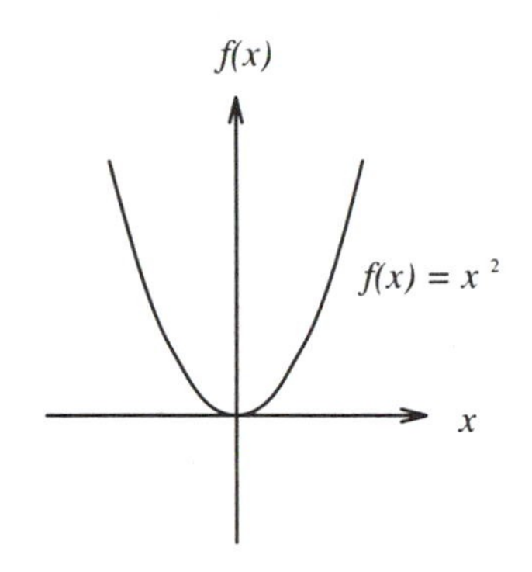

Figure 3.15: The Graph of a Function that Is Not Injective

> **Definition 3.17** *A function* $f : A \to B$ *is* **bijective** *if and only if* f *is surjective and injective.*

It is now possible to solve the problem in Example 3.24, which is repeated here.

> **Example 3.27 – The Problem of Solving the Equation** $f(x) = y$
>
> Given a function $f : A \to B$ and any point $y \in B$, find a point $x \in A$ such that $f(x) = y$. That is, given any value for $y \in B$, you want to find a value for $x \in A$ that satisfies the following equation:
>
> $$f(x) = y. \qquad (3.12)$$

One approach to solving this problem involves finding a new function g that you can apply to both sides of (3.12) to obtain

$$g(f(x)) = g(y). \qquad (3.13)$$

In the event that g "cancels" f, you obtain from (3.13) that

$$x = g(y). \qquad (3.14)$$

Finally, to verify that the value of $x = g(y)$ in (3.14) *is* a solution to equation (3.12) in Example 3.27, you need to check that

$$f(x) = f(g(y)) = y. \qquad (3.15)$$

In other words, you also need to know that f "cancels" g.

From these observations, the key to solving the problem in Example 3.27 is to find this function g, so begin by identifying what the domain and codomain of g should be. Because you want to apply g to both sides of

(3.12) and because both $f(x)$ and y are elements in B, the *domain* of g must be B. Also, from (3.14), you can see that $g(y) = x$ and because $x \in A$, it follows that the *codomain* of g must be A. In other words, the function g must satisfy $g : B \to A$.

What *additional* properties must g satsify? To go from (3.13) to (3.14), you need to know that $g(f(x)) = x$. Also to satisfy (3.15), you need $f(g(y)) = y$. A function with both of these desirable properties is described in the following definition, in which you will find the use of the previous concepts of the identity function (see Section 3.2.1), the composition of two functions (see Definition 3.12), and the comparison of two functions for equality (see Definition 3.11).

> **Definition 3.18** *Suppose that $f : A \to B$. A function $g : B \to A$ is the* **inverse function** *of f if and only if $g \circ f = i_A$ and $f \circ g = i_B$.*

The use of Definition 3.18 is that, if a function $f : A \to B$ has an inverse function g, then you can always find a unique solution to equation (3.12) in Example 3.27, as stated next.

> **Solution to Example 3.27**
>
> Suppose that $f : A \to B$. If $g : B \to A$ is an inverse function of f, then for any $y \in B$, the unique solution to the equation
>
> $$f(x) = y$$
>
> is
>
> $$x = g(y).$$

The final question, then, is: Under what conditions does f *have* an inverse function? One answer is provided in the following theorem.

> **Theorem 3.1** *If $f : A \to B$ is bijective, then there is a unique inverse function of f, denoted by $f^{-1} : B \to A$.*

Proof. You will find it helpful to refer to the picture of a bijective function in Figure 3.16. Proceeding with the proof, it is first necessary to construct the function $f^{-1} : B \to A$. Thus, a rule is needed that associates to each $y \in B$, a unique value of $x \in A$. From Figure 3.16, you can define the function f^{-1} as follows:

$$f^{-1}(y) = \text{that unique value of } x \in A \text{ for which } f(x) = y.$$

The fact that f is surjective insures that for each $y \in B$, *there is* a point $x \in A$ such that $f(x) = y$. Also, because f is injective, there is only one such point $x \in A$.

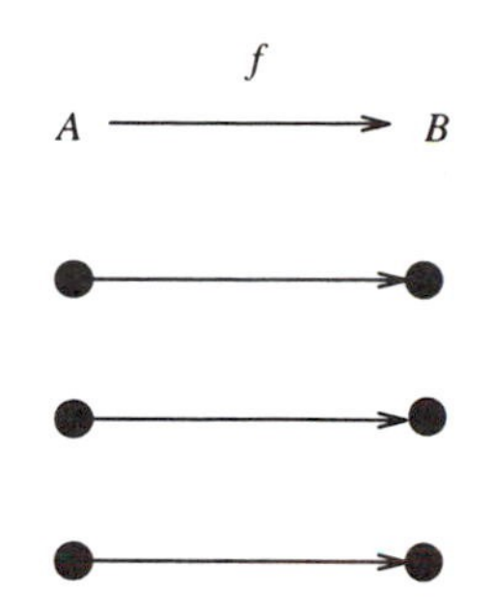

Figure 3.16: An Example of a Bijective Function

Having constructed f^{-1}, it must now be shown that $f^{-1} \circ f = i_A$. To show these two functions are equal, according to Definition 3.11, you must show that

for all elements $x \in A$, $(f^{-1} \circ f)(x) = i_A(x) = x$.

Because of the quantifier *for all*, choose an element $x \in A$ and show that $(f^{-1} \circ f)(x) = x$. But, letting $y = f(x)$, it follows that

$$\begin{aligned}(f^{-1} \circ f)(x) &= f^{-1}(f(x)) \\ &= f^{-1}(y) \\ &= \text{that unique element } z \in A \text{ such that } f(z) = y \\ &= x.\end{aligned}$$

Next, it must be shown that $f \circ f^{-1} = i_B$. To show these two functions are equal, according to Definition 3.11, you must show that

for all elements $y \in B$, $(f \circ f^{-1})(y) = i_B(y) = y$.

Because of the quantifier *for all*, choose an element $y \in B$ and show that $(f \circ f^{-1})(y) = y$. But, letting x be the unique element of A such that $f(x) = y$, it follows that

$$\begin{aligned}(f \circ f^{-1})(y) &= f(f^{-1}(y)) \\ &= f(\text{the unique element } z \in A \text{ such that } f(z) = y) \\ &= f(x) \\ &= y.\end{aligned}$$

Finally, it must be shown that this function f^{-1} is unique. Following proof techniques, the approach is to assume that $g : B \to A$ is *any* function for which $g \circ f = i_A$ and $f \circ g = i_B$. It is then necessary to show that $g = f^{-1}$, and this part is left to the exercises.

QED

The Inverse Image of a Set

You have just seen that when $f : A \to B$ is a bijective function, you can find an inverse function $f^{-1} : B \to A$. This inverse function is used to solve equation (3.12) in Example 3.27, that is, given any value for $y \in B$, you can use f^{-1} to find a value for $x \in A$ for which

$$f(x) = y. \tag{3.16}$$

Even when f is *not* bijective, you can still try to find a value of $x \in A$ that satisfies (3.16). However, as shown in Figure 3.11, there may be *no* such x or there may be *several different values* of x that satisfy (3.16).

In any event, for each $y \in B$, you can create the set of all points $x \in A$ that *do* satisfy (3.16). That is, let

$$\widehat{f^{-1}}(y) = \{x \in A : f(x) = y\}.$$

This set may be empty or it may contain one or more elements. For example, if $f(x) = x^2$ and $y = 9$, then

$$\begin{aligned} \widehat{f^{-1}}(9) &= \{\text{real numbers } x : x^2 = 9\} \\ &= \{-3, 3\}. \end{aligned}$$

As a visual example, in Figure 3.11, $\widehat{f^{-1}}(y) = \{a, b\}$ when $y = v$ and $\widehat{f^{-1}}(y) = \emptyset$ when $y = u$. Do not confuse $\widehat{f^{-1}}(y)$, which is a *set*, with $f^{-1}(y)$, which is an *element* in the domain of f.

A generalization of $\widehat{f^{-1}}(y)$ arises when you replace the element y with a *subset* Y of B, as shown in the following definition.

Definition 3.19 *Suppose that $f : A \to B$ and $Y \subseteq B$. The* **inverse image of Y under f**, *denoted by $\widehat{f^{-1}}(Y)$, is the following set:*

$$\begin{aligned} \widehat{f^{-1}}(Y) &= \{x \in A : f(x) = y \text{ for some } y \in Y\} \\ &= \{x \in A : f(x) \in Y\}. \end{aligned}$$

Example 3.28 – The Inverse Image of a Set

If $A = \{1, 2, 3\}$ and $B = \{a, b, c\}$ and $f : A \to B$ is defined by

Continued

$$f(1) = a, \ f(2) = a, \ f(3) = b,$$

then the following are $\widehat{f}^{-1}(Y)$ for various sets Y:

Y	$\widehat{f}^{-1}(Y)$
$\{a\}$	$\{1,2\}$
$\{b\}$	$\{3\}$
$\{c\}$	$\emptyset$
$\{a,b\}$	$\{1,2,3\}$
$\{a,c\}$	$\{1,2\}$
$\{b,c\}$	$\{3\}$
$\{a,b,c\}$	$\{1,2,3\}$

In this section you have learned the visual and symbolic representation for the abstract concept of a function. You have seen how to compare two functions for equality and how to combine two such functions by composition. You have also learned of the important properties of a surjective, injective, and bijective function. In the next section, you will learn another concept in discrete mathematics that is useful in problem solving.

3.3 Graphs

The concept of a *graph* was introduced through the process of abstraction in Section 1.5.1 as a problem-solving tool in which circles, called *vertices*, are used to represent certain physical items in a problem (such as cities, people, and so on). Selected pairs of those vertices are then connected by a line, called an *edge*, to indicate a relationship between the objects represented by the two connected vertices (such as a nonstop flight between two cities, or the fact that two people know each other, and so on). In this section, you will learn more about graphs and see several new problems that you can solve using graphs. As usual, you need to develop visual images, translate those images to symbolic form, create definitions, discover useful properties of graphs, and more.

3.3.1 Graphs and Their Representations

The concept of a graph is repeated in the following definition.

Definition 3.20 *An* **undirected graph** *(or, more simply, a* **graph***) consists of a nonempty finite set V of vertices and a finite set E of edges, each of which connects two vertices in V. The graph is written $G = (V, E)$.*

The first issue is how to represent a graph so that you can both think about and work with that concept. There is one standard visual representation

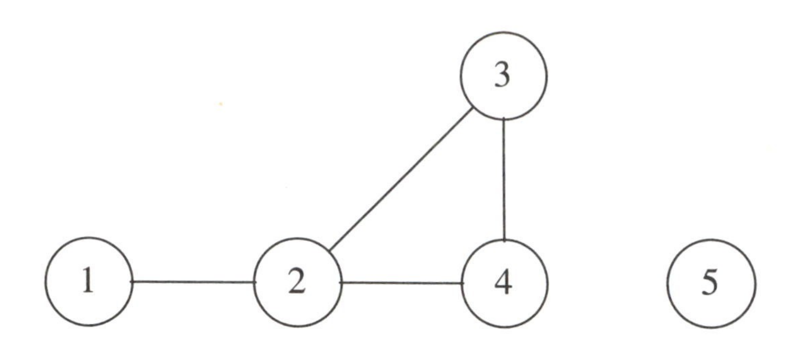

Figure 3.17: A Visual Representation of a Graph

Figure 3.18: A Loop that Connects a Vertex to Itself

and several different symbolic forms that are useful when solving problems on graphs, all of which are described next.

A Visual Representation of a Graph

The most natural way to picture a graph is to use circles (dots, squares, or similar shapes) to represent the vertices of a graph. Each edge is represented by a line that connects the corresponding pair of vertices. An example of a graph is shown in Figure 3.17. To simplify the subsequent discussion, from here on it is assumed that

1. The graph has no **loops**, that is, no edges that connect a vertex to itself (see Figure 3.18).

2. The graph has no **multiple edges**, that is, no more than one edge connects each pair of vertices (see Figure 3.19).

Suppose that a graph $G = (V, E)$ has n vertices and p edges. It is common to label the vertices of a graph for reference purposes. Any symbolic name will do but, for convenience, from here on, the vertices are numbered sequentially from 1 to n, unless otherwise indicated. You can also label the edges. Alternatively, because there is at most one edge connecting a vertex i to a vertex j, you can refer to that edge by the symbol ij, as is done here. Some other important aspects of graphs include the following:

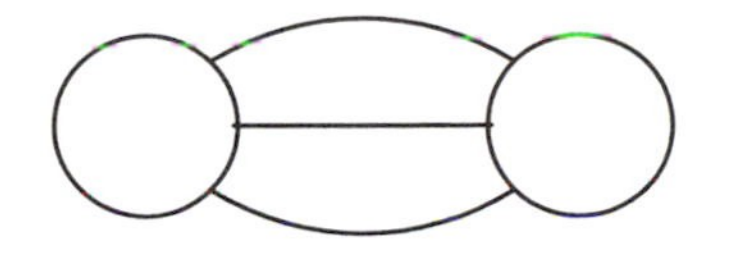

Figure 3.19: A Graph with Multiple Edges

1. Two vertices i and j connected by an edge e are **adjacent** to each other, or equivalently, are **neighbors** of each other. Each vertex is an **endpoint of the edge** e and e is said to be **incident** to i and j.

2. The **degree of a vertex** i, denoted by $d_G(i)$, is the number of edges incident to i, that is, the number of edges having vertex i as an endpoint. If there is only one graph under consideration, then the degree of a vertex is written as $d(i)$. For example, the degree of vertex 2 in Figure 3.17 is $d(2) = 3$ and the degree of vertex 5 is $d(5) = 0$. Because a graph has no loops or multiple edges, $0 \leq d(i) \leq n - 1$.

The visual representation of a graph is helpful in developing methods for solving numerous problems, as you will see later in this section. However, computers cannot "see" or "work with" this visual representation. More suitable *computational* representations are described next.

Symbolic Representations of a Graph

Two different ways to represent a graph for computational purposes are now presented. Each method has its advantages and disadvantages.

The Adjacency Matrix. One way to represent a graph symbolically is to use a rectangular table of numbers, called the **adjacency matrix**, denoted by the symbol A, in which there are n rows and n columns – one row and one column corresponding to each vertex. The number A_{ij} in row i and column j of this matrix is either 0 or 1, according to the following:

$$A_{ij} = \begin{cases} 1, & \text{if there is an edge from vertex } i \text{ to vertex } j \\ 0, & \text{if there is } \textit{no} \text{ edge from vertex } i \text{ to vertex } j. \end{cases}$$

The word *matrix* means a rectangular table of values arranged in rows and columns. The adjacency matrix for a graph is illustrated in the next example.

Example 3.29 – The Adjacency Matrix for a Given Graph

The adjacency matrix corresponding to the graph in Figure 3.17 is

	1	2	3	4	5
1	0	1	0	0	0
2	1	0	1	1	0
3	0	1	0	1	0
4	0	1	1	0	0
5	0	0	0	0	0

For each graph you draw, you can create an associated adjacency matrix. Similarly, for each adjacency matrix, you can draw a corresponding graph, as shown in the next example.

Example 3.30 – The Graph for a Given Adjacency Matrix

The graph corresponding to the following adjacency matrix is shown in Figure 3.20:

	1	2	3	4	5	6
1	0	1	0	0	0	1
2	1	0	1	0	0	0
3	0	1	0	1	0	0
4	0	0	1	0	1	0
5	0	0	0	1	0	1
6	1	0	0	0	1	0

The Edge List. Another symbolic representation of a graph is the **edge list**, in which you first provide the number of vertices and edges in the graph and then make a list of all the *edges*. Each edge is specified by the numbers of the two vertices connected by that edge, as illustrated in the following examples.

Example 3.31 – The Edge List of a Given Graph

The edge list for the graph in Figure 3.17 is as follows:

5 vertices and 4 edges:
$(1, 2)$; $(2, 3)$; $(2, 4)$; $(3, 4)$

Likewise, for each edge list, you can draw a corresponding graph, as shown in the next example.

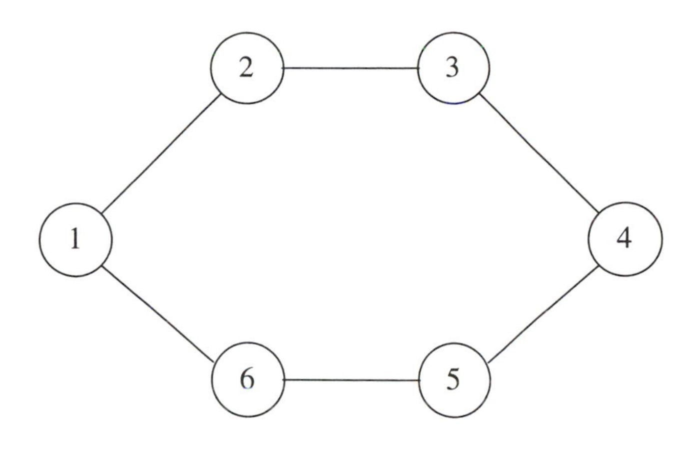

Figure 3.20: The Graph Corresponding to the Adjacency Matrix in Example 3.30

Example 3.32 – The Graph for a Given Edge List

The graph for the following edge list is given in Figure 3.20:

6 vertices and 6 edges:
$(1,2)$; $(2,3)$; $(3,4)$; $(4,5)$; $(5,6)$; $(6,1)$

Comparing the Adjacency Matrix with the Edge List. To compare these two foregoing symbolic representations, consider translating the operation of finding the degree of a vertex i by "counting the edges incident to i" to corresponding operations on the adjacency matrix and the edge list. You will then discover some differences between these two symbolic representations.

To obtain the degree of vertex i from the adjacency matrix, you add the n numbers in the row (or column) corresponding to vertex i. For the edge-list representation, you scan the p edges in the list and count the number of times vertex i appears. If the number of vertices (n) is smaller than the number of edges (p), then using the adjacency matrix is more efficient than using the edge list. In contrast, the adjaceny matrix consists of n^2 numbers, but the edge list has only $2p$ numbers. Here you can see a trade-off between the amount of storage space needed and the computational efficiency of these two symbolic representations of a graph.

3.3.2 Operations on Graphs

In this section, you will learn several operations you can perform on graphs. The most common ones are *unary* operations, in which you create a new

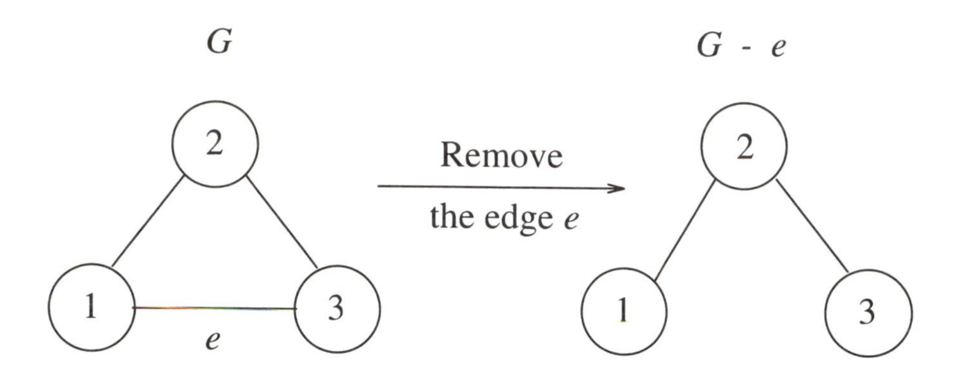

Figure 3.21: Deleting an Edge from a Graph

graph G' from an original graph G, as described next. Throughout, let $G = (V, E)$ be a graph.

Deleting and Adding Edges

One common operation is to **delete an edge** $e \in E$ **from** G. This is accomplished visually by removing the edge e from the picture. For example, Figure 3.21 illustrates the graph obtained by removing from G the edge from vertex 1 to vertex 3. The translation of "removing an edge e of a graph" to symbolic form is to create a new graph, denoted by $G - e$, having the same vertex set as G but whose edge set is $E - \{e\}$. That is, you create from $G = (V, E)$ the new graph $G' = (V', E')$ in which $V' = V$ and $E' = E - \{e\}$.

The opposite operation is to **add an edge** $e \notin E$ **to** G. This is accomplished visually by drawing the new edge e. For example, Figure 3.22 illustrates the graph obtained by adding to G the edge from vertex 2 to vertex 4. Symbolically, you create from $G = (V, E)$ the new graph $G' = (V', E')$ in which $V' = V$ and $E' = E \cup \{e\}$. This new graph is written as $G + e$.

Deleting and Adding Vertices

Another common operation is to **delete a vertex** $v \in V$ **from** G. Unlike deleting an edge, you *cannot* simply "remove" a vertex v, for what happens to all the edges incident to that vertex? Those edges cannot remain in the graph, so, when you remove a vertex v, you must remove simultaneously all edges incident to v. For example, Figure 3.23 illustrates the graph obtained by removing from G vertex 2 and the four edges incident to vertex 2. The process of "removing a vertex v from a graph" translates to the symbolic operation of creating from $G = (V, E)$ the new graph $G' = (V', E')$ in which $V' = V - \{v\}$ and $E' = E - \{e \in E : e \text{ is incident to } v\}$. This new graph is written as $G - v$.

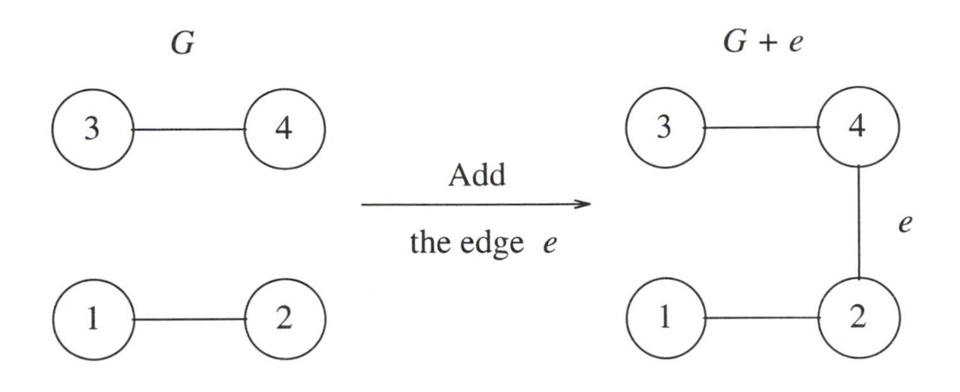

Figure 3.22: Adding an Edge to a Graph

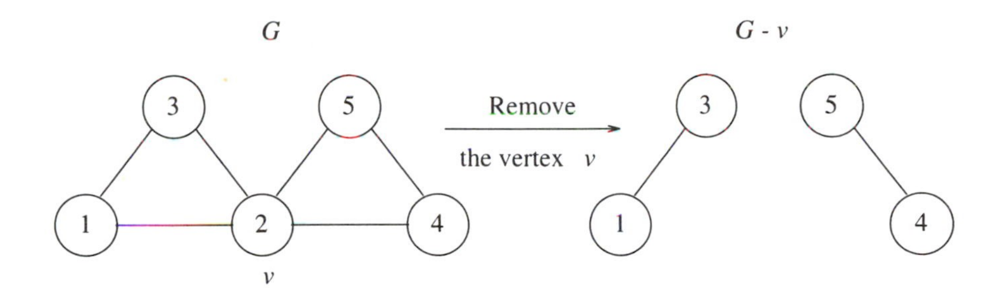

Figure 3.23: Deleting a Vertex from a Graph

The opposite operation is to **add a vertex** $v \notin V$ **to** G. This is accomplished visually by drawing the new vertex v. Symbolically, you create from $G = (V, E)$ the new graph $G' = (V', E')$ in which $V' = V \cup \{v\}$ and $E' = E$. This new graph is written as $G + v$.

Creating a Subgraph

A generalization of the concept of deleting edges and vertices arises when you need to work with some *portion* of a graph $G = (V, E)$. For example, look at the graph G in Figure 3.24. You might only want to work with the portion of G consisting of vertices 1 and 2 together with that one connecting edge. The objective now is to create an appropriate definition to describe the concept of a "portion of a graph G," referred to as a *subgraph* of G.

Notice that to create a subgraph, you select a *subset* of the vertices and edges from the original graph $G = (V, E)$. Thus, the concept of "creating a subgraph" translates to the symbolic operation of "selecting a subset" of the vertices and edges of G, leading you to the following first attempt:

> A *subgraph* of $G = (V, E)$ is a graph $G' = (V', E')$ in which $V' \subseteq V$ and $E' \subseteq E$.

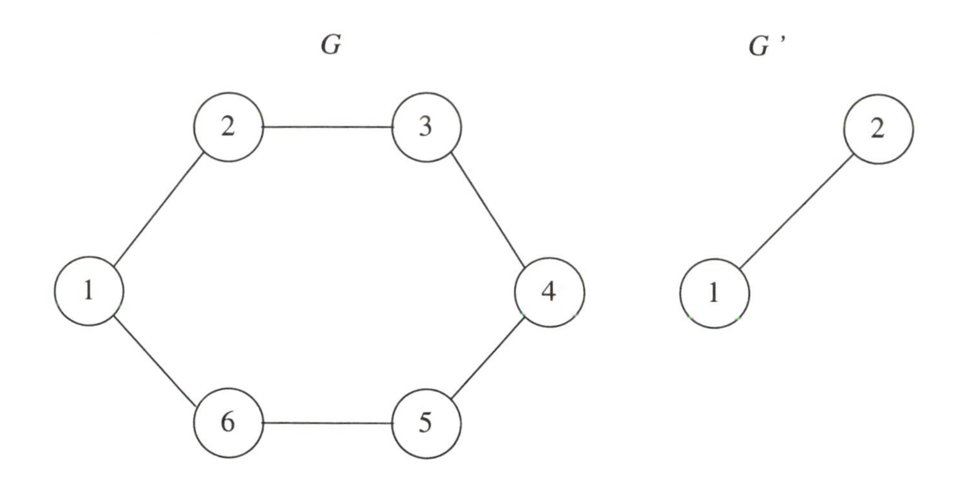

Figure 3.24: A Subgraph G' of a Graph G

However, as you learned in Section 2.2.2, after creating a definition, you must *verify* that definition. Although it *is true* that the subgraph G' in Figure 3.24 is obtained by choosing $V' \subseteq V$ and $E' \subseteq E$, is it *true* that when you *arbitrarily* choose subsets V' of V and E' of E, you always get what you think of as a subgraph? The answer is *no*. For example, if from G in Figure 3.24 you choose only vertices 1, 2, and 3 to be in the subgraph, then you *cannot* select the edge from vertex 3 to vertex 4 to be in the subgraph. On the basis of this example, modify the preceding attempt at the definition of a subgraph to obtain the following correct definition.

> **Definition 3.21** *A* **subgraph** *of a graph $G = (V, E)$ is a graph $G' = (V', E')$ in which $V' \subseteq V$, $E' \subseteq E$, and every edge of E' connects two vertices in V'.*

A special case of Definition 3.21 is the operation of deleting an edge e from a graph $G = (V, E)$ to create the subgraph $G' = (V', E')$ in which $V' = V$ and $E' = E - \{e\}$. Similarly, the operation of deleting a vertex v from G is a special case of Definition 3.21 in which $V' = V - \{v\}$ and $E' = E - \{e \in E : e \text{ is incident to } v\}$.

Special Graphs

After working with graphs for a while, you will discover that several special graphs arise frequently. Some of these are described next.

The Trivial Graph. The **trivial graph** is the graph that consists of a single vertex and no edges.

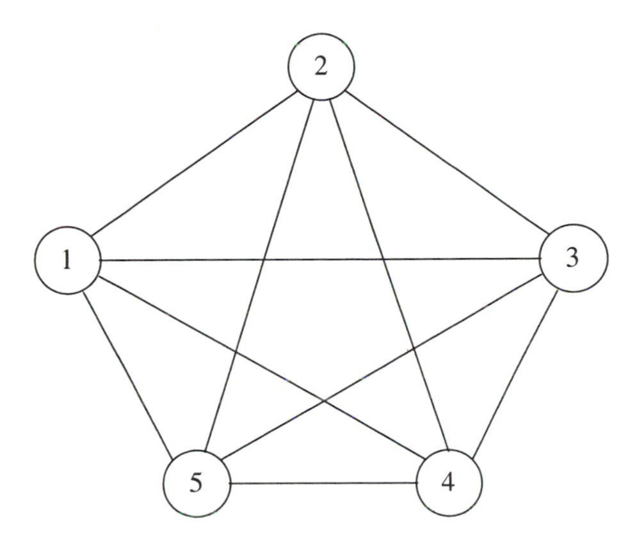

Figure 3.25: The Complete Graph on 5 Vertices (K_5)

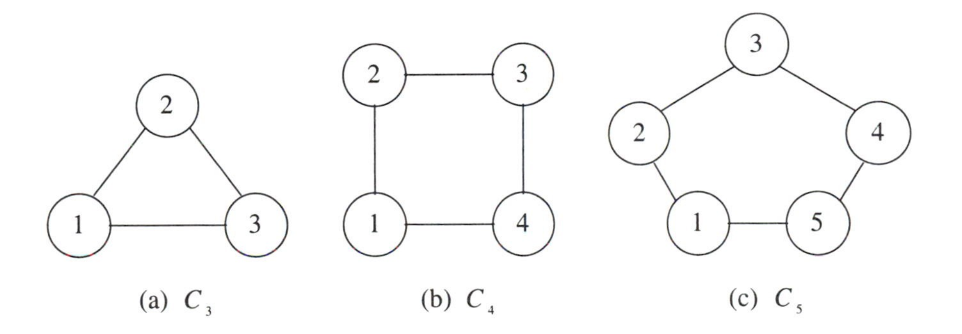

Figure 3.26: The Cycles C_3, C_4, and C_5

The Complete Graph. The **complete graph** on n vertices, denoted by K_n, is the graph in which every possible pair of vertices are connected with an edge. For example, K_5 is shown in Figure 3.25. Also note that K_1 is the trivial graph. The number of edges in K_n is "n choose 2":

$$\binom{n}{2} = \frac{n(n-1)}{2}.$$

A Cycle. Another useful graph is a *cycle* on n vertices, denoted by C_n. The cycles C_3, C_4, and C_5 are shown in Figure 3.26. The cycle C_6 is illustrated in Figure 3.24. The objective now is to create a formal definition of a cycle.

You might at first try to use the fact that the number of edges and vertices in a cycle is the same. However, Figure 3.27 illustrates a graph

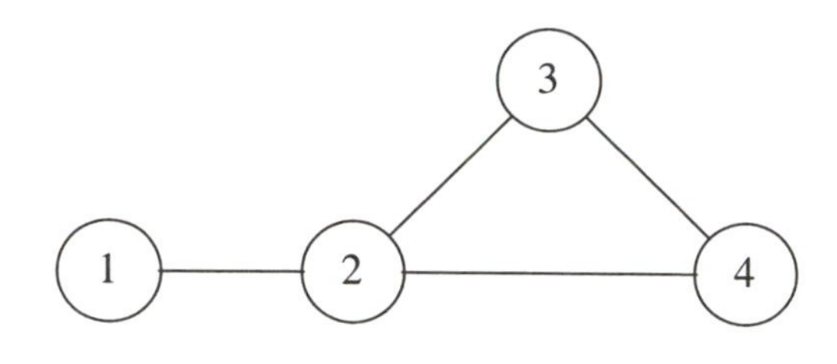

Figure 3.27: A Graph with the Same Number of Vertices and Edges that Is Not a Cycle

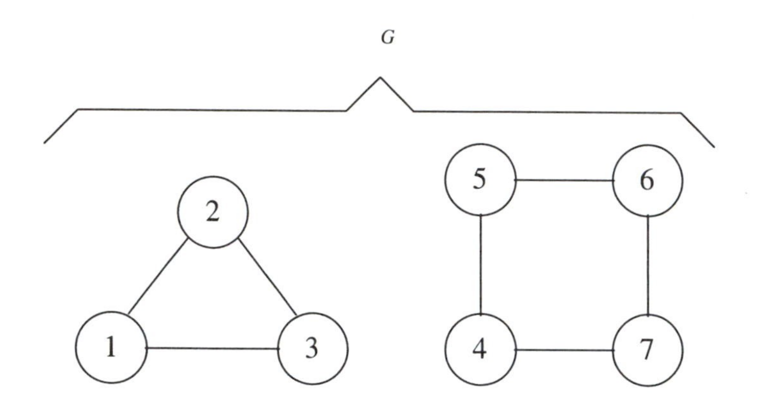

Figure 3.28: A Graph in Which Each Vertex Has Degree 2 but that Is Not a Cycle

with 4 vertices and 4 edges that is *not* a cycle. What, then, is a cycle?

Another approach is to use the concept of the degrees of the vertices to describe a cycle, leading you to the following:

A *cycle* is a graph in which each vertex has degree 2.

Verifying this definition, it *is true* that all the *cycles* in Figure 3.26 satisfy the property that each vertex has degree 2. But, is it also *true* that every graph in which each vertex has degree 2 is a cycle? The answer is *no* – look at the graph in Figure 3.28. Each vertex in this graph has degree 2 and yet that graph is *not* what you think of as a cycle.

The problem in Figure 3.28 is that the graph has two "pieces" – one piece consisting of the vertices 1, 2, and 3 and the other piece consisting of the vertices 4, 5, 6, and 7. If you rule out more than one piece, then you obtain the following correct definition of a cycle (in which the word "connected" means "one piece"):

Definition 3.22 *A* **cycle** *is a connected graph in which each vertex has degree 2.*

The problem with Definition 3.22 is that the meaning of "connected" is not yet precise. Again, you have a visual image of what a connected graph is – namely, a graph consisting of one *piece* – but you must translate that image to symbolic form. This task is accomplished in the next section.

3.3.3 The Property of a Graph Being Connected

The objective of this section is to create a definition of what it means for a graph to be *connected*, that is, to consist of one piece. For example, the graph in Figure 3.27 *is* connected and the graph in Figure 3.28 is *not* connected. What property does the graph in Figure 3.27 have that the graph in Figure 3.28 does not?

One way to describe this desirable property is to realize that in a connected graph, you can "go from any vertex to any other vertex by using the edges of the graph." For example, in Figure 3.27, you can go from vertex 1 to vertex 3 by using the edge that connects vertex 1 to vertex 2 and then the edge from vertex 2 to vertex 3. In contrast, in Figure 3.28, you *cannot* go from vertex 1 to vertex 4 by using the edges of the graph. This concept of "going from one vertex to another by using the edges" is made precise in the following definition.

Definition 3.23 *A* **walk** *in a graph $G = (V, E)$ is an alternating sequence of vertices and edges, say,*

$$v_0,\ e_1,\ v_1,\ e_2,\ v_2, \ldots,\ v_{k-1},\ e_k,\ v_k$$

in which $v_0, \ldots, v_k \in V$ and $e_1, \ldots, e_k \in E$ and each edge e_i connects vertex v_{i-1} to vertex v_i, for $i = 1, \ldots, k$. This is said to be **a walk from** v_0 **to** v_k *and is sometimes written as $v_0 - v_k$. The* **length of a walk** *is the number of edges in the walk (k, in this case).*

Observe that because there are no multiple edges, you can represent a walk unambiguously by the sequence of vertices in the walk, as shown in the following example.

Example 3.33 – Walks in a Graph

Each of the following are walks in the graph in Figure 3.28:

(a)	4	a walk from vertex 4 to vertex 4 of length 0.
(b)	4-5	a walk from vertex 4 to vertex 5 of length 1.
(c)	4-5-4	a walk from vertex 4 to vertex 4 of length 2.
(d)	4-5-6	a walk from vertex 4 to vertex 6 of length 2.
(e)	5-6-7-4-5	a walk from vertex 5 to vertex 5 of length 4.

You can distinguish the following two kinds of walks in Example 3.33:

1. A **trail**, in which no *edge* is repeated. (Walks (a), (b), (d), and (e) in Example 3.33 are trails.)

2. A **path**, in which no *vertex* is repeated. (Walks (a), (b), and (d) in Example 3.33 are paths.)

You can now use the concept of a walk together with the quantifier *for all* to describe a connected graph, as shown in the following definition.

> **Definition 3.24** *A graph $G = (V, E)$ is* **connected** *if and only if for each pair of vertices $i, j \in V$, there is a walk in G from i to j.*

The meaning of a cycle given in Definition 3.22 is now precise – a cycle is a connected graph in which each vertex has degree 2.

When G contains few vertices and edges, you can draw the graph and determine *visually* whether G is connected or not. How would you do so if the graph has too many vertices and edges to draw? A method for doing so is presented next.

3.3.4 Solving Problems Using Graphs

In this section, you will see a general approach to solving problems involving graphs. The first such problem is that of determining if a graph is connected or not.

Determining if a Graph Is Connected

From Definition 3.24, a graph G is connected if and only if there is a walk between every pair of vertices. The objective now is to solve the following problem.

> **Example 3.34 – The Connectivity Problem**
>
> Determine whether a given graph $G = (V, E)$ is connected or not.

Recall from Chapter 1 that the process of solving a problem involves using information you know (the *input* data) to determine information you do not know (the *outputs*). What data do you have for the problem in Example 3.34? The answer is the graph G consisting of its vertices and edges. The desired output is: *yes*, the graph is connected or *no*, the graph is not connected.

As discussed in Section 1.1.1, you would ideally like to solve this problem in closed form. In this case, that means developing a simple rule or formula that determines whether the graph G is connected or not by using the vertices and edges of G. Unfortunately, no such closed-form solution is known, so, instead, a *numerical method* (see Section 1.1.2) is needed. That numerical method is described in the form of an algorithm for obtaining the solution.

A Solution Procedure for Determining if a Graph is Connected. Throughout the rest of this section, let G be a graph on n vertices and i and j be two vertices in G. Recall that a graph is connected if there is a walk between *every* pair of vertices. You already know that there is a walk from every vertex i to itself – namely, the walk of length 0 obtained by standing at vertex i. Also, from the graph, you know that whenever there is an edge from i to j there is also a walk – namely, the walk consisting of that edge. The issue, then, is to determine if there is an $i - j$ walk between those vertices $i \neq j$ that are not connected by an edge. To find out, a sequence of n matrices of numbers is created, say, $A^{(0)}, \ldots, A^{(n)}$, in which for each $k = 0, \ldots, n$, $A_{ij}^{(k)}$ has the following meaning:

$$A_{ij}^{(k)} = \begin{cases} 1, & \text{if there is an } i - j \text{ walk that uses } \textit{only} \text{ the} \\ & \text{vertices whose numbers are } 1, \ldots, k \\ 0, & \text{otherwise.} \end{cases}$$

When you obtain $A_{ij}^{(n)}$, you have tried to use all the n vertices in finding a walk from i to j. Thus, you will know that the original graph $G = (V, E)$ is connected if and only if every value of $A_{ij}^{(n)}$ is 1.

Starting with $A^{(0)}$ defined by

$$A_{ij}^{(0)} = \begin{cases} 1, & \text{if } i = j \text{ or } ij \in E \\ 0, & \text{if } i \neq j \text{ and } ij \notin E, \end{cases}$$

you sequentially create the next matrix $A^{(k)}$ from $A^{(k-1)}$ by using the following idea:

$$A_{ij}^{(k)} = \begin{cases} 1, & \text{if there already is an } i - j \text{ walk in } A^{(k-1)} \text{ or} \\ & \text{if there is an } i - k \text{ walk and a } k - j \text{ walk in } A^{(k-1)} \\ 0, & \text{otherwise.} \end{cases}$$

Can you translate these ideas into specific formulas using $A^{(k-1)}$? The answer is given in the following solution.

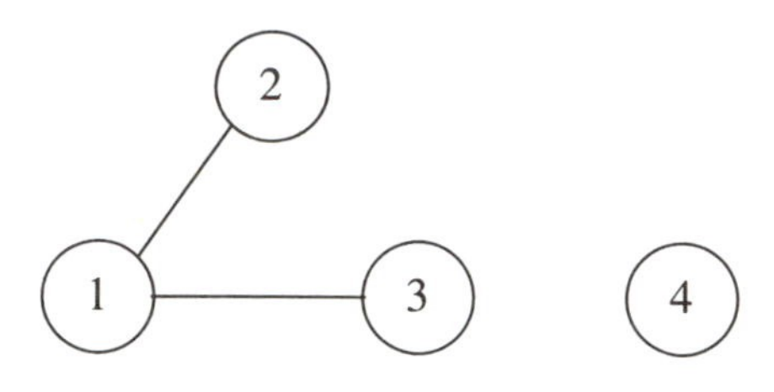

Figure 3.29: An Example of a Graph that Is Not Connected

Solution to Example 3.34

To determine if a graph $G = (V, E)$ is connected, create the matrix $A^{(0)}$ as follows:

$$A_{ij}^{(0)} = \begin{cases} 1, & \text{if } i = j \text{ or } ij \in E \\ 0, & \text{if } i \neq j \text{ and } ij \notin E. \end{cases}$$

Then, for each $k = 1, \ldots, n$, compute the next matrix $A^{(k)}$ from $A^{(k-1)}$ by using the following formula:

$$A_{ij}^{(k)} = \begin{cases} 1, & \text{if } A_{ij}^{(k-1)} = 1 \text{ or} \\ & \text{if } A_{ik}^{(k-1)} = 1 \text{ and } A_{kj}^{(k-1)} = 1 \\ 0, & \text{otherwise.} \end{cases}$$

The graph G is connected if and only if for each $i, j = 1, \ldots, n$, $A_{ij}^{(n)} = 1$.

By applying these computations to the graph in Figure 3.29, you obtain the following sequence of matrices:

$$A^{(0)} = \begin{array}{c|c|c|c|c|} & 1 & 2 & 3 & 4 \\ \hline 1 & 1 & 1 & 1 & 0 \\ \hline 2 & 1 & 1 & 0 & 0 \\ \hline 3 & 1 & 0 & 1 & 0 \\ \hline 4 & 0 & 0 & 0 & 1 \\ \hline \end{array}$$

$$A^{(1)} = \begin{array}{c|c|c|c|c|} & 1 & 2 & 3 & 4 \\ \hline 1 & 1 & 1 & 1 & 0 \\ \hline 2 & 1 & 1 & 1 & 0 \\ \hline 3 & 1 & 1 & 1 & 0 \\ \hline 4 & 0 & 0 & 0 & 1 \\ \hline \end{array}$$

$$A^{(4)} = A^{(3)} = A^{(2)} = A^{(1)}.$$

The fact that $A_{14}^{(4)} = 0$ indicates that the graph in Figure 3.29 is *not* connected because there is no walk from vertex 1 to vertex 4.

The proof that this algorithm does in fact determine whether a graph is connected or not follows.

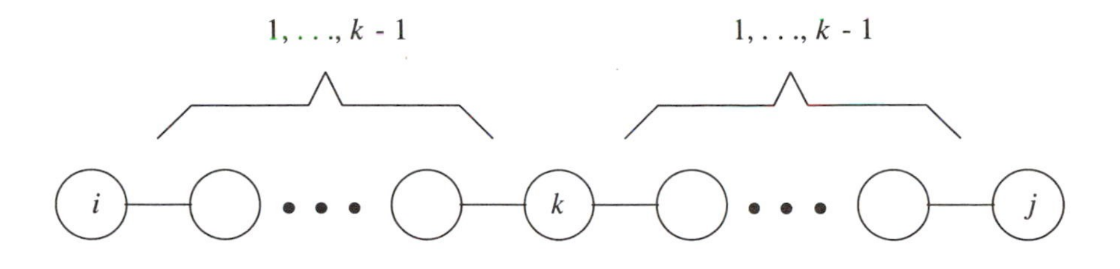

Figure 3.30: Combining Two Walks that Use Only the Vertices Whose Numbers Are $1, \ldots, k-1$

Theorem 3.2 *Let $G = (V, E)$ be a graph on n vertices. If the matrices $A^{(0)}, \ldots, A^{(n)}$ are generated according to the Solution to Example 3.34, then for all vertices i and j, and for all $k = 0, \ldots, n$,*

$$A_{ij}^{(k)} = \begin{cases} 1, & \text{if there is an } i-j \text{ walk that uses only the} \\ & \text{vertices whose numbers are } 1, \ldots, k \\ 0, & \text{otherwise.} \end{cases} \tag{3.17}$$

In particular, G is connected if and only if for all vertices i and j, $A_{ij}^{(n)} = 1$.

Proof. In view of the fact that $A^{(k)}$ is defined in terms of $A^{(k-1)}$, a proof by induction on k is used. So, for $k = 0$, the formula for $A_{ij}^{(0)}$ in the Solution to Example 3.34 yields a value of 1, if there is a walk from i to j using no vertices and 0, otherwise. Thus, (3.17) is correct for $k = 0$.

Now assume that $A_{ij}^{(k-1)}$ has the meaning indicated in (3.17). To obtain the desired result for $A^{(k)}$, recall that $A_{ij}^{(k)}$ is determined by the following formula:

$$A_{ij}^{(k)} = \begin{cases} 1, & \text{if } A_{ij}^{(k-1)} = 1 \text{ or} \\ & \text{if } A_{ik}^{(k-1)} = 1 \text{ and } A_{kj}^{(k-1)} = 1 \\ 0, & \text{otherwise.} \end{cases}$$

Therefore, if $A_{ij}^{(k)} = 1$, then either $A_{ij}^{(k-1)} = 1$, in which case, by the induction hypothesis, there is a walk from i to j that uses only the vertices whose numbers are $1, \ldots, k-1$, or, alternatively, $A_{ik}^{(k-1)} = 1$ and $A_{kj}^{(k-1)} = 1$, in which case, by the induction hypothesis, there are walks from i to k and from k to j that use only the vertices whose numbers are $1, \ldots, k-1$. As seen in Figure 3.30, combining these two walks sequentially results in a walk from i to j that uses only the vertices whose numbers are $1, \ldots, k$.

Finally, if $A_{ij}^{(k)} = 0$, then there is no walk from i to j that uses only the vertices whose numbers are $1, \ldots, k$. This is because if there *were* such a walk – say, W – then either W uses k or not. If W does not use k, then $A_{ij}^{(k-1)}$ would be 1 and hence so would $A_{ij}^{(k)}$. If W uses k then, as seen in Figure 3.31, the portion of W from i to the *first* occurrence of k is a walk

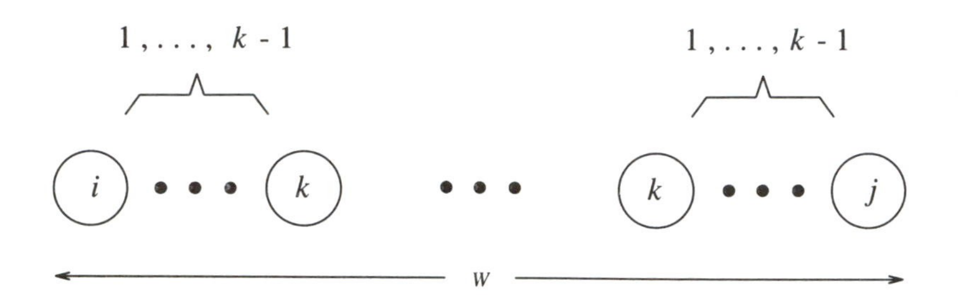

Figure 3.31: A Walk W from i to j that Uses k Several Times

from i to k that uses only the vertices whose numbers are $1, \ldots, k-1$, so $A_{ik}^{(k-1)} = 1$. Likewise, that portion of W in Figure 3.31 from the *last* occurrence of k to j is a walk from k to j that uses only the vertices whose numbers are $1, \ldots, k-1$, so $A_{kj}^{(k-1)} = 1$ also. But then, because $A_{ik}^{(k-1)} = 1$ and $A_{kj}^{(k-1)} = 1$, it follows that $A_{ij}^{(k)}$ would be 1. However, because $A_{ij}^{(k)} = 0$, there can be no such walk W for i to j. Thus, (3.17) is correct when $A_{ij}^{(k)} = 0$.

By induction, it now follows that for each $k = 0, \ldots, n$, $A_{ij}^{(k)}$ has the meaning in (3.17). Consequently, G is connected if and only if for all vertices i and j, $A_{ij}^{(n)} = 1$. QED

Although you now know that the algorithm in the Solution to Example 3.34 is correct, recall from Section 1.1.3 that one of the uses of mathematics is to solve problems as *efficiently* as possible. In the exercises, you will see how to develop a more efficient method for determining if a graph is connected or not. For now, you will learn how to solve another problem involving graphs.

Finding Shortest Paths in a Graph

As another example of using graphs, consider the following problem faced by the management of Florida Airways.

Example 3.35 – The Shortest-Path Problem of Florida Airways

Florida Airways will start providing service between the following cities: Miami (M), Daytona (D), Orlando (O), Tampa (T), and Jacksonville (J). The nonstop fares in dollars between certain pairs of these cities are given in the following matrix:

Continued

From/To	M	D	O	T	J
M	–	89	120	–	–
D	89	–	55	75	–
O	120	55	–	59	69
T	–	75	59	–	–
J	–	–	69	–	–

Management wants to use these nonstop fares as the basis for determining the fares between *all* pairs of cities. Specifically, they need to know the least-cost route among all the possible ways to fly between two cities.

Problem Formulation. To solve these types of problems using a graph, you should first **formulate the problem** by following these steps (which were demonstrated with the traveling-salesperson problem in Section 1.5.1).

Steps for Formulating a Problem Using a Graph

Step 1. Use vertices to represent certain objects in the problem. Explain the meaning of those vertices in the context of the problem. (For Example 3.35, use 5 vertices, one for each of the 5 cities.)

Step 2. Use edges to connect certain pairs of these vertices to reflect an associated relationship between the objects being connected. Explain the meaning of those edges in the context of the problem. (For Example 3.35, use an edge to connect two vertices if there is a nonstop flight between the corresponding cities.)

Step 3. Identify any additional data and determine whether those values are associated with the vertices and/or the edges of the graph. (For Example 3.35, the data are the nonstop fares in the given matrix, each of which is associated with an *edge* of the graph.)

Step 4. Draw the resulting graph together with the data identified in Step 3. (The graph for Example 3.35 is shown in Figure 3.32.)

Step 5. State the objective of the problem in terms of a *general graph and its data* rather than the specific problem. (For Example 3.35, this objective is stated below.)

Example 3.36 – The Shortest-Path Problem

Given a graph $G = (V, E)$ and a nonnegative cost associated with each edge of G, find the cost of the shortest path between all pairs of vertices in G, where the cost of a path is the sum of the individual costs of the edges in the path.

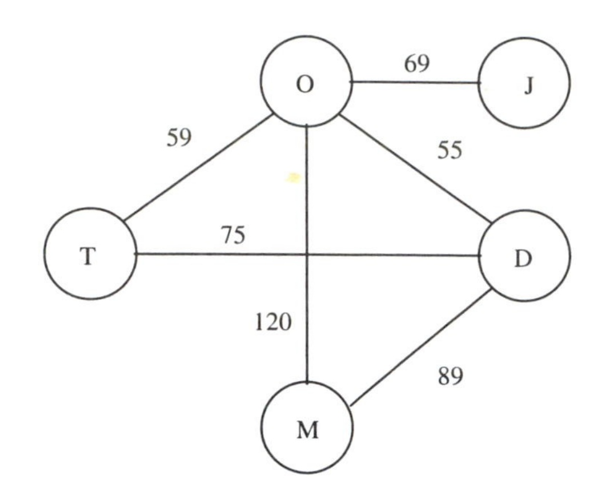

Figure 3.32: The Graph for the Problem of Florida Airways

Developing a Solution Procedure. The input data in Example 3.36 are the graph and the costs of the edges in a given matrix, say, C. The desired output is a matrix of least costs between all pairs of vertices. One approach to solving the shortest-path problem is similar in nature to the method you learned for determining if a graph is connected or not. That is, you create a sequence of matrices – say, $C^{(0)}, \ldots, C^{(n)}$ – in which for each $k = 0, \ldots, n$, $C^{(k)}$ has the following meaning:

$$C_{ij}^{(k)} = \text{the cost of the shortest path from } i \text{ to } j \text{ when using } \textit{only} \text{ those vertices whose numbers are } 1, \ldots, k.$$

Thus, when you obtain $C^{(n)}$, you will know that $C_{ij}^{(n)}$ is the cost of the shortest path from vertex i to vertex j because you are using all n vertices.

The specific way in which you compute $C^{(k)}$ from $C^{(k-1)}$ is based on the observation that in computing $C_{ij}^{(k)}$, you can use or not use the vertex whose number is k when going from i to j. (Note that because the costs of the edges are *nonnegative*, the vertex k is used at most once in a shortest path from i to j.) If you choose to use k, then the least-cost way to get from i to j is to go from i to k first (at a least cost of $C_{ik}^{(k-1)}$) and then go from k to j (at a least cost of $C_{kj}^{(k-1)}$). Alternatively, if you choose *not* to use vertex k in going from i to j, then the least cost for doing so is $C_{ij}^{(k-1)}$. The choice of whether to use k or not depends on whichever alternative is less expensive. That is,

$$C_{ij}^{(k)} = \min\{C_{ij}^{(k-1)}, C_{ik}^{(k-1)} + C_{kj}^{(k-1)}\}.$$

The solution procedure for the shortest-path problem is summarized as follows.

Solution to Example 3.36

The solution to the shortest-path problem on a graph $G = (V, E)$ in which there are nonnegative costs associated with the edges of G provided in the matrix C is to start with

$$C_{ij}^{(0)} = \begin{cases} C_{ij}, & \text{if } ij \in E \\ 0, & \text{if } ij \notin E \text{ and } i = j \\ \infty, & \text{if } ij \notin E \text{ and } i \neq j. \end{cases}$$

Then, for each $k = 1, \ldots, n$, sequentially create the next matrix $C^{(k)}$ from $C^{(k-1)}$ by using the following formula:

$$C_{ij}^{(k)} = \min\{C_{ij}^{(k-1)}, C_{ik}^{(k-1)} + C_{kj}^{(k-1)}\}.$$

The final matrix, $C^{(n)}$, contains the costs of the shortest paths.

By applying this solution procedure to the specific problem in Example 3.35, you obtain the following sequence of matrices:

$C^{(0)} =$

	M	D	O	T	J
M	0	89	120	∞	∞
D	89	0	55	75	∞
O	120	55	0	59	69
T	∞	75	59	0	∞
J	∞	∞	69	∞	0

$C^{(1)} =$

	M	D	O	T	J
M	0	89	120	164	∞
D	89	0	55	75	∞
O	120	55	0	59	69
T	164	75	59	0	∞
J	∞	∞	69	∞	0

$C^{(2)} =$

	M	D	O	T	J
M	0	89	120	164	189
D	89	0	55	75	124
O	120	55	0	59	69
T	164	75	59	0	128
J	189	124	69	128	0

$$C^{(5)} = C^{(4)} = C^{(3)} = C^{(2)}.$$

From $C^{(5)}$, you now know all of the least-cost fares. For example, because $C_{MJ}^{(5)} = 189$, the least-cost fare from Miami to Jacksonville is \$189.

The proof that this algorithm does, in fact, provide the costs of the shortest paths between all pairs of vertices is provided in the next theorem.

Theorem 3.3 *Suppose that $G = (V, E)$ is a graph that has a nonnegative number C_{ij} associated with each edge $ij \in E$. If for all vertices i and j and for each $k = 0, \ldots, n$, $C_{ij}^{(k)}$ is defined by the formula in the Solution to Example 3.36, then*

$$C_{ij}^{(k)} = \text{the cost of the shortest path from } i \text{ to } j \text{ when using only the vertices } 1, \ldots, k. \tag{3.18}$$

In particular, $C^{(n)}$ contains the costs of the shortest paths from i to j in G.

Proof. In view of the fact that $C^{(k)}$ is defined in terms of $C^{(k-1)}$, a proof by induction is used. So, for $k = 0$, note that the formula for $C_{ij}^{(0)}$ in the Solution to Example 3.36 is the cost of the shortest path from i to j when using no other vertices. Thus, (3.18) is correct for $k = 0$.

Now assume that $C^{(k-1)}$ has the meaning given in (3.18). To obtain the desired result for $C^{(k)}$, recall that $C_{ij}^{(k)}$ is computed by the following formula:

$$C_{ij}^{(k)} = \min\{C_{ij}^{(k-1)}, C_{ik}^{(k-1)} + C_{kj}^{(k-1)}\}.$$

To see that the foregoing formula is correct, let P be a shortest path from i to j that uses only the vertices whose numbers are $1, \ldots, k$. Now either P use k or not. If P does not use k, then P uses only vertices whose numbers are $1, \ldots, k-1$. In this case, by the induction hypothesis, $C_{ij}^{(k-1)}$ is the cost of P.

Alternatively, if P uses k then, because all the edge-costs are nonnegative, k appears in P only *once* (otherwise, you can remove the cycle from k to k and have a path of less cost). Now, as seen in Figure 3.33, the least-cost way to go from i to j is to go first from i to k by a path that uses only the vertices whose numbers are $1, \ldots, k-1$ and then go from k to j by a path that also uses only the vertices whose numbers are $1, \ldots, k-1$. The total cost of P is the sum of these two portions which, by the induction hypothesis, is $C_{ik}^{(k-1)} + C_{kj}^{(k-1)}$.

Therefore, the cost of the shortest path from i to j that uses only the vertices whose numbers are $1, \ldots, k$ is the smaller of $C_{ij}^{(k-1)}$ (if you do *not* use k) and $C_{ik}^{(k-1)} + C_{kj}^{(k-1)}$ (if you *do* use k). That is,

$$C_{ij}^{(k)} = \min\{C_{ij}^{(k-1)}, C_{ik}^{(k-1)} + C_{kj}^{(k-1)}\}.$$

This completes the induction step and shows that for each $k = 0, \ldots, n$, $C_{ij}^{(k)}$ computed by (3.18) is indeed the cost of the shortest path from i to j when using only the vertices whose numbers are $1, \ldots, k$. QED

The shortest-path problem is a *generalization* of the problem of determining whether a graph is connected or not. This is because you can use the data from the special case of determining if a graph is connected to

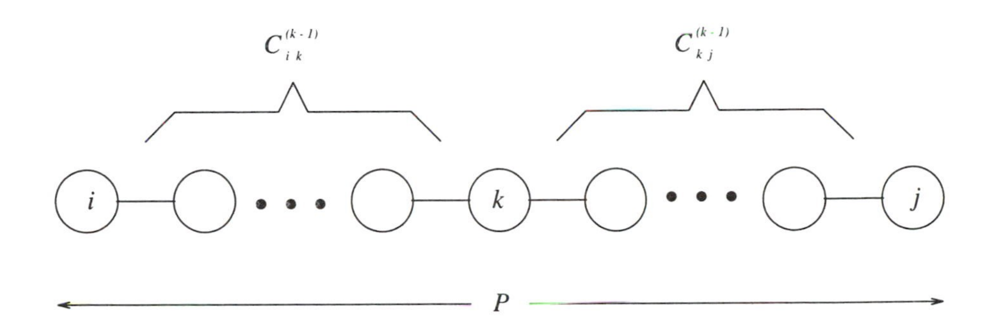

Figure 3.33: A Path P from i to j that Uses k Exactly Once

create data for the general problem of determining the costs of the shortest paths in such a way that when you obtain the solution to the general problem, you have solved the special case. This is done as follows.

Suppose you want to determine if the graph $G = (V, E)$ is connected or not. Create an assoicated shortest-path problem in which the cost is 1 for each edge of G. Now, suppose you find the costs of the shortest paths between all pairs of vertices in this problem. These final costs represent the number of edges in the shortest paths because all edge-costs are 1. Thus, if these final costs are all *finite*, then the graph is connected because *there is* a path between all pairs of vertices. On the other hand, if the cost of the shortest path between two vertices i and j is ∞, then the graph is *not* connected because there is no walk from i to j – that is, there is no way to get from i to j by using the edges of G. Thus, you can use the shortest-path solution procedure to determine if a graph $G = (V, E)$ is connected or not. However, recall from Section 1.4 that using the general solution procedure (finding the shortest paths, in this case) is less efficient than solving the special case directly (determining if a graph is connected or not, in this case).

Further Generalizations of Graphs

Throughout this section, it has been assumed that a graph has no loops or multiple edges. For solving some problems, you need to work with a generalization of a graph that allows for loops and multiple edges.

Another generalization of a graph arises when there is a *direction* associated with each edge of the graph. In this case, there is a difference between the edge that starts at vertex i and ends at vertex j and the edge that starts at vertex j and ends at vertex i. This type of graph is defined in Section 1.5.1 and is repeated here.

Definition 3.25 *A* **directed graph** *consists of a nonempty finite set V of vertices and a finite set $\overrightarrow{E}$ of* **directed edges**, *each of which connects a vertex i to a vertex j. The directed graph is written $\overrightarrow{G} = (V, \overrightarrow{E})$.*

In the exercises, you are asked to generalize many of the concepts in this section to directed graphs. For now, it is time to bring together your knowledge of sets, functions, and graphs.

3.4 Putting It All Together

The objective of this section is to illustrate the techniques of unification, generalization, and abstraction in an example that uses sets, functions, and graphs. To that end, recall from Section 3.3.4, the connectivity problem of determining whether a graph is connected or not (see Example 3.34) and the shortest-path problem (see Example 3.36). In describing the solution procedure for the shortest-path problem, it was mentioned that "One approach for solving the shortest-path problem is similar in nature to the method you learned for determining if a graph is connected or not." If indeed you see some similarities and if you think like a mathematician, then you should investigate this relationship more carefully in the hope of *unifying* these two problems into a single, more general problem that includes both problems as special cases. The reason for doing so is that you can then develop a *single* solution procedure. This unification is accomplished through the process of abstraction, as seen in what follows.

3.4.1 Identifying Similarities and Differences

The first step in unification is to identify as many similarities and differences as you can between the connectivity and shortest-path problems and their solution procedures. Review that material from Section 3.3.4 at this time. How many similarities and differences can you find? Here is a partial list.

Similarities in the Connectivity and Shortest-Path Problems

1. Both problems involve a graph $G = (V, E)$ on n vertices.

2. Both solution procedures involve creating a sequence of $(n + 1)$ matrices, say, $L^{(0)}, \ldots, L^{(n)}$, the last of which provides the solution to the problem.

3. The data for each problem are used to create the initial matrix $L^{(0)}$.

4. The meaning of each $L_{ij}^{(k)}$ is similar in that paths between the vertices i and j use *only* those vertices whose numbers are $1, \ldots, k$.

5. In both cases, $L^{(k)}$ is computed from $L^{(k-1)}$. (There are even similarities in the specific way $L^{(k-1)}$ is used in each case.)

Differences Between the Connectivity and Shortest-Path Problems

1. The data for the shortest-path problem include a nonnegative number (cost) associated with each edge of the graph whereas the connectivity problem does not.

2. The objectives of the two problems are different in nature. The objective of the connectivity problem is a *yes* or *no* answer – *yes*, the graph *is* connected or *no*, the graph is *not* connected. The objective of the shortest-path problem is to determine a value for each pair of vertices – the cost of the shortest path between those vertices.

3. The specific way in which the matrices $L^{(0)}, \ldots, L^{(n)}$ are created is different in each case.

Additional similarities and differences are described in the subsequent development.

Overcoming the Differences

The objective now is to examine the foregoing differences with the goal of finding a way to make the two problems more similar. For example, the first apparent difference mentioned above is that the shortest-path problem has a nonnegative number associated with each edge of the graph but the connectivity problem does not. However, at the end of Section 3.3.4, it is shown that the connectivity problem is a special case of the shortest-path problem. This is accomplished by considering a cost of 1 associated to each edge of the graph in the connectivity problem. In other words, just like a shortest-path problem, you can assume that there is a nonnegative number (1) associated to each edge of the graph of the connectivity problem.

Now consider the second identified difference. Although the two objectives are different in nature, both problems are solved by using the final matrix, $L^{(n)}$. Thus, for all practical purposes, you can consider the objectives of both problems to be that of obtaining $L^{(n)}$.

The final difference identified above is that the specific way in which the sequence of matrices is computed is different. Finding a common way to describe those computations is a challenge that is addressed subsequently in this section.

3.4.2 Developing an Abstract System

The next step is to develop an *abstract system*, as described in Section 1.5. One component of that system is a graph $G = (V, E)$ on n vertices. Each problem also has some data that you can describe symbolically by a function c that associates to each edge of G, a nonnegative number, that is, $c : E \to \mathbb{R}_+$, where

$$\mathbb{R}_+ = \{\text{real numbers } x : x \geq 0\}.$$

The following, then, constitutes your first attempt at the abstract system.

First Attempt at an Abstract System
Consider an abstract system consisting of a graph $G = (V, E)$ together with a function $c : E \to \mathbb{R}_+$.

Now consider what you *do* with this information. According to the solution procedures, the first step is to use these data to create the first matrix, $L^{(0)}$. In the case of the connectivity problem, that first matrix is

$$L_{ij}^{(0)} = A_{ij}^{(0)} = \begin{cases} 1, & \text{if } ij \in E \text{ or } i = j \\ 0, & \text{if } ij \notin E \text{ and } i \neq j \end{cases} \tag{3.19}$$

and for the shortest-path problem whose edge costs are given as C_{ij}, that first matrix is

$$L_{ij}^{(0)} = C_{ij}^{(0)} = \begin{cases} C_{ij}, & \text{if } ij \in E \\ 0, & \text{if } ij \notin E \text{ and } i = j \\ \infty, & \text{if } ij \notin E \text{ and } i \neq j. \end{cases} \tag{3.20}$$

To make the similarities between the two matrices $L_{ij}^{(0)}$ in (3.19) and (3.20) even more clear, you can rewrite the computations in (3.19) as follows:

$$L_{ij}^{(0)} = A_{ij}^{(0)} = \begin{cases} 1, & \text{if } ij \in E \\ 1, & \text{if } ij \notin E \text{ and } i = j \\ 0, & \text{if } ij \notin E \text{ and } i \neq j. \end{cases} \tag{3.21}$$

From (3.20) and (3.21), you can see that when $ij \in E$, the values of $L_{ij}^{(0)}$ in both cases are obtained from the data consisting of the function c. However, *when $ij \notin E$, two special values are assigned to $L_{ij}^{(0)}$*: 0 and ∞, in the shortest-path problem and 1 and 0, in the connectivity problem. For unification purposes, those special values are denoted hereafter by α and β.

By using α and β together with the cost function c, you can now construct the first matrix, $L^{(0)}$, from the abstract system, as follows:

$$L_{ij}^{(0)} = \begin{cases} c(ij), & \text{if } ij \in E \\ \alpha, & \text{if } ij \notin E \text{ and } i = j \\ \beta, & \text{if } ij \notin E \text{ and } i \neq j. \end{cases} \tag{3.22}$$

This computation is so straightforward that you might just as well consider the information in $L^{(0)}$ as the data of the problem. This data, then, associates to *each* pair of vertices, a number from $\mathbb{R}_+$ or one of the special

Connectivity Problem	Shortest-Path Problem
$S = \{0, 1\}$	$S = \mathbb{R}_+ \cup \{\infty\}$
$\alpha = 1$	$\alpha = 0$
$\beta = 0$	$\beta = \infty$

Table 3.2: The Values of S, α, and β for the Two Special Cases of the Abstract System

values, α or β. Symbolically, you can now think of the data as a new function $c : V \times V \to \mathbb{R}_+ \cup \{\alpha, \beta\}$.

Generalizing even further by using abstraction, instead of working with the specific values in $\mathbb{R}_+ \cup \{\alpha, \beta\}$, why not allow these values to belong to an *arbitrary* set, say, S? That is, $c : V \times V \to S$, with the understanding that S contains two special elements, α and β. This now constitutes the second attempt at creating an abstract system.

Second Attempt at an Abstract System

Consider an abstract system consisting of a graph $G = (V, E)$ together with a set S containing the special elements α and β and a function $c : V \times V \to S$.

The values of S, α, and β for the two special cases of the connectivity and shortest-path problems are summarized in Table 3.2. The first matrix is then computed by the formulas in (3.22).

The next step is to use the values of $L^{(0)}$ to compute the values of $L^{(1)}, \ldots, L^{(n)}$. However, keep in mind that the values in each of these matrices pertain to a *path*. More specifically, the values of $L_{ij}^{(k)}$ are values associated with a path from vertex i to vertex j in which only the vertices having the numbers $1, \ldots, k$ are used. Specifically, then, how do you associate a value to a *path*?

One answer arises by thinking of the shortest-path problem. The "cost" of a path is the *sum* of the cost of the edges in the path. For the abstract system, then, you might reason as follows: a path contains a sequence of edges, each edge consists of a pair of vertices (say, s and t), and the given function c associates to each such pair of vertices, a value $c(s, t)$. Why, then, not add the c-values of the edges in the path? That is, suppose you have a path P consisting of the following vertices:

$$P = v_0, v_1, \ldots, v_r.$$

Why not define the cost of P by

$$\text{cost}(P) = c(v_0, v_1) + \cdots + c(v_{r-1}, v_r)? \tag{3.23}$$

There are two problems with doing so.

Connectivity Problem	Shortest-Path Problem
$S = \{0, 1\}$	$S = \mathbb{R}_+ \cup \{\infty\}$
$\alpha = 1$	$\alpha = 0$
$\beta = 0$	$\beta = \infty$
$\oplus$ = multiplication	$\oplus$ = addition

Table 3.3: The Values of S, α, β, and $\oplus$ for the Two Special Cases of the Abstract System

1. In the abstract system, the values of $c(v_{k-1}, v_k)$ are not *numbers*, but rather, elements of an arbitrary set S, because $c : V \times V \to S$. Thus, you cannot add the c-values in (3.23).

2. A second problem is that, although you *add* the cost of the edges in the shortest-path problem to obtain the cost of a path P, it is *not* the case that you add the cost of the edges in the connectivity problem. Try to determine how you use the c-values of 0 and 1 in the connectivity problem to compute the cost of a path – which is to be 1, if all the edges in the path have a c-value of 1 and 0, otherwise. One way to do so is to *multiply* the c-values of the edges in the path, as follows:

$$\text{cost}(P) = c(v_0, v_1) \cdot \ldots \cdot c(v_{r-1}, v_r). \tag{3.24}$$

One way to overcome both of these obstacles is to replace the *arithmetic* operation of addition (+) in (3.23) that combines two *numbers*, with a *closed binary operation* (say, $\oplus$) that combines two elements of S to produce a new element of S. With this binary operator $\oplus$, it is now possible to define the cost of a path $P = v_0, \ldots, v_r$ by

$$\text{cost}(P) = c(v_0, v_1) \oplus \cdots \oplus c(v_{r-1}, v_r). \tag{3.25}$$

From (3.23), you can see that for the shortest-path problem, the operator $\oplus$ is addition (with the understanding that for any real number x, $\infty + x = \infty$) and for the connectivity problem, from (3.24), the operator $\oplus$ is multiplication. You therefore want this binary operation of $\oplus$ to have the same useful properties as arithmetic addition and multiplication. In any event, you have added the operator $\oplus$ to the abstract system, which now becomes your next attempt.

Third Attempt at an Abstract System

Consider an abstract system consisting of a graph $G = (V, E)$ together with a set S containing the special elements α and β, a function $c : V \times V \to S$, and a closed binary operator $\oplus$ on S.

The values of S, α, β, and $\oplus$ for the two special cases of the connectivity and shortest-path problems are summarized in Table 3.3. You use these values to construct the first matrix, $L^{(0)}$, from the formulas in (3.22).

Finally, consider the way in which $L^{(k)}$ is computed from $L^{(k-1)}$. Again, look at the special cases of the connectivity problem:

$$L_{ij}^{(k)} = A_{ij}^{(k)} = \begin{cases} 1, & \text{if } A_{ij}^{(k-1)} = 1 \text{ or} \\ & \text{if } A_{ik}^{(k-1)} = 1 \text{ and } A_{kj}^{(k-1)} = 1 \\ 0, & \text{otherwise} \end{cases} \tag{3.26}$$

and the shortest-path problem:

$$L_{ij}^{(k)} = C_{ij}^{(k)} = \min\{C_{ij}^{(k-1)},\ C_{ik}^{(k-1)} + C_{kj}^{(k-1)}\}. \tag{3.27}$$

Try to find similarities in these computations. Perhaps the most important similarity is that, in both cases, the values of $L_{ij}^{(k-1)}$, $L_{ik}^{(k-1)}$, and $L_{kj}^{(k-1)}$ from the *previous* matrix are used to determine $L_{ij}^{(k)}$. Of course the *specific* way in which these values are used is different. The objective is to find a *common* way to describe these operations in the context of the abstract system.

Again, using the shortest-path problem as a guide, you can replace the arithmetic operation of + in (3.27) with the binary operation of $\oplus$ and the matrix C with L. The result is the following:

$$L_{ij}^{(k)} = \min\{L_{ij}^{(k-1)},\ L_{ik}^{(k-1)} \oplus L_{kj}^{(k-1)}\}. \tag{3.28}$$

Now consider the process of taking the minimum in (3.28). A syntax error arises because the values of $L_{ij}^{(k-1)}$ and $L_{ik}^{(k-1)} \oplus L_{kj}^{(k-1)}$ are not numbers, but rather, elements of the set S. One way to correct this syntax error is to replace the operation of *min*, that applies to *numbers*, with a closed binary operation (say, $\odot$) that combines two elements of S – namely, $L_{ij}^{(k-1)}$ and $L_{ik}^{(k-1)} \oplus L_{kj}^{(k-1)}$ – to produce a new value in S. Thus, you can rewrite (3.28) as follows:

$$L_{ij}^{(k)} = L_{ij}^{(k-1)} \odot (L_{ik}^{(k-1)} \oplus L_{kj}^{(k-1)}). \tag{3.29}$$

In the special case of the shortest-path problem, $\odot$ is the act of computing the minimum of two values. Can you figure out what operation $\odot$ is for the special case of the connectivity problem in (3.26)? In other words, can you define $\odot$ so that (3.26) can be written in exactly the same way as (3.29)?

The first step is to rewrite "$A_{ik}^{(k-1)} = 1$ and $A_{kj}^{(k-1)} = 1$" in (3.26) by using multiplication, as follows:

$$A_{ij}^{(k)} = \begin{cases} 1, & \text{if } A_{ij}^{(k-1)} = 1 \text{ or} \\ & \text{if } A_{ik}^{(k-1)} \cdot A_{kj}^{(k-1)} = 1 \\ 0, & \text{otherwise.} \end{cases} \tag{3.30}$$

Connectivity Problem	Shortest-Path Problem
$S = \{0, 1\}$	$S = \mathbb{R}_+ \cup \{\infty\}$
$\alpha = 1$	$\alpha = 0$
$\beta = 0$	$\beta = \infty$
$\oplus$ = multiplication	$\oplus$ = addition
$\odot$ = max	$\odot$ = min

Table 3.4: The Values of S, α, β, $\oplus$, and $\odot$ for the Two Special Cases of the Abstract System

To express (3.30) as an operation involving $A_{ij}^{(k-1)}$ and $A_{ik}^{(k-1)} \cdot A_{kj}^{(k-1)}$ requires creativity. One way to do so is as follows:

$$A_{ij}^{(k)} = \max\{A_{ij}^{(k-1)},\ A_{ik}^{(k-1)} \cdot A_{kj}^{(k-1)}\}. \tag{3.31}$$

Verify for yourself that the computations in (3.31) provide the same results as those in (3.30).

To make (3.31) look more like (3.29), replace the multiplication symbol in (3.31) with the operator $\oplus$ and the matrix A with the matrix L to obtain the following:

$$L_{ij}^{(k)} = \max\{L_{ij}^{(k-1)},\ L_{ik}^{(k-1)} \oplus L_{kj}^{(k-1)}\}. \tag{3.32}$$

Finally, think of the operation of taking the maximum in (3.32) as a *new* operation denoted by $\odot$. Thus, you can rewrite (3.32) as follows:

$$L_{ij}^{(k)} = L_{ij}^{(k-1)} \odot (L_{ik}^{(k-1)} \oplus L_{kj}^{(k-1)}), \tag{3.33}$$

which is precisely the same as (3.29).

With this unification, the final abstract system is as follows.

The Abstract System

Consider an abstract system consisting of a graph $G = (V, E)$ together with a set S containing the special elements α and β, a function $c : V \times V \to S$, and two closed binary operators $\oplus$ and $\odot$ on S.

The values of $S, \alpha, \beta, \oplus$, and $\odot$ for the two special cases of the connectivity and shortest-path problems are summarized in Table 3.4.

3.4.3 Stating the Problem in Terms of the Abstract System

Now that you have an abstract system, the question is what do you want to *do* with that system? That is, what is the problem you want to solve?

Using the special case of the shortest-path problem as a guide, you want to find the costs of the shortest-paths between all pairs of vertices in the graph. In terms of the abstract system, consider, then, a list of *all* paths between the two vertices i and j in the graph G, say, $P^1, \ldots, P^r$. There is a cost associated with each of these paths, namely, the cost defined in (3.25) using the operator $\oplus$. Consider, therefore, the following list:

Path	Cost
P^1	cost(P^1)
P^2	cost(P^2)
$\vdots$	$\vdots$
P^r	cost(P^r)

To solve the shortest-path problem, you want to select the *smallest* of these costs, that is, the *minimum* cost. In terms of the abstract system, recall that the operation of finding the minimum in the shortest-path problem is represented by the binary operator $\odot$. Thus, the objective of the shortest-path problem in terms of the abstract system is to compute

$$\text{cost}(P^1) \odot \text{cost}(P^2) \odot \cdots \odot \text{cost}(P^r). \tag{3.34}$$

The expression in (3.34) is *also* the objective of the connectivity problem. This is because, for the connectivity problem, you want to determine if *any* of the paths $P^1, \ldots, P^r$ has a cost of 1. Another way to say this is that you want to see if

$$\max\{\text{cost}(P^1), \ldots, \text{cost}(P^r)\} = 1. \tag{3.35}$$

For the connectivity problem, the operation of taking the maximum is the same as using the binary operator $\odot$ and hence, (3.35) is the same as (3.34).

The objective in (3.34) together with the abstract system becomes the following abstract problem.

The Abstract Problem

Consider an abstract system consisting of a graph $G = (V, E)$ together with a set S containing the special elements α and β, a function $c : V \times V \to S$, and two closed binary operators $\oplus$ and $\odot$ on S. For each pair of vertices i and j, find the value of

$$L_{ij}^{(*)} = \text{cost}(P^1) \odot \text{cost}(P^2) \odot \cdots \odot \text{cost}(P^r),$$

where $P^1, \ldots, P^r$ are all paths from i to j in G.

3.4.4 Developing a Solution Procedure

You now need to develop a method for solving the abstract problem. On the basis of the algorithms for solving the connectivity and the shortest-path problems, the algorithm for solving the abstract problem is as follows.

Algorithm for Solving the Abstract Problem

Given the abstract system, start by computing

$$L_{ij}^{(0)} = \begin{cases} c(i,j), & \text{if } ij \in E \\ \alpha, & \text{if } ij \notin E \text{ and } i = j \\ \beta, & \text{if } ij \notin E \text{ and } i \neq j. \end{cases} \tag{3.36}$$

Then, for each $k = 1, \ldots, n$, sequentially create the next matrix $L^{(k)}$ from $L^{(k-1)}$ by using the following formula:

$$L_{ij}^{(k)} = L_{ij}^{(k-1)} \odot (L_{ik}^{(k-1)} \oplus L_{kj}^{(k-1)}). \tag{3.37}$$

The solution is contained in the final matrix, $L^{(n)}$.

3.4.5 Developing an Axiomatic System

To prove that the foregoing algorithm solves the abstract problem, you need the abstract system to satisfy certain properties, or *axioms*. The specific axioms are those that enable you to prove the following theorem.

Theorem 3.4 *Suppose that the abstract system consisting of a graph $G = (V, E)$ on n vertices together with a set S containing the special elements α and β, a function $c : V \times V \to S$, and two closed binary operators $\oplus$ and $\odot$ satisfy the following properties:*

(1) *For all $x, y, z \in S$, $x \odot (y \odot z) = (x \odot y) \odot z$.*

(2) *For all $x \in S$, $x \odot \beta = \beta \odot x = x$.*

(3) *For all $x, y, z \in S$, $x \oplus (y \oplus z) = (x \oplus y) \oplus z$.*

(4) *For all $x \in S$, $x \oplus \alpha = \alpha \oplus x = x$.*

(5) *For all $x \in S$, $x \oplus \beta = \beta \oplus x = \beta$.*

(6) *For all $x, y \in S$, $x \odot y = y \odot x$ and $x \odot x = x$.*

(7) *For all $x, y, z \in S$, $x \oplus (y \odot z) = (x \oplus y) \odot (x \oplus z)$ and $(x \odot y) \oplus z = (x \odot z) \oplus (y \odot z)$.*

(8) *For all $x \in S$, $\alpha \odot x = \alpha \odot (x \oplus x)$.*

If $L^{(0)}$ is computed according to (3.36) and if, for each $k = 1, \ldots, n$, $L^{(k)}$ is computed by (3.37), then for all pairs of vertices i and j,

$$L_{ij}^{(n)} = cost(P^1) \odot cost(P^2) \odot \cdots \odot cost(P^r),$$

where $P^1, P^2, \ldots, P^r$ are all the paths from i to j in G.

You can verify that these eight axioms are satisfied by the special cases of the connectivity and shortest-path problems using the information in Table 3.4; however, the *development* of these axioms and the *proof* of Theorem 3.4 are beyond the scope of this book but can be found in *The Design and Analysis of Computer Algorithms* by Aho, Hopcroft, and Ullman, Addison-Wesley Publishing Company, 1974, starting on page 195.

Summary

In this section, you have seen how the unification of the connectivity and shortest-path problems is accomplished through the use of abstraction. The following two points are worth noting.

1. The special cases are used as a guide when developing the abstract system. Each time a statement is derived for the abstract system, that statement is verified against all of the special cases. This is analogous to testing a newly developed computer program (the abstract system) on all available data (the special cases).
2. It may be necessary to modify and rewrite the way you think about a particular aspect of a special case so that unification is possible with the other special cases. This rewriting is done in this section for the computations of the matrices $A^{(0)}, \ldots A^{(n)}$ in the connectivity problem and $C^{(0)}, \ldots, C^{(n)}$ in the shortest-path problem.

Chapter Summary

In this chapter, you have seen how unification, generalization, and abstraction are used in studying sets, functions, and graphs in discrete mathematics. In each case, you have seen how to develop visual images, translate those images to symbolic form, create definitions, and so on. A summary of these topics follows.

Sets

A set is a collection of objects that is used as a problem-solving tool, especially in the process of abstraction. A set is represented visually by a Venn diagram or symbolically using a list or set-builder notation. You can compare two sets A and B by checking if A is contained in B ($A \subseteq B$) or if $A = B$.

Unary operations on sets include finding the complement and the power set. In each case, you start with a set A and use the unary operator to create another set. In a similar manner, the binary operations of union, intersection, and Cartesian product combine *two* sets to create a new set.

Functions

A function is a useful problem-solving tool because a function associates to each object in its domain, a unique object in its codomain. Care is needed to insure that you apply the function *only* to an element in the domain because it may not be possible to perform the function on other elements.

A function is represented visually by a black box in which you put in an element from the domain and, after performing some operations, obtain an element in the codomain. Alternatively, you can visualize the graph of a function. Symbolically, a function is represented by a rule in the form of a list, a closed-form expression, or a numerical method that specifies how to use the function input to obtain the output.

Typical operations include comparing two functions for equality, combining them with the arithmetic operators $+$, $-$, $\cdot$, and $/$, when appropriate, or composing them by performing the functions sequentially. You have also learned about the properties of a surjective, injective, and bijective function and how a bijective function is used to solve a specific type of equation.

Graphs

A graph is a problem-solving tool that arises through the process of abstraction when you identify a finite number of vertices to represent certain objects in a problem. Pairs of these vertices are connected by an edge to represent a relationship of interest between the two objects corresponding to the conected vertices.

A graph is represented visually by drawing a circle for each vertex and a line for each edge. The adjacency matrix and the edge list are two symbolic ways to represent a graph.

You also learned the operations of deleting and adding edges and vertices as well as creating subgraphs. Through the process of identifying similarities and differences, you saw how to develop the formal definitions of a cycle and a connected graph. You also saw numerical methods for determining if a graph is connected or not and for finding the shortest path between all pairs of vertices.

Exercises

Exercise 3.1 Suppose that A and B are subsets of a universal set U. Draw a Venn diagram that illustrates each of the following:

(a) The universal set U and the fact that $A \cap B = \emptyset$.

(b) The universal set U and the fact that $A \cap B$ has only one element.

(c) The universal set U and the fact that $A \cap B$ has many elements.

Exercise 3.2 Suppose that A, B, and C are sets. Draw a Venn diagram that illustrates each of the following:

(a) $A \cap C = \emptyset$, $A \cap B \neq \emptyset$, and $B \cap C \neq \emptyset$.

(b) Suppose that $A \cap B \neq \emptyset$. Illustrate the set $A - (A \cap B)$.

Exercise 3.3 Create a visual image for each of the following sets of points. Then translate that image to symbolic form using set-builder notation.

(a) The region on and above the graph of the function $y = x^2 - 2x + 2$.

(b) The straight line $y = 4 - 2x$ together with the region on the same side of the line as the origin.

Exercise 3.4 Generalize in symbolic form each of the sets in Exercise 3.3 according to the following guidelines. Introduce appropriate notation as necessary.

(a) Generalize the set in Exercise 3.3(a) to an arbitrary real-valued function of one variable instead of $x^2 - 2x + 2$.

(b) (i) Generalize the set in Exercise 3.3(b) to an arbitrary line in the plane.

(ii) Generalize the set in part (i) to an arbitrary plane in three dimensions.

(iii) Generalize the set in part (ii) to an arbitrary plane in n dimensions.

Exercise 3.5 Consider the following universal set of countries:

AllCountries = {America, Canada, England, France, Germany, Japan, Korea, Malaysia, Mexico, Taiwan}

together with the following subsets:

NorthAmerican = {America, Canada, Mexico},
European = {England, France, Germany},
Asian = {Japan, Korea, Malaysia, Taiwan},
Leaders = {America, Germany, Japan}.

Translate the following statements regarding the foregoing sets to symbolic form. Use *only* the names of the sets and the operators union ($\cup$), intersection ($\cap$), and set difference ($-$). Do *not* use the names of the individual countries.

(a) Those countries that are either European or Asian.

(b) Those leaders that are either Asian or North American.

(c) Those countries that are Asian or European but not leaders.

Exercise 3.6 Translate the following statements regarding the sets in Exercise 3.5 to symbolic form. Use *only* the names of the sets and the operators union ($\cup$), intersection ($\cap$), and set difference ($-$). Do *not* use the names of the individual countries.

(a) Those countries that are not North American.

(b) Those leaders that are neither European nor Asian.

Exercise 3.7 Suppose that a and b are real numbers with $a \leq b$ and recall that a *closed interval* from a to b is the following set:

$$[a, b] = \{\text{real numbers } x : a \leq x \leq b\}.$$

(a) By using set-builder notation, generalize a closed interval to a *closed rectangle in the plane* whose bottom leftmost point has coordinates (a, c) and whose top rightmost point has coordinates (b, d).

(b) By using set-builder notation, generalize the set in part (a) to a *closed rectangle in n dimensions.* Introduce appropriate notation.

Exercise 3.8 Repeat parts (a) and (b) of Exercise 3.7 using the Cartesian product. That is, express a closed rectangle in the plane and a closed rectangle in n dimensions in terms of Cartesian products of other sets.

Exercise 3.9 Complete the Solution to Example 3.6 in Section 3.1.2 by proving that $B \subseteq A$ for the following sets A and B:

$$A = \{x \in \mathbb{R} : -1 \leq x \leq 2\},$$
$$B = \{x \in \mathbb{R} : x^2 - x - 2 \leq 0\}.$$

Exercise 3.10 Prove that if A is a set with $n \geq 1$ elements, then the power set, 2^A, has 2^n elements. (Hint: Use induction.)

Exercise 3.11 Prove each of the following properties from Table 3.1 in Section 3.1.4 pertaining to the two sets A and B.

(a) $A \cup (B \cap C) = (A \cup B) \cap (A \cup C)$.

(b) $(A \cup B)^c = A^c \cap B^c$.

Exercise 3.12 Prove each of the following properties from Table 3.1 in Section 3.1.4 pertaining to the two sets A and B.

(a) $A \cap (B \cup C) = (A \cap B) \cup (A \cap C)$.

(b) $(A \cap B)^c = A^c \cup B^c$.

Exercise 3.13 Explain why the following is *not* a real-valued function: associate to the real numbers a, b, and c for which $b^2 - 4ac \geq 0$, a solution to the equation

$$ax^2 + bx + c = 0.$$

Exercise 3.14 Suppose you want to associate to each positive integer n, the following value:

$$\begin{cases} 1, & \text{if the digits of } n \text{ appear consecutively somewhere} \\ & \text{in the decimal expansion of } \pi \\ 0, & \text{otherwise.} \end{cases}$$

Discuss the reasons why you might and might not want to consider this association to be a function.

Exercise 3.15 Draw a visual representation similar to the one in Figure 3.7 in Section 3.2.1 for each of the following functions that clearly indicates the *number* of inputs and outputs.

(a) $f : \mathbb{R}^1 \to \mathbb{R}^1$.

(b) $g : \mathbb{R}^n \to \mathbb{R}^1$.

(c) $h : \mathbb{R}^n \to \mathbb{R}^p$.

Exercise 3.16 Draw a visual representation of the function that associates to two sets A and B, the Cartesian product $A \times B$. (Hint: To simplify, think of A and B as intervals of real numbers and consider Exercise 3.8.)

Exercise 3.17 For each of the following associations, identify a suitable domain and codomain so that you can compute the resulting function at every point in the domain.

(a) A closed unary operator u on a set A.

(b) The association to each real number x, the value $1/x$.

(c) The association to each group of six real numbers a, b, c, d, e, and f, the *unique* values of the real numbers x and y that satisfy the following two equations:

$$\begin{aligned} ax + by &= e, \\ cx + dy &= f. \end{aligned}$$

(d) The association to each group of three real numbers a, b, and c, the *set* of all real solutions to the equation

$$ax^2 + bx + c = 0.$$

Exercise 3.18 For each of the following associations, identify a suitable domain and codomain so that you can compute the resulting function at every point in the domain.

(a) A closed binary operator β on a set A.

(b) The association to each real number x, the value of $\tan(x)$.

(c) The association to each integrable function $f : \mathbb{R} \to \mathbb{R}$, a differentiable function $F : \mathbb{R} \to \mathbb{R}$ such that for every $x \in \mathbb{R}$, $F'(x) = f(x)$.

Exercise 3.19 Create a definition in symbolic form for each of the following concepts. Introduce appropriate notation as necessary.

(a) A *fixed point of a real-valued function of one variable*, which is any point where the graph of this function crosses the line through the origin having a slope of 1.

(b) An *increasing function*, which is a real-valued function of one variable whose value increases as you move from left to right along the x-axis.

Exercise 3.20 Create a definition in symbolic form for each of the following concepts. Introduce appropriate notation as necessary.

(a) The *set of fixed points of a function* whose codomain is the same as its domain, which are those points where the value of the function is unchanged.

(b) A *convex function* $f : \mathbb{R} \to \mathbb{R}$, which is a function with the property that between any two points x and z on the x-axis, the graph of f between those two points lies below the line segment connecting $(x, f(x))$ and $(z, f(z))$. (Hint: Draw a picture and observe that you can represent a point between x and z as $tx + (1 - t)z$ for some value of t between 0 and 1. Also, the point on the line segment connecting $(x, f(x))$ and $(z, f(z))$ that corresponds to $tx + (1 - t)z$ is $tf(x) + (1 - t)f(z)$.)

Exercise 3.21 Suppose that $f, g : \mathbb{R}^1 \to \mathbb{R}^1$.

(a) Does a syntax error arise when you generalize the operation $f + g$ to functions $f, g : \mathbb{R}^n \to \mathbb{R}^1$? Why or why not? Explain.

(b) Does a syntax error arise when you generalize the operation f/g to functions $f, g : \mathbb{R}^n \to \mathbb{R}^1$? Why or why not? Explain.

Exercise 3.22 Suppose that $f, g : \mathbb{R}^1 \to \mathbb{R}^1$.

(a) Does a syntax error arise when you generalize the operation $f + g$ to functions $f, g : \mathbb{R}^n \to \mathbb{R}^p$? Why or why not? Explain.

(b) Does a syntax error arise when you generalize the operation f/g to functions $f, g : \mathbb{R}^n \to \mathbb{R}^p$? Why or why not? Explain.

Exercise 3.23 Complete the proof of Proposition 3.1 in Section 3.2.3 by showing that if A and B are sets with $f : A \to B$ and range $f = B$, then f is surjective.

Exercise 3.24 Complete the proof of Theorem 3.1 in Section 3.2.3 by showing that if A and B are sets, $f : A \to B$ is bijective, $f^{-1} : B \to A$ is the inverse function of f, and $g : B \to A$ is any function for which $g \circ f = i_A$, and $f \circ g = i_B$, then $g = f^{-1}$.

Exercise 3.25 Identify each of the following functions as being surjective, injective, and/or bijective. Explain your reasoning.

(a) $f : \mathbb{R} \to [-1, 1]$ defined by $f(x) = \sin(x)$.

(b) $f : \{x \in \mathbb{R} : x \geq 0\} \to \mathbb{R}$ defined by $f(x) = \sqrt{x}$.

(c) $f : \{x \in \mathbb{R} : x \geq 0\} \to \{x \in \mathbb{R} : x \geq 0\}$ defined by $f(x) = \sqrt{x}$.

Exercise 3.26 If possible, draw an example of a linear function $f(x) = ax + b$ that has the following properties and give a specific numerical example in the form of values for the real numbers a and b. If you cannot do so, then prove that there is no such linear function.

(a) A linear function $f : \mathbb{R} \to \mathbb{R}$ that is bijective.

(b) A linear function $f : \mathbb{R} \to \mathbb{R}$ that is injective but not surjective.

(c) A linear function $f : \mathbb{R} \to \mathbb{R}$ that is surjective but not injective.

(d) On the basis of your results in parts (a), (b), and (c), find conditions on the values of a and b so that the function $f(x) = ax + b$ is bijective. In other words, fill in the underlined portion in the following proposition and then prove the statement:

> The linear function $f : \mathbb{R} \to \mathbb{R}$ defined by $f(x) = ax + b$ is bijective if and only if *some condition on a and b is true*.

Note: In the remaining exercises, assume that the graph $G = (V, E)$ is undirected and has no loops or multiple edges, unless otherwise stated.

Exercise 3.27 Formulate each of the following problems by using a graph. Explain the meaning of the vertices and the edges in the context of the problem. Then state the objective of the problem in terms of the graph.

(a) At a party of six people, you will always find three people all of whom know each other or else three people none of whom know each other.

(b) Suppose that six microchips on a circuit board are arranged in two groups of three each. You want to show that it is impossible to connect each microchip in one group to all three microchips in the other group without at least two of the wires crossing each other.

Exercise 3.28 Formulate each of the following problems by using a graph. Explain the meaning of the vertices and the edges in the context of the problem and draw the associated graph. Then state the problem and objective in terms of a *general* graph with associated data. You need *not* solve the problem.

(a) A company has five employees, each of whom can perform one or more of five different jobs, as shown in the following table:

Person	Jobs They Can Perform
Bob	1, 5
Mary	1, 2
Ted	1, 3, 4
Sarah	2, 5
John	2, 5

Is it possible to accomplish all jobs by assigning one and only one person to each job?

(b) You need to perform eight jobs on a certain type of machine that can work on only one job at a time. Each job has the following known starting time and ending time ($[a_i, b_i]$): [5, 10], [11, 12], [11, 13], [4, 12], [1, 6], [17, 19], [18, 20], [14, 16]. You want to determine the fewest number of machines needed to complete all eight jobs.

Exercise 3.29 Translate each of the following operations on a graph $G = (V, E)$ to an operation on the adjacency matrix A associated with G.

(a) Adding an edge from vertex i to vertex j.

(b) Deleting the edge from vertex i to vertex j.

(c) Adding a new vertex to G.

(d) Deleting vertex i from G.

Exercise 3.30 Repeat Exercise 3.29 for the edge-list representation.

Exercise 3.31 Generalize the concept of an adjacency matrix to allow for loops and multiple edges. In other words, how does the presence of loops and multiple edges affect the values in the adjacency matrix?

Exercise 3.32 Generalize the concept of an adjacency matrix to allow for a *directed* graph. In other words, how does the presence of directed edges affect the values in the adjacency matrix?

Exercise 3.33 By introducing appropriate notation, create a definition with symbols for the following unary operation on a graph:

> The *complement* of a graph is the graph in which two vertices are adjacent if and only if those vertices are not adjacent in the original graph.

Exercise 3.34 Define a cycle of a graph in terms of a *walk* that has certain properties. Verify your definition by checking that K_1 and K_2 do *not* satisfy the property you develop, but K_3 does. (Recall that K_n is the complete graph on n vertices.)

Exercise 3.35 Each of the following graphs is *bipartite*:

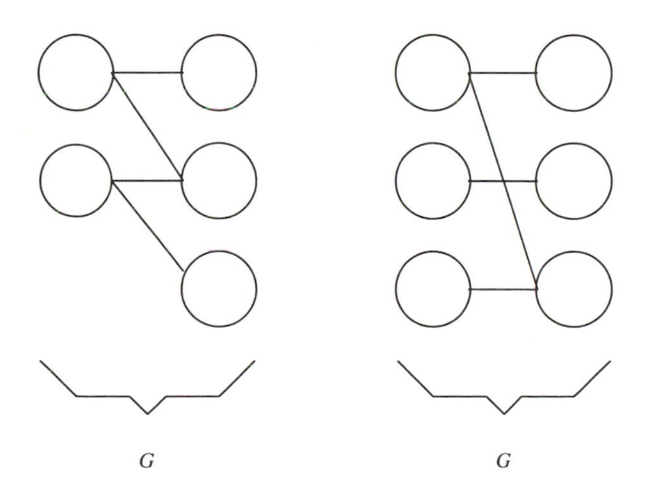

In contrast, *none* of the following graphs is bipartite:

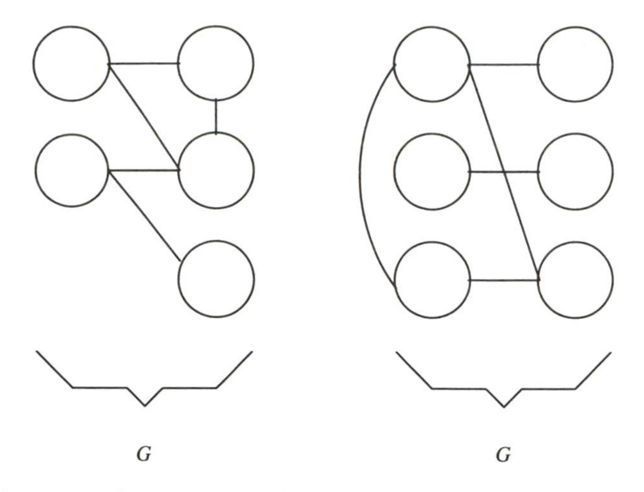

By identifying similarities and differences, create a definition of a bipartite graph. (Hint: Notice that the vertices are divided into two groups. Which vertices are connected by edges in the bipartite graphs?)

Exercise 3.36 Each of the following graphs is a *tree*:

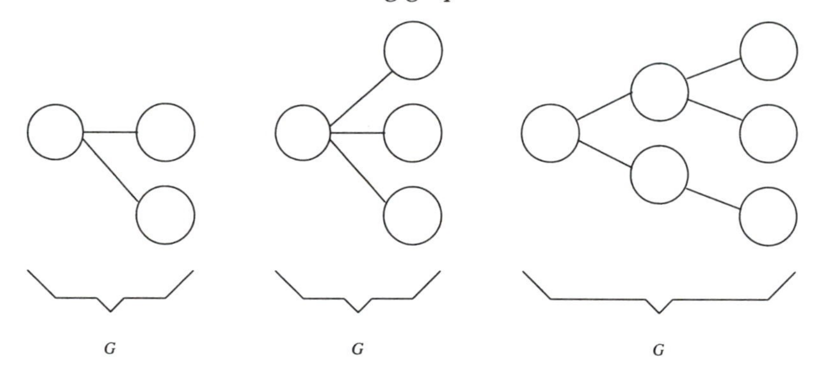

In contrast, *none* of the following graphs is a tree:

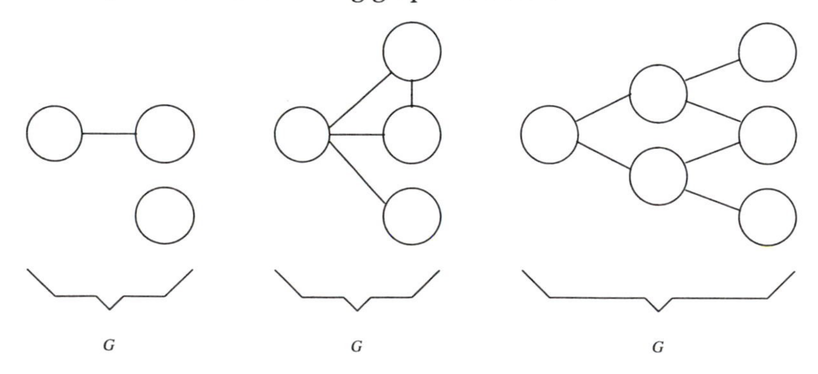

By identifying similarities and differences, create a definition of a tree.

Exercise 3.37 Recall that a connected graph is a graph consisting of one "piece." When a graph is *not* connected, it consists of *more than* one piece.

Each piece is called a *component* of the graph. For example, the following graph consist of *three* components, labeled G_1, G_2, and G_3:

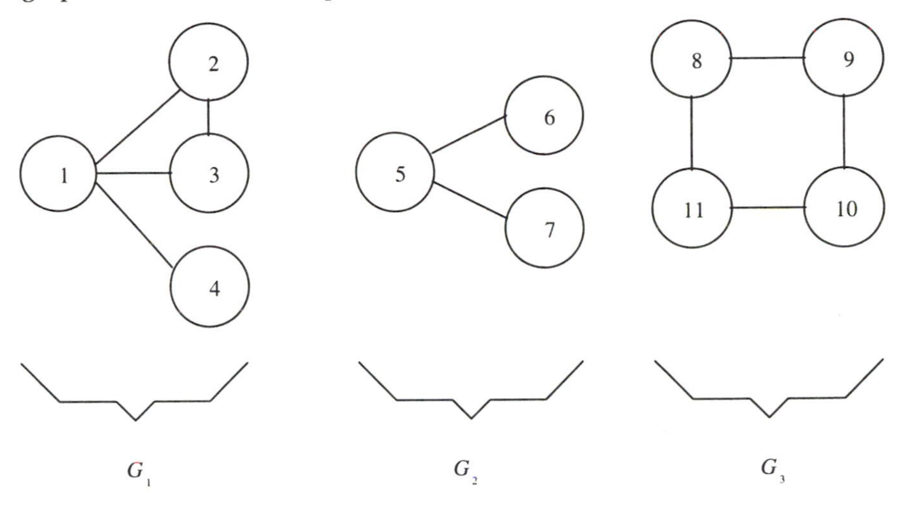

A student suggested the following definition for a component:

> A *component* of a graph $G = (V, E)$ is a connected subgraph of G.

Verify this definition. Is it *true* that every component of G satisfies the property of being a connected subgraph of G? Is it also *true* that *every* connected subgraph of G is what you think of as a component? Why or why not? Explain using the foregoing example of a graph.

Exercise 3.38 On the basis of your results in Exercise 3.37, modify the definition of a component so that it is correct.

Exercise 3.39 Suppose that i and j are vertices of a graph G.

(a) In the statement "i is adjacent to j," the words "is adjacent to" constitute a binary relation on the vertices of G. Is this relation reflexive? Symmetric? Transitive? Why or why not? Explain. (See Definitions 2.16, 2.17, and 2.18 in Section 2.3.3.)

(b) Repeat part (a) for a *directed* graph.

Exercise 3.40 Suppose that i and j are vertices of a graph G.

(a) In the statement "i is connected by a walk to j," the words "is connected by a walk" constitute a binary relation on the vertices of G. Is this relation reflexive? Symmetric? Transitive? Why or why not? Explain. (See Definitions 2.16, 2.17, and 2.18 in Section 2.3.3.)

(b) Repeat part (a) for a *directed* graph, using the binary relation of a *directed walk* from i to j, that is, a walk in which all edges point from i toward j.

Exercise 3.41 Consider the problem of determining if a graph G is connected or not.

(a) Give a reason for trying to prove the following proposition:

> A graph G is connected if and only if there is a walk from vertex 1 to each other vertex in G.

(b) Prove the proposition in part (a).

Exercise 3.42 On the basis of the result in Exercise 3.41, develop an algorithm along the following lines to determine if a graph G having n vertices is connected or not. Let S be the set of all vertices adjacent to vertex 1. Then determine the set S of vertices reachable from vertex 1 by walks of length at most 2. Continue in this way until you obtain the set S of vertices reachable from vertex 1 by walks of length at most n. How do you use this final set S to determine if G is connected or not? Explain.

Chapter 4

SELECTED TOPICS IN LINEAR ALGEBRA

In this chapter, you will see how the concepts of Chapters 1 and 2 are applied to selected topics in **linear algebra** – the study of ordered lists of numbers. The various different lists you learn about in this chapter are a sequential generalization of real numbers in the set $\mathbb{R}$. You should review the material on unary operators in Section 2.3.1, on binary operators in Section 1.5.2, on sets in Section 3.1, and on functions in Section 3.2.

4.1 Vectors

You already know how to work with individual numbers, a generalization of which is a list of *two* numbers, for example,

$$(1, 2).$$

Continuing this generalization, you can work with a list of three numbers, for example,

$$(-1.5,\ 0,\ 4.3).$$

The next definition presents a generalization to a list of n real numbers.

Definition 4.1 *An n-**vector** $\mathbf{x} = (x_1, \ldots, x_n)$ (also called a **vector**) is an ordered list of n real numbers. The positive integer n is the **dimension** of $\mathbf{x}$ and, for each $i = 1, \ldots, n$, the number x_i is called **component** (or **coordinate**) i of $\mathbf{x}$.*

Throughout this book, real numbers are denoted by italicized lowercase letters – for example, a, t, and x_1. In contrast, n-vectors are denoted by

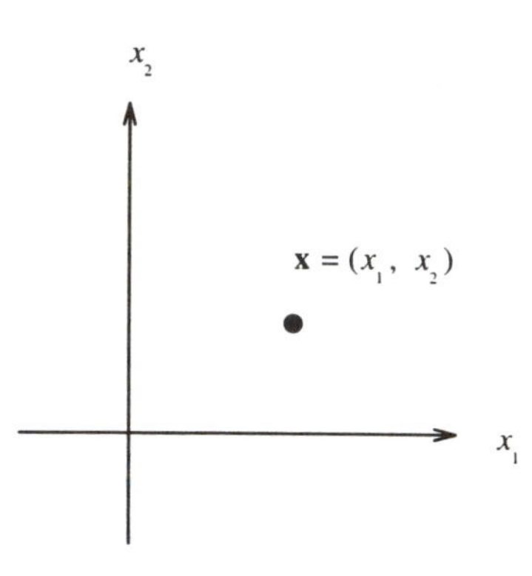

Figure 4.1: Visualizing a Vector as a Point

lowercase letters in boldface – for example, **x**, **y**, and **d**. The set of all n-vectors is written as $\mathbb{R}^n$, that is,

$$\mathbb{R}^n = \overbrace{\mathbb{R} \times \mathbb{R} \times \cdots \times \mathbb{R}}^{n \text{ times}},$$

which, as you will recall from Section 3.1.4, is the Cartesian product of the set $\mathbb{R}$ of real numbers, n times. An n-vector is often written as $\mathbf{x} \in \mathbb{R}^n$.

As you learned in Section 2.1.1, it is helpful to develop an image associated with a definition. For example, to visualize a vector $\mathbf{x} = (x_1, x_2)$ that has two components, consider a horizontal axis for the x_1 value and a vertical axis for the x_2 value. There are two standard ways to picture the vector $\mathbf{x} = (x_1, x_2)$:

1. As a point in the plane whose coordinates are (x_1, x_2), as seen in Figure 4.1.

2. As an arrow in the plane whose tail is at the origin and whose head is at the coordinates (x_1, x_2), as seen in Figure 4.2(a). The same vector results if you move the original vector parallel to itself, as shown in Figure 4.2(b). Unless otherwise stated, assume that a vector is drawn from the origin.

You can extend these visualizations to a vector having three components, but you *cannot* do so when the vector has four or more components. Nevertheless, you can work with vectors having any number of components. Herein lies one of the advantages of generalization: the ability to create and to work with concepts that you cannot visualize.

Now you will learn about certain special n-vectors and various comparisons and operations you can perform on n-vectors. Throughout this section, you will need to convert images to symbolic form and vice versa, as you learned to do in Section 2.1.1.

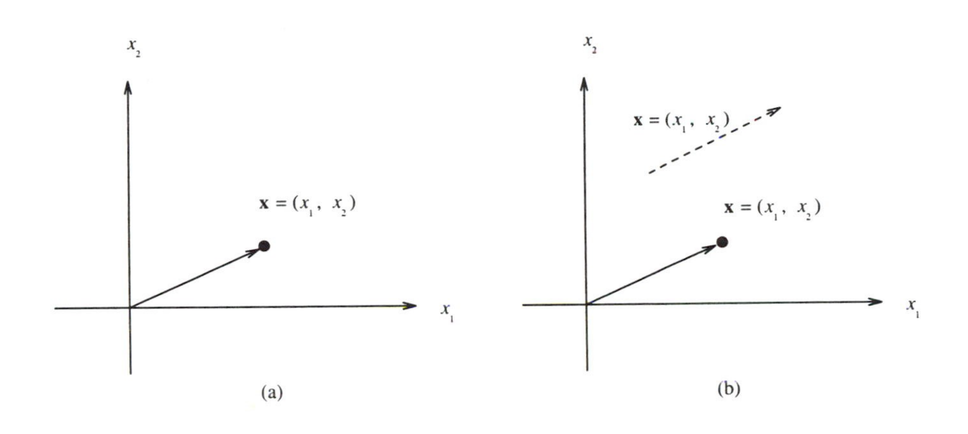

Figure 4.2: Visualizing a Vector as an Arrow

4.1.1 Special Vectors

One special n-vector that arises frequently is the **zero vector**, denoted by $\mathbf{0}$, which is the n-vector each of whose components is 0. For example, the zero vector in two dimensions is

$$\mathbf{0} = (0, 0)$$

and the zero vector in three dimensions is

$$\mathbf{0} = (0, 0, 0).$$

The dimension of the zero vector is generally understood from the context. Geometrically, the zero vector is a single point located at the origin of the coordinate system and, as such, has no length.

Other n-vectors that arise in applications are the **standard unit vectors**, each of which is the zero vector except in one component, whose value is 1. The following two standard unit vectors in two dimensions are illustrated in Figure 4.3:

$$(1, 0) \text{ and } (0, 1).$$

The three standard unit vectors in three dimensions are

$$(1, 0, 0),\ (0, 1, 0),\ \text{and } (0, 0, 1).$$

In n dimensions, the n standard unit vectors, denoted by $\mathbf{e}^1, \ldots, \mathbf{e}^n$, are:

$$\begin{aligned}
\mathbf{e}^1 &= (1, 0, 0, \ldots, 0) \\
\mathbf{e}^2 &= (0, 1, 0, \ldots, 0) \\
\vdots\ &\ \vdots \qquad\quad \vdots \\
\mathbf{e}^n &= (0, 0, \ldots, 0, 1).
\end{aligned}$$

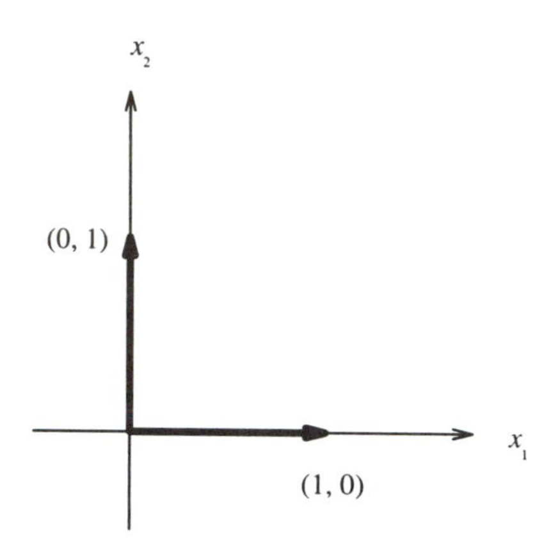

Figure 4.3: The Standard Unit Vectors in Two Dimensions

4.1.2 Comparing Two Vectors

As mentioned previously, the two n-vectors in Figure 4.2(b) are really the *same*, the meaning of which is made precise in the following definition.

> **Definition 4.2** *Two n-vectors* $\mathbf{x} = (x_1, \ldots, x_n)$ *and* $\mathbf{y} = (y_1, \ldots, y_n)$ *are* **equal**, *written* $\mathbf{x} = \mathbf{y}$, *if and only if for each* $i = 1, \ldots, n$, $x_i = y_i$.

This symbol = is a *binary relation* that is used to compare two n-vectors for equality. Thus, the statement

$$\mathbf{x} = \mathbf{y}$$

means that $\mathbf{x}$ *is* equal to $\mathbf{y}$, or equivalently, that the result of checking whether $\mathbf{x}$ is equal to $\mathbf{y}$ is *true*. Analogously, the statement

$$\mathbf{x} \neq \mathbf{y}$$

means that $\mathbf{x}$ is *not* equal to $\mathbf{y}$, or equivalently, that the result of checking whether $\mathbf{x}$ is equal to $\mathbf{y}$ is *false*.

You can also compare two n-vectors for inequality, as follows.

> **Definition 4.3** *The n-vector* $\mathbf{x} = (x_1, \ldots, x_n)$ *is* **less than or equal to** *the n-vector* $\mathbf{y} = (y_1, \ldots, y_n)$, *written* $\mathbf{x} \leq \mathbf{y}$, *if and only if for each* $i = 1, \ldots, n$, $x_i \leq y_i$. *Also,* $\mathbf{x} < \mathbf{y}$ *means that for each* $i = 1, \ldots, n$, $x_i < y_i$.

Definition 4.4 *The* n*-vector* $\mathbf{x} = (x_1, \ldots, x_n)$ *is* **greater than or equal to** *the* n*-vector* $\mathbf{y} = (y_1, \ldots, y_n)$*, written* $\mathbf{x} \geq \mathbf{y}$*, if and only if for each* $i = 1, \ldots, n,\ x_i \geq y_i$*. Also,* $\mathbf{x} > \mathbf{y}$ *means that for each* $i = 1, \ldots, n,\ x_i > y_i$*.*

The symbols $\leq, <, \geq$, and $>$ are *binary relations* that are used to compare two n-vectors for inequality. Thus, for example, the statement

$$\mathbf{x} \geq \mathbf{y}$$

means that $\mathbf{x}$ *is* greater than or equal to $\mathbf{y}$, or equivalently, that the result of checking whether $\mathbf{x}$ is greater than or equal to $\mathbf{y}$ is *true*. Analogously, the statement

$$\mathbf{x} \not\geq \mathbf{y}$$

means that $\mathbf{x}$ is *not* greater than or equal to $\mathbf{y}$, or equivalently, that the result of checking whether $\mathbf{x}$ is greater than or equal to $\mathbf{y}$ is *false*.

Recall from Section 2.2 that one of the advantages of creating a symbolic definition with quantifiers is that you can write the negation of the statement more easily. For example, by applying the rules for negating a statement containing the quantifier *for all* (see Section 1.6.10), the statement that $\mathbf{x} \not\geq \mathbf{y}$ means that *there is* an integer i with $1 \leq i \leq n$ such that $x_i < y_i$. Observe that there is a significant difference between using the symbol $\geq$ to compare two *numbers* and using that same symbol to compare two n*-vectors*. If x and y are numbers for which $x \not\geq y$, then you know that $x < y$. This statement is *not* necessarily *true* if $\mathbf{x}$ and $\mathbf{y}$ are n-vectors. This is because

$\mathbf{x} \not\geq \mathbf{y}$ means that there is at least one integer i such that $x_i < y_i$,

whereas

$\mathbf{x} < \mathbf{y}$ means that for each $i = 1, \ldots, n,\ x_i < y_i$.

Always think about what objects are being compared with the symbols $=, \leq, <, \geq$, and $>$. The use of these symbols for comparing two n-vectors is illustrated in the following example.

Example 4.1 – Comparing Two Vectors

The following comparisons are all *true*:

$$\begin{array}{rcll} (2,4) & = & (2,4) & \\ (2,4) & \neq & (4,2) & \text{(the order is important)} \\ (2,4) & \geq & (0,0) & \text{(because } 2 \geq 0 \text{ and } 4 \geq 0\text{)} \\ (-1,4,2) & \not\geq & (-1,3,5) & \text{(because } 2 < 5\text{).} \end{array}$$

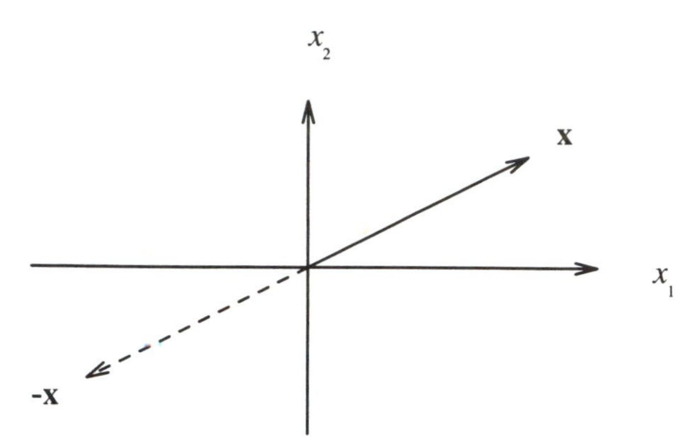

Figure 4.4: Reversing a Vector

4.1.3 Operations on Vectors

The operations on n-vectors described in this section fall into one of the following four categories:

1. Unary operations on an n-vector that result in an *n-vector*.
2. Unary operations on an n-vector that result in a *real number*.
3. Binary operations that combine two n-vectors to produce another *n-vector*.
4. Binary operations that combine two n-vectors to produce a *real number*.

Unary Operations on a Vector that Result in a Vector

One useful operation is to "reverse" an n-vector $\mathbf{x}$ to produce the new n-vector $-\mathbf{x}$ that has the same length as $\mathbf{x}$ but points in the opposite direction, as illustrated in Figure 4.4. To translate the visual image in Figure 4.4 to symbolic form, compare the components of $\mathbf{x}$ to those of $-\mathbf{x}$ for numerous specific examples. Doing so should lead you to the conclusion that when $\mathbf{x} = (x_1, \ldots, x_n)$,

$$-\mathbf{x} = (-x_1, \ldots, -x_n). \tag{4.1}$$

A generalization of the unary operation in (4.1) arises when you consider multiplying each component of an n-vector $\mathbf{x} = (x_1, \ldots, x_n)$ by an *arbitrary* real number, t, to obtain the new n-vector $t\mathbf{x}$. Algebraically,

$$t\mathbf{x} = t(x_1, \ldots, x_n) = (tx_1, \ldots, tx_n). \tag{4.2}$$

Observe that (4.2) is a generalization of $-\mathbf{x}$ because, when you replace t with the specific value of -1 in (4.2), you obtain (4.1).

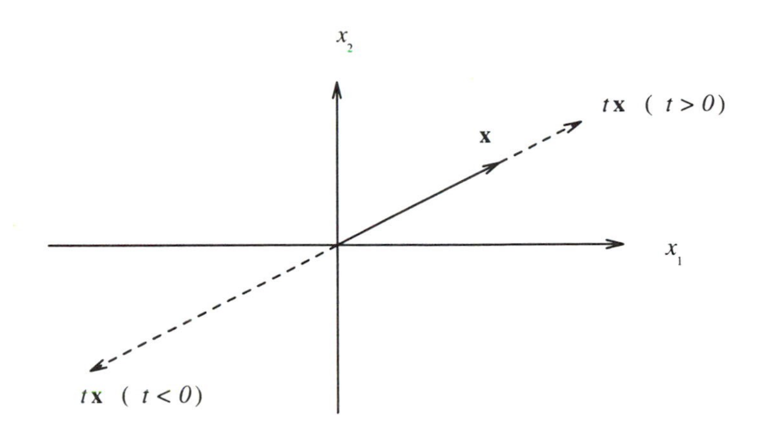

Figure 4.5: The Vectors $t\mathbf{x}$ When $t > 0$ and When $t < 0$

The next step is to create a geometric interpretation of $t\mathbf{x}$. By taking several specific values of t and drawing the vectors $\mathbf{x}$ and $t\mathbf{x}$ in two dimensions, you should conclude that the operation in (4.2) has the effect of creating an n-vector whose length is $|t|$ times the length of $\mathbf{x}$. The direction of $t\mathbf{x}$ is the same as that of $\mathbf{x}$, if $t > 0$ and opposite to that of $\mathbf{x}$, if $t < 0$, as shown in Figure 4.5. Of course, if $t = 0$, then $t\mathbf{x}$ is the zero vector, that is, $0\mathbf{x} = \mathbf{0}$.

Unary Operations on a Vector that Result in a Real Number

As you have seen geometrically, associated with each n-vector is a direction and a length. The length indicates how far the head of the vector is from the origin. For example, if $\mathbf{x} = (3, 4)$, then, as seen in Figure 4.6, the distance from (3, 4) to the origin is given by the Pythagorean theorem:

$$\sqrt{(3-0)^2 + (4-0)^2} = 5.$$

For a general vector $\mathbf{x} = (x_1, x_2)$ in two dimensions, the distance from (x_1, x_2) to the origin is

$$\sqrt{(x_1 - 0)^2 + (x_2 - 0)^2} = \sqrt{x_1^2 + x_2^2}.$$

Generalizing to a vector $\mathbf{x} = (x_1, x_2, x_3)$ in three dimensions, you have

$$\text{length}(\mathbf{x}) = \sqrt{x_1^2 + x_2^2 + x_3^2}.$$

Finally, generalizing to a vector in n dimensions results in the following definition for the length of an n-vector.

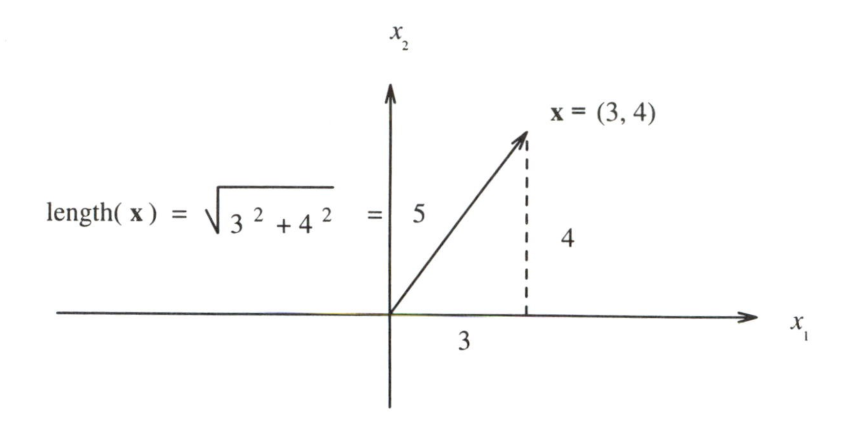

Figure 4.6: Finding the Length of a Vector

Definition 4.5 *The* **length** *of an n-vector $\mathbf{x} = (x_1, \ldots, x_n)$, denoted by $||\mathbf{x}||$, is computed by the following formula:*

$$||\mathbf{x}|| = \sqrt{x_1^2 + \cdots + x_n^2}. \tag{4.3}$$

The length of a vector is also called the **norm** *of the n-vector.*

The norm is a unary operation on an n-vector that results in a *real number*. A numerical example follows.

Example 4.2 – Computing the Length of a Vector

If $\mathbf{x} = (1, -2, 3)$, then

$$||\mathbf{x}|| = \sqrt{1^2 + (-2)^2 + 3^2} = \sqrt{14}.$$

You can also use the norm to compute the *distance* between two vectors. To see how, consider the two vectors $\mathbf{x} = (x_1, x_2)$ and $\mathbf{y} = (y_1, y_2)$ as two points in two dimensions. As seen in Figure 4.7, from the Pythagorean theorem, the distance from $\mathbf{x}$ to $\mathbf{y}$ is

$$d = \sqrt{(x_1 - y_1)^2 + (x_2 - y_2)^2}.$$

Generalizing to n-vectors $\mathbf{x} = (x_1, \ldots, x_n)$ and $\mathbf{y} = (y_1, \ldots, y_n)$,

$$\text{Distance from } \mathbf{x} \text{ to } \mathbf{y} = \sqrt{(x_1 - y_1)^2 + \cdots + (x_n - y_n)^2}. \tag{4.4}$$

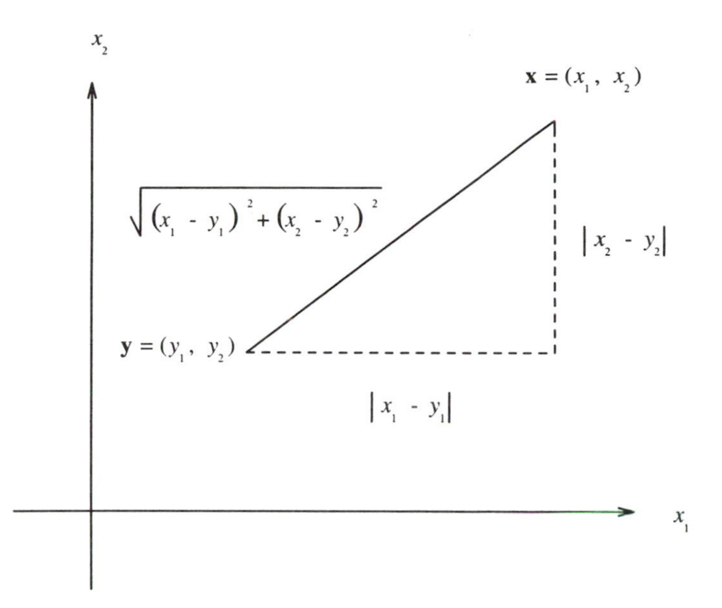

Figure 4.7: Computing the Distance Between Two Vectors

Binary Operations on a Vector that Result in a Vector

Two useful binary operations that combine two n-vectors to produce an n-vector are addition and subtraction, as you will now see.

Adding Vectors. For the vectors **x** and **y** in two dimensions, the operation of addition results in the new vector **x** + **y** whose tail is at the origin and whose head is that of **y** *after* moving **y** parallel to itself until its tail coincides with the head of **x**, as seen in Figure 4.8.

Look at the coordinates of several specific examples of vectors **x**, **y**, and **x** + **y** in two dimensions and then translate this geometric operation to algebraic form. Specifically, express the *unknown* coordinates of **x** + **y** in terms of the *known* coordinates of **x** and **y**. You should come to the following conclusion:

$$\mathbf{x} + \mathbf{y} = (x_1 + y_1, x_2 + y_2).$$

Generalizing to n-vectors $\mathbf{x} = (x_1, \ldots, x_n)$ and $\mathbf{y} = (y_1, \ldots, y_n)$ results in

$$\mathbf{x} + \mathbf{y} = (x_1 + y_1, \ldots, x_n + y_n).$$

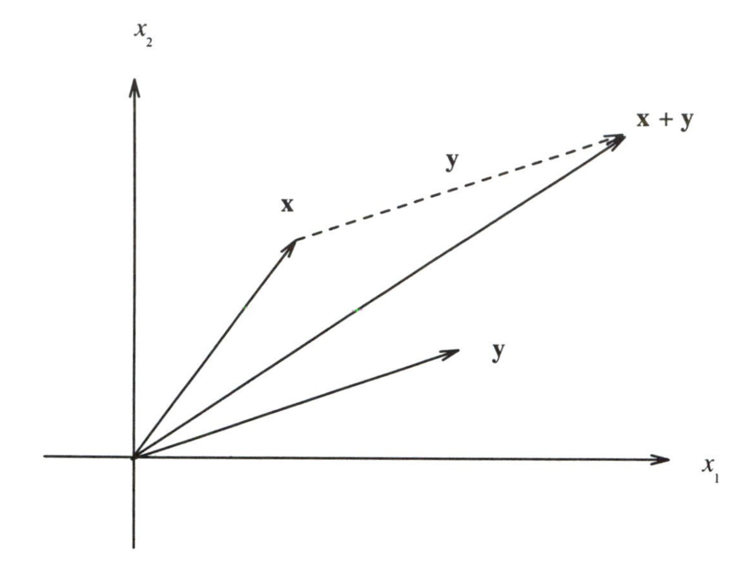

Figure 4.8: The Geometry of Adding Two Vectors

Example 4.3 – Adding Two Vectors

If $\mathbf{x} = (1, -2, 4)$ and $\mathbf{y} = (0, 2, -5)$, then

$$\begin{aligned}\mathbf{x} + \mathbf{y} &= (1, -2, 4) + (0, 2, -5)\\ &= (1 + 0,\ -2 + 2,\ 4 + (-5))\\ &= (1, 0, -1).\end{aligned}$$

Subtracting Vectors. In a similar manner, subtracting n-vectors is defined as follows:

$$\mathbf{x} - \mathbf{y} = (x_1 - y_1, \ldots, x_n - y_n).$$

Example 4.4 – Subtracting Two Vectors

If $\mathbf{x} = (1, -2, 4)$ and $\mathbf{y} = (0, 2, -5)$, then

$$\begin{aligned}\mathbf{x} - \mathbf{y} &= (1, -2, 4) - (0, 2, -5)\\ &= (1 - 0,\ -2 - 2,\ 4 - (-5))\\ &= (1, -4, 9).\end{aligned}$$

One way to create a visual image of $\mathbf{x} - \mathbf{y}$ is to multiply $\mathbf{y}$ by -1 to create the vector $-\mathbf{y}$ (thus reversing the direction of $\mathbf{y}$) and then add $-\mathbf{y}$ to

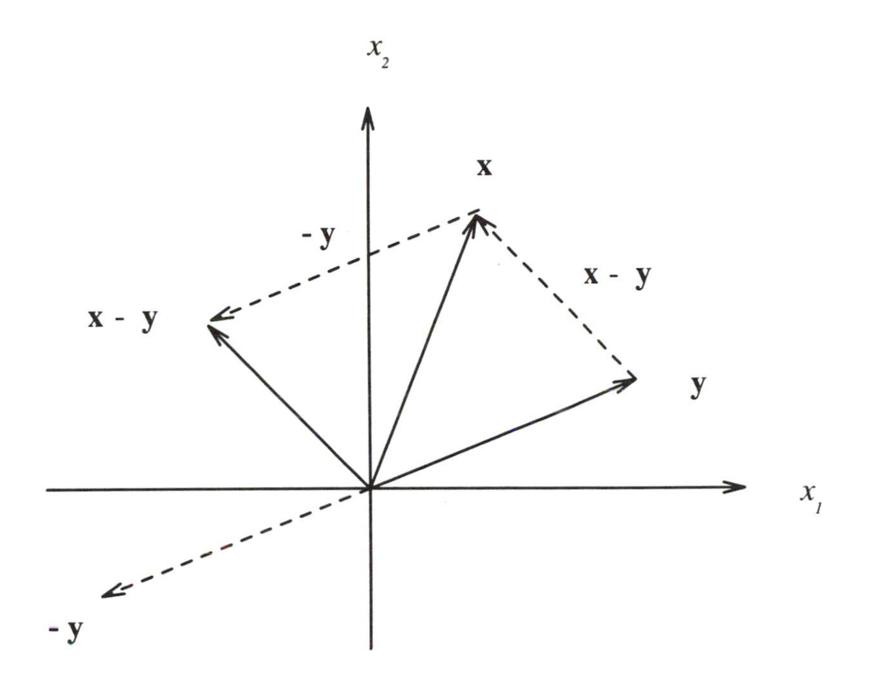

Figure 4.9: The Geometry of Subtracting Two Vectors

x, as shown in Figure 4.9. Alternatively, from Figure 4.9, you can see that the vector $\mathbf{x} - \mathbf{y}$ is the vector that points from the head of **y** to the head of **x**. That is, $\mathbf{x} - \mathbf{y}$ is the vector whose tail is at the head of **y** and whose head is at the head of **x**.

You might also notice that the formula for computing the distance from $\mathbf{x} = (x_1, \ldots, x_n)$ to $\mathbf{y} = (y_1, \ldots, y_n)$, as given in (4.4), is really the norm of $\mathbf{x} - \mathbf{y}$, that is,

$$\text{Distance from } \mathbf{x} \text{ to } \mathbf{y} = ||\mathbf{x} - \mathbf{y}||.$$

Binary Operations on a Vector that Result in a Real Number

There is one important binary operation that combines two n-vectors **x** and **y** to produce a *real number*. To illustrate, consider three items whose unit prices are, say, x_1, x_2, and x_3, respectively. If you purchase y_1 units of the first item, y_2 units of the second item, and y_3 units of the third item, then the total cost is $x_1y_1 + x_2y_2 + x_3y_3$. By thinking of x_1, x_2, and x_3 as the three components of a vector **x** and y_1, y_2, and y_3 as the three components of a vector **y**, the total cost is obtained by combining **x** and **y** in a special way to create a single number. This particular binary operation on n-vectors is common in applications and is formalized in the following definition.

Definition 4.6 *The* **dot product** *of the n-vectors* $\mathbf{x} = (x_1, \ldots, x_n)$ *and* $\mathbf{y} = (y_1, \ldots, y_n)$, *denoted by* $\mathbf{x} \cdot \mathbf{y}$ *or* $\mathbf{xy}$, *is the real number computed by the following formula:*

$$\mathbf{x} \cdot \mathbf{y} = \sum_{i=1}^{n} x_i y_i = x_1 y_1 + \cdots + x_n y_n.$$

This value is also called the **inner product** *of* $\mathbf{x}$ *and* $\mathbf{y}$.

Example 4.5 – Computing the Dot Product of Two Vectors

The following are examples of the dot product of different vectors in two dimensions:

$\mathbf{x}$	$\mathbf{y}$	$\mathbf{x} \cdot \mathbf{y}$	$= x_1y_1 + x_2y_2$	
$(1,2)$	$(0,3)$	$(1,2) \cdot (0,3)$	$= 1(0) + 2(3)$	$= 6$
$(1,2)$	$(0,-3)$	$(1,2) \cdot (0,-3)$	$= 1(0) + 2(-3)$	$= -6$
$(1,2)$	$(4,-2)$	$(1,2) \cdot (4,-2)$	$= 1(4) + 2(-2)$	$= 0$

There is a useful geometric interpretation of the *sign* of $\mathbf{x}{\cdot}\mathbf{y}$. To understand this in two dimensions, first draw the vector $\mathbf{x}$ pointing out of the origin. Then draw a line through the origin that is perpendicular to $\mathbf{x}$. As seen in Figure 4.10, this line divides the plane into two parts, each called a *half space*. The sign of $\mathbf{x}{\cdot}\mathbf{y}$ has the following meaning (see Figure 4.10):

1. $\mathbf{x}{\cdot}\mathbf{y} > 0$ if and only if $\mathbf{y}$ lies in the *same* half space as $\mathbf{x}$.

2. $\mathbf{x}{\cdot}\mathbf{y} < 0$ if and only if $\mathbf{y}$ lies in the *opposite* half space as $\mathbf{x}$.

3. $\mathbf{x}{\cdot}\mathbf{y} = 0$ if and only if $\mathbf{y}$ lies *on* the line and is perpendicular to $\mathbf{x}$. Another way to say this is that $\mathbf{x}$ and $\mathbf{y}$ are **orthogonal**.

4.1.4 Properties of Vectors and Their Operations

After working with n-vectors for some time, you will discover that the unary and binary operations in Section 4.1.3 satisfy certain properties, some of which are summarized in the following proposition.

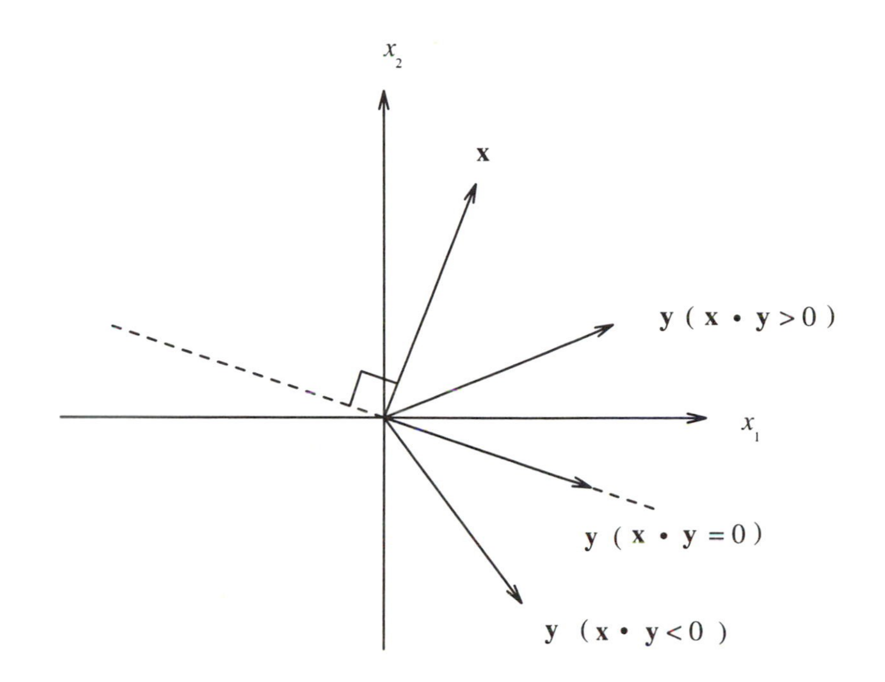

Figure 4.10: The Meaning of the Sign of $\mathbf{x} \cdot \mathbf{y}$

Proposition 4.1 *If* $\mathbf{x}$, $\mathbf{y}$, *and* $\mathbf{z}$ *are* n*-vectors and* t *and* u *are real numbers, then the following properties hold:*

(a) $\mathbf{x} + \mathbf{y} = \mathbf{y} + \mathbf{x}$.

(b) $(\mathbf{x} + \mathbf{y}) + \mathbf{z} = \mathbf{x} + (\mathbf{y} + \mathbf{z})$.

(c) $\mathbf{x} + \mathbf{0} = \mathbf{0} + \mathbf{x} = \mathbf{x}$.

(d) $\mathbf{x} + (-\mathbf{x}) = \mathbf{0}$.

(e) $t(u\mathbf{x}) = (tu)\mathbf{x}$.

(f) $t(\mathbf{x} + \mathbf{y}) = (t\mathbf{x}) + (t\mathbf{y})$.

(g) $(t + u)\mathbf{x} = (t\mathbf{x}) + (u\mathbf{x})$.

(h) $1\,\mathbf{x} = \mathbf{x}$.

Proof. You will find it helpful to use visual images of vectors and their operations in two dimensions to understand these properties and to motivate the following proofs.

(a) To see that the n-vectors $\mathbf{x} + \mathbf{y}$ and $\mathbf{y} + \mathbf{x}$ are equal, use Definition 4.2

to show that

$$\text{for each } i = 1, \ldots, n, \ (\mathbf{x}+\mathbf{y})_i = (\mathbf{y}+\mathbf{x})_i.$$

So, let i be an integer with $1 \leq i \leq n$. Then

$$\begin{aligned}
(\mathbf{x}+\mathbf{y})_i &= x_i + y_i && \text{(definition of adding vectors)} \\
&= y_i + x_i && \text{(by the commutative property of adding real numbers)} \\
&= (\mathbf{y}+\mathbf{x})_i && \text{(definition of adding vectors).}
\end{aligned}$$

(b-d) The proofs of these properties are left to the exercises.

(e) Again, use Definition 4.2 to show that the n-vector $t(u\mathbf{x})$ is equal to the n-vector $(tu)\mathbf{x}$. So, as in the proof of part (a), let i be an integer with $1 \leq i \leq n$ and show that

$$[t(u\mathbf{x})]_i = [(tu)\mathbf{x}]_i.$$

But,

$$\begin{aligned}
[t(u\mathbf{x})]_i &= t(ux_i) && \text{(definition of multiplying a vector } u\mathbf{x} \text{ by a real number } t) \\
&= (tu)x_i && \text{(associative property of multiplying real numbers)} \\
&= [(tu)\mathbf{x}]_i && \text{(definition of multiplying a vector } \mathbf{x} \text{ by a real number } tu).
\end{aligned}$$

(f-h) The proofs of these properties are left to the exercises.

This completes the proof.

QED

Properties of the Norm and Dot Product of Vectors

Recall that for n-vectors $\mathbf{x} = (x_1, \ldots, x_n)$ and $\mathbf{y} = (y_1, \ldots, y_n)$,

$$||\mathbf{x}|| = \sqrt{x_1^2 + \cdots + x_n^2}, \quad (4.5)$$

$$\mathbf{x} \cdot \mathbf{y} = x_1 y_1 + \cdots + x_n y_n. \quad (4.6)$$

The next proposition establishes numerous relationships between (4.5) and (4.6). Keep in mind that, geometrically, $||\mathbf{x}||$ is the length of $\mathbf{x}$.

Proposition 4.2 *Suppose that* $\mathbf{x}$, $\mathbf{y}$, *and* $\mathbf{z}$ *are* n*-vectors and that* t *is a real number. Then the following properties hold:*

(a) $\mathbf{x} \cdot \mathbf{x} \geq 0$.

Continued

(b) $\mathbf{x} \cdot \mathbf{x} = 0$ if and only if $\mathbf{x} = \mathbf{0}$.

(c) $||t\mathbf{x}|| = |t|\ ||\mathbf{x}||$.

(d) $\mathbf{x} \cdot \mathbf{x} = ||\mathbf{x}||^2$, that is, $||\mathbf{x}|| = \sqrt{\mathbf{x} \cdot \mathbf{x}}$.

(e) $\mathbf{x} \cdot \mathbf{y} = \mathbf{y} \cdot \mathbf{x}$.

(f) $\mathbf{x} \cdot (\mathbf{y} + \mathbf{z}) = (\mathbf{x} \cdot \mathbf{y}) + (\mathbf{x} \cdot \mathbf{z})$.

(g) $t(\mathbf{x} \cdot \mathbf{y}) = (t\mathbf{x}) \cdot \mathbf{y} = \mathbf{x} \cdot (t\mathbf{y})$.

Proof. Throughout, let

$$\begin{aligned}
\mathbf{x} &= (x_1, \ldots, x_n), \\
\mathbf{y} &= (y_1, \ldots, y_n), \\
\mathbf{z} &= (z_1, \ldots, z_n).
\end{aligned}$$

(a) The fact that $\mathbf{x} \cdot \mathbf{x} \geq 0$ is evident from the formula in (4.6).

(b) Suppose first that $\mathbf{x} \cdot \mathbf{x} = 0$. From (4.6), this means that

$$x_1 x_1 + \cdots + x_n x_n = x_1^2 + \cdots + x_n^2 = 0.$$

The only way this sum of nonnegative numbers can be 0 is if each number is 0, that is, for each $i = 1, \ldots, n,\ x_i = 0$, so, $\mathbf{x} = \mathbf{0}$.

Now assume that $\mathbf{x} = \mathbf{0}$, that is, for each $i = 1, \ldots, n,\ x_i = 0$. Then

$$\mathbf{x} \cdot \mathbf{x} = x_1 x_1 + \cdots + x_n x_n = 0(0) + \cdots + 0(0) = 0.$$

(c, d) The proofs of these properties are left to the exercises.

(e) To see that the two numbers $\mathbf{x} \cdot \mathbf{y}$ and $\mathbf{y} \cdot \mathbf{x}$ are equal, use (4.6) and observe that

$$\begin{aligned}
\mathbf{x} \cdot \mathbf{y} &= x_1 y_1 + \cdots + x_n y_n && \text{(definition of } \mathbf{x} \cdot \mathbf{y}\text{)} \\
&= y_1 x_1 + \cdots + y_n x_n && \text{(commutative property of multiplying real numbers)} \\
&= \mathbf{y} \cdot \mathbf{x} && \text{(definition of } \mathbf{y} \cdot \mathbf{x}\text{).}
\end{aligned}$$

(f, g) The proofs of these properties are left to the exercises.

This completes the proof.

QED

In this section, you have worked with n-vectors and seen how to perform various unary and binary operations on n-vectors. In some cases those operations result in n-vectors but in other cases the result is a real number. You have also seen some of the properties these operations satisfy. In Section 4.2, you will see another type of ordered list of numbers that serves as a generalization of n-vectors.

4.2 Matrices

You now know how to work with n-vectors, which are ordered lists of numbers. A useful generalization is a rectangular *table* of numbers, such as:

$$A = \begin{bmatrix} 2 & -3 & 1 \\ 0 & 4 & -2 \end{bmatrix}.$$

This particular table consist of six numbers organized in two *rows* of three *columns* each. The next definition presents a generalization to a table of m rows and n columns.

Definition 4.7 *An* $(m \times n)$ **matrix** A *(read as "an m by n matrix A") is a rectangular table of real numbers organized in* m *rows of* n *columns. The* **dimension** *of* A *is the values of* m *and* n*. The number in row* i $(1 \le i \le m)$ *and column* j $(1 \le j \le n)$ *of* A *is called* **element** A_{ij} *of the matrix.*

In contrast to vectors – which are denoted by lowercase boldface letters such as **d**, **x**, **y** – matrices are denoted in this book by uppercase italicized letters – for example, A, B, C, X. Also, the first subscript of a matrix always refers to a *row* and the second subscript refers to a *column*. The set of all $(m \times n)$ matrices is denoted by $\mathbb{R}^{m \times n}$. An $(m \times n)$ matrix is often written as $A \in \mathbb{R}^{m \times n}$. The following is another example of a matrix.

Example 4.6 – An Example of a Matrix

The following (4×7) matrix represents a calendar for the month of February:

$$A = \begin{bmatrix} 1 & 2 & 3 & 4 & 5 & 6 & 7 \\ 8 & 9 & 10 & 11 & 12 & 13 & 14 \\ 15 & 16 & 17 & 18 & 19 & 20 & 21 \\ 22 & 23 & 24 & 25 & 26 & 27 & 28 \end{bmatrix}.$$

4.2.1 Representations of Matrices

From Example 4.6, you can see that a matrix

$$A = \begin{bmatrix} A_{11} & A_{12} & \dots & A_{1n} \\ A_{21} & A_{22} & \dots & A_{2n} \\ \vdots & \vdots & \vdots & \vdots \\ A_{m1} & A_{m2} & \dots & A_{mn} \end{bmatrix}$$

is a generalization of a vector in that a matrix consists of an ordered list of *vectors*. In fact, there are two different ways to visualize a matrix in terms of vectors.

Row Representation of a Matrix

Realizing that each *row* of a matrix $A \in \mathbb{R}^{m \times n}$ is an n-vector, you can think of A as consisting of its m rows, stacked on top of each other:

$$A = \begin{bmatrix} A_{11} & A_{12} & \dots & A_{1n} \\ A_{21} & A_{22} & \dots & A_{2n} \\ \vdots & \vdots & \vdots & \vdots \\ A_{m1} & A_{m2} & \dots & A_{mn} \end{bmatrix} \begin{matrix} \longleftarrow \text{row } 1 \\ \longleftarrow \text{row } 2 \\ \vdots \\ \longleftarrow \text{row } m \end{matrix}$$

For each $i = 1, \dots, m$, row i of A is denoted in this book by $A_{i.}$. For instance, for the matrix in Example 4.6,

$$\begin{aligned} A_{1.} &= [1 \;\; 2 \;\; 3 \;\; 4 \;\; 5 \;\; 6 \;\; 7], \\ A_{2.} &= [8 \;\; 9 \;\; 10 \;\; 11 \;\; 12 \;\; 13 \;\; 14], \\ A_{3.} &= [15 \;\; 16 \;\; 17 \;\; 18 \;\; 19 \;\; 20 \;\; 21], \\ A_{4.} &= [22 \;\; 23 \;\; 24 \;\; 25 \;\; 26 \;\; 27 \;\; 28]. \end{aligned}$$

In this case, $A_{i.}$ represents the dates of the seven days in week i of February. For example, $A_{3.}$ represents the dates of the seven days in the the *third* week of February.

Column Representation of a Matrix

Alternatively, realizing that each *column* of a matrix $A \in \mathbb{R}^{m \times n}$ is an m-vector, you can think of A as consisting of its n columns, written next to each other:

$$\begin{matrix} \text{Col.} & 1 & 2 & \dots & n \\ & \downarrow & \downarrow & \cdots & \downarrow \end{matrix}$$

$$A = \begin{bmatrix} A_{11} & A_{12} & \dots & A_{1n} \\ A_{21} & A_{22} & \dots & A_{2n} \\ \vdots & \vdots & \vdots & \vdots \\ A_{m1} & A_{m2} & \dots & A_{mn} \end{bmatrix}.$$

For each $j = 1, \dots, n$, column j of A is denoted in this book by $A_{.j}$. For instance, for the matrix in Example 4.6,

$$A_{.1} = \begin{bmatrix} 1 \\ 8 \\ 15 \\ 22 \end{bmatrix}, \; A_{.2} = \begin{bmatrix} 2 \\ 9 \\ 16 \\ 23 \end{bmatrix}, \; \dots, \; A_{.7} = \begin{bmatrix} 7 \\ 14 \\ 21 \\ 28 \end{bmatrix}.$$

In this case, $A_{.j}$ represents those dates in February that fall on the same day of the week. For example, the dates in $A_{.1}$ might all be Saturdays.

Vectors as a Special Case of Matrices

As you just saw, a matrix is a collection of its row (or column) vectors. Analogously, a given n-vector $\mathbf{x} = (x_1, \ldots, x_n)$ is a special type of matrix that has two possible representations, depending on whether you think of $\mathbf{x}$ as a column or a row, as described next.

Column Vectors. You can think of the n-vector $\mathbf{x} = (x_1, \ldots, x_n)$ as an $(n \times 1)$ matrix whose one *column* consists of the entries in $\mathbf{x}$:

$$\mathbf{x} = \begin{bmatrix} x_1 \\ x_2 \\ \vdots \\ x_n \end{bmatrix}.$$

In this case, the vector $\mathbf{x}$ is referred to as a **column vector**.

Row Vectors. You can also think of the n-vector $\mathbf{x} = (x_1, \ldots, x_n)$ as a $(1 \times n)$ matrix whose one *row* consists of the entries in $\mathbf{x}$:

$$\mathbf{x}^t = [x_1, x_2, \ldots, x_n].$$

In this case, the vector is referred to as a **row vector** and is denoted by $\mathbf{x}^t$ to differentiate the row vector from the column vector.

Now you will learn about certain special matrices and about various comparisons and operations that you can perform on matrices in general.

4.2.2 Special Matrices

After working with matrices in various applications, you will find that certain matrices arise frequently and play a special role. Some of those matrices are described now.

The Zero Matrix

One special matrix is the **zero matrix**, denoted by 0, in which each element is 0. For example, the (2×3) zero matrix is

$$0 = \begin{bmatrix} 0 & 0 & 0 \\ 0 & 0 & 0 \end{bmatrix}.$$

The dimension of the zero matrix is generally understood from the context.

Square Matrices

Another special type of matrix is one in which the number of rows is the same as the number of columns. Such a matrix with n rows and n columns is a **square matrix of order** n. The elements $A_{11}, A_{22}, \ldots, A_{nn}$ of a square matrix A of order n are called the **diagonal elements**. Several examples of square matrices are described next.

Diagonal Matrices. A **diagonal matrix** is a square matrix of order n in which all elements, except possibly for the diagonal elements, are 0. The $(n \times n)$ zero matrix is a diagonal matrix, as are the following matrices:

$$A = \begin{bmatrix} 1 & 0 \\ 0 & -2 \end{bmatrix}, \ B = \begin{bmatrix} 4 & 0 & 0 \\ 0 & 0 & 0 \\ 0 & 0 & \frac{1}{2} \end{bmatrix}.$$

The Identity Matrix. One particular diagonal matrix that arises for instance when solving a system of linear equations (as described in Section 4.3) is the $(n \times n)$ **identity matrix**, denoted by I, in which the diagonal elements are all 1, that is:

$$I = \begin{bmatrix} 1 & 0 & \dots & 0 \\ 0 & 1 & \dots & 0 \\ \vdots & \vdots & \ddots & \vdots \\ 0 & 0 & \dots & 1 \end{bmatrix}.$$

You might notice that row i and column i of I are the standard unit vectors, $\mathbf{e}^i$. So, another way to write the standard unit vector $\mathbf{e}^i$ is $I_{i.}$, which is a row vector, or $I_{.i}$, which is a column vector.

Permutation Matrices. A **permutation** in mathematics is a rearrangement. For example, if you start with the n-vector $\mathbf{x} = (1, 2, 3, 4)$, then a permutation of $\mathbf{x}$ is the n-vector (4, 2, 1, 3). Generalizing to matrices, you can rearrange their columns. For example, rearranging the columns of the $(n \times n)$ identity matrix I results in a permutation matrix of I. More formally, a **permutation matrix** is a square matrix in which each row and each column has exactly one 1 and all other entries in that row and column are 0. The following are examples of permutation matrices:

$$P = \begin{bmatrix} 0 & 1 \\ 1 & 0 \end{bmatrix}, \ Q = \begin{bmatrix} 0 & 1 & 0 \\ 0 & 0 & 1 \\ 1 & 0 & 0 \end{bmatrix}.$$

Eta Matrices. An **eta matrix** is the identity matrix, except possibly for one column, say, column k. The identity matrix is an eta matrix, as are the following matrices:

$$A = \begin{bmatrix} 1 & 0 \\ 0 & -2 \end{bmatrix}, \ B = \begin{bmatrix} 1 & 0 & 0 \\ 0 & 1 & 0 \\ \frac{1}{2} & 0 & 1 \end{bmatrix}, \ C = \begin{bmatrix} 1 & 0 & 1 & 0 \\ 0 & 1 & -1 & 0 \\ 0 & 0 & 2 & 0 \\ 0 & 0 & 3 & 1 \end{bmatrix}.$$

In this example, column 2 of A differs from column 2 of the identity matrix; for B, it is column 1 that differs from column 1 of the identity matrix; and for C, it is column 3 that is different.

4.2.3 Comparing Two Matrices

The primary comparison between two matrices is determining whether or not they are the *same*, the precise meaning of which is made clear in the following definition.

> **Definition 4.8** *Two* $(m \times n)$ *matrices* A *and* B *are* **equal**, *written* $A = B$, *if and only if for each* $i = 1, \ldots, m$ *and for each* $j = 1, \ldots, n$, $A_{ij} = B_{ij}$.

Alternatively, thinking of a matrix in terms of its *rows*, $A = B$ if and only if for each row $i = 1, \ldots, n$, $A_{i.} = B_{i.}$. Equivalently, in terms of the *columns*, $A = B$ if and only if for each column $j = 1, \ldots, n$, $A_{.j} = B_{.j}$.

Because the symbol = is used to compare real numbers, vectors, and matrices, always be careful to determine *what* objects are being compared. For example, the first occurrence of the symbol = in Definition 4.8 is used to compare two *matrices* (A and B) but the last occurrence of = compares two *real numbers* (A_{ij} and B_{ij}).

When the symbol = is used to compare two matrices of the same dimension for equality, the statement

$$A = B$$

means that A *is* equal to B, or equivalently, that the result of checking whether A is equal to B is *true*. Analogously, the statement

$$A \neq B$$

means that A is *not* equal to B, or equivalently, that the result of checking whether A is equal to B is *false*. By using the rules for negating a statement that contains the quantifier *for all*, observe that

$$A \neq B$$

means that

there exist $1 \leq i \leq m$ and $1 \leq j \leq n$ such that $A_{ij} \neq B_{ij}$.

The use of these symbols for comparing matrices is illustrated in the following example.

Example 4.7 – Comparing Two Matrices for Equality

It is *true* that

$$\begin{bmatrix} -1 & 0 & 2 \\ 0 & 3 & -2 \end{bmatrix} = \begin{bmatrix} -1 & 0 & 2 \\ 0 & 3 & -2 \end{bmatrix}.$$

Also, for the $(n \times n)$ identity matrix I and the $(n \times n)$ zero matrix 0, it is *true* that $I \neq 0$.

4.2.4 Operations on Matrices

Several unary and binary operations on matrices are described in this section. Each operation always results in a matrix.

Unary Operations on Matrices

A unary operation transforms a given matrix to a new matrix. Two such unary operations are described next.

Multiplying a Matrix by a Real Number. One useful unary operation is to multiply a matrix $A \in \mathbb{R}^{m \times n}$ by a real number t to obtain an $(m \times n)$ matrix, denoted by tA. The elements of tA are those of A multiplied by t, that is:

$$(tA)_{ij} = tA_{ij}, \text{ for each } i = 1, \dots, m \text{ and for each } j = 1, \dots, n.$$

Example 4.8 – Multiplying a Matrix by a Real Number

If

$$A = \begin{bmatrix} -1 & 0 & 2 \\ 0 & 3 & -2 \end{bmatrix}$$

and $t = 2$, then

$$tA = 2\begin{bmatrix} -1 & 0 & 2 \\ 0 & 3 & -2 \end{bmatrix} = \begin{bmatrix} -2 & 0 & 4 \\ 0 & 6 & -4 \end{bmatrix}.$$

You can see that for the special case of $t = 0$, tA is the zero matrix, that is, $0A = 0$. Also, for the special case of $t = 1$, $1A = A$.

The Transpose of a Matrix. Another unary operation is to **transpose** a matrix $A \in \mathbb{R}^{m \times n}$, in which you create a new matrix, denoted by A^t, that has n rows and m columns. Specifically, row i of A becomes column i of A^t, that is,

$$(A^t)_{ji} = A_{ij}, \text{ for each } i = 1, \dots, m \text{ and for each } j = 1, \dots, n.$$

You have already seen the use of the transpose to distinguish a *column* vector, $\mathbf{x}$, from the corresponding *row* vector, $\mathbf{x}^t$. Another example of the transpose of a matrix follows.

Example 4.9 – The Transpose of a Matrix

If $A = \begin{bmatrix} -1 & 0 & 2 \\ 0 & 3 & -2 \end{bmatrix}$, then $A^t = \begin{bmatrix} -1 & 0 \\ 0 & 3 \\ 2 & -2 \end{bmatrix}$.

Binary Operations on Matrices

The remaining arithmetic operations of adding, subtracting, and multiplying matrices constitute binary operations that combine two matrices to produce another matrix.

Adding and Subtracting Matrices. You can add two matrices of the same dimension to create a new matrix of that dimension. Specifically, **matrix addition** is defined as follows: If $A, B \in \mathbb{R}^{m \times n}$, then $A + B$ is the $(m \times n)$ matrix obtained by adding the corresponding elements of A and B, that is:

$$(A + B)_{ij} = A_{ij} + B_{ij}, \text{ for each } i = 1, \ldots, m \text{ and } j = 1, \ldots, n.$$

Example 4.10 – Adding Two Matrices

If

$$A = \begin{bmatrix} -1 & 0 & 2 \\ 0 & 3 & -2 \end{bmatrix} \text{ and } B = \begin{bmatrix} 2 & 3 & -2 \\ -1 & 0 & 4 \end{bmatrix},$$

then

$$\begin{aligned} A + B &= \begin{bmatrix} -1 & 0 & 2 \\ 0 & 3 & -2 \end{bmatrix} + \begin{bmatrix} 2 & 3 & -2 \\ -1 & 0 & 4 \end{bmatrix} \\ &= \begin{bmatrix} -1 + 2 & 0 + 3 & 2 + (-2) \\ 0 + (-1) & 3 + 0 & -2 + 4 \end{bmatrix} \\ &= \begin{bmatrix} 1 & 3 & 0 \\ -1 & 3 & 2 \end{bmatrix}. \end{aligned}$$

Matrix subtraction is defined analogously, that is, if $A, B \in \mathbb{R}^{m \times n}$, then $A - B$ is the $(m \times n)$ matrix in which

$$(A - B)_{ij} = A_{ij} - B_{ij}, \text{ for each } i = 1, \ldots, m \text{ and } j = 1, \ldots, n.$$

Multiplying Matrices. The operation of multiplying two matrices also results in a matrix. One way to multiply two matrices $A, B \in \mathbb{R}^{m \times n}$ is similar to addition and subtraction in that you multiply the corresponding elements of A and B. However, as with multiplying two n-vectors, much experience with matrices in applications has led to a different and more useful method for multiplying two matrices.

The more practical method of multiplying the matrix A by the matrix B results in a new matrix, denoted by AB, in which the element in row i and column j of AB is the dot product of the row vector $A_{i.}$ and the column vector $B_{.j}$, that is,

$$(AB)_{ij} = A_{i.} \cdot B_{.j}. \tag{4.7}$$

Before proceeding, observe that to compute the dot product in (4.7), both the vectors $A_{i.}$ and $B_{.j}$ must have the same dimension. In terms of the matrices A and B, this means that the number of *columns* of A must be the same as the number of *rows* of B. In other words, with the method in (4.7), it is only possible to multiply a matrix A by a matrix B if $A \in \mathbb{R}^{m \times p}$ and $B \in \mathbb{R}^{p \times n}$, that is, if A has p columns and B has p rows. In this case, **matrix multiplication** is defined as follows. If $A \in \mathbb{R}^{m \times p}$ and $B \in \mathbb{R}^{p \times n}$, then AB is the $(m \times n)$ matrix in which

$$(AB)_{ij} = A_{i.} \cdot B_{.j}, \text{ for each } i = 1, \dots, m \text{ and } j = 1, \dots, n. \tag{4.8}$$

Alternatively, using the definition of the dot product (see Definition 4.6 in Section 4.1.3), you can rewrite each element in (4.8) as follows:

$$(AB)_{ij} = A_{i.} \cdot B_{.j} = \sum_{k=1}^{p} A_{ik} B_{kj}. \tag{4.9}$$

Example 4.11 – Multiplying Two Matrices

If

$$A = \begin{bmatrix} 1 & 2 & 0 \\ 0 & -1 & 3 \end{bmatrix} \text{ and } B = \begin{bmatrix} 1 & 0 & -1 & 2 \\ 2 & -1 & 3 & 0 \\ 1 & -1 & 0 & 4 \end{bmatrix},$$

then $A \in \mathbb{R}^{2 \times 3}$ and $B \in \mathbb{R}^{3 \times 4}$, so $AB \in \mathbb{R}^{2 \times 4}$ and

$$\begin{aligned}
(AB)_{11} &= A_{1.} \cdot B_{.1} = (1,2,0) \cdot (1,2,1)^t &&= 5, \\
(AB)_{12} &= A_{1.} \cdot B_{.2} = (1,2,0) \cdot (0,-1,-1)^t &&= -2, \\
(AB)_{13} &= A_{1.} \cdot B_{.3} = (1,2,0) \cdot (-1,3,0)^t &&= 5, \\
(AB)_{14} &= A_{1.} \cdot B_{.4} = (1,2,0) \cdot (2,0,4)^t &&= 2, \\
\\
(AB)_{21} &= A_{2.} \cdot B_{.1} = (0,-1,3) \cdot (1,2,1)^t &&= 1, \\
(AB)_{22} &= A_{2.} \cdot B_{.2} = (0,-1,3) \cdot (0,-1,-1)^t &&= -2, \\
(AB)_{23} &= A_{2.} \cdot B_{.3} = (0,-1,3) \cdot (-1,3,0)^t &&= -3, \\
(AB)_{24} &= A_{2.} \cdot B_{.4} = (0,-1,3) \cdot (2,0,4)^t &&= 12.
\end{aligned}$$

By putting each of the foregoing numbers in the correct position of the matrix AB, you obtain

$$AB = \begin{bmatrix} 5 & -2 & 5 & 2 \\ 1 & -2 & -3 & 12 \end{bmatrix}.$$

Another example of matrix multiplication is the use of a permutation matrix to interchange two rows or columns of an $(n \times n)$ matrix, A. In particular, to interchange rows i and j of A, you multiply A on the *left* by the $(n \times n)$ permutation matrix P obtained by interchanging rows i and j of the $(n \times n)$ identity matrix, I. This process is illustrated now.

Example 4.12 – Interchanging Two Rows of a Matrix

To interchange rows 2 and 3 of the matrix

$$A = \begin{bmatrix} 1 & -1 & 0 \\ 0 & 0 & 1 \\ 0 & -1 & 3 \end{bmatrix},$$

construct the following permutation matrix P by interchanging rows 2 and 3 of I:

$$P = \begin{bmatrix} 1 & 0 & 0 \\ 0 & 0 & 1 \\ 0 & 1 & 0 \end{bmatrix}.$$

When you now multiply A on the left by P, you interchange rows 2 and 3 of A because

$$PA = \begin{bmatrix} 1 & 0 & 0 \\ 0 & 0 & 1 \\ 0 & 1 & 0 \end{bmatrix} \begin{bmatrix} 1 & -1 & 0 \\ 0 & 0 & 1 \\ 0 & -1 & 3 \end{bmatrix} = \begin{bmatrix} 1 & -1 & 0 \\ 0 & -1 & 3 \\ 0 & 0 & 1 \end{bmatrix}.$$

In the exercises, you are asked to show that to interchange *columns* i and j of A, you multiply A on the *right* by the $(n \times n)$ permutation matrix P obtained by interchanging columns i and j of the $(n \times n)$ identity matrix.

4.2.5 Properties of Matrices and Their Operations

After working with matrices for some time, you will discover that the operations of addition and subtraction satisfy the same properties as do those operations on real numbers, as summarized in the following proposition.

Proposition 4.3 *If $A, B, C, 0 \in \mathbb{R}^{m \times n}$, then the following properties hold:*

(a) $A + B = B + A$.

(b) $(A + B) + C = A + (B + C)$.

(c) $A + 0 = 0 + A = A$.

(d) $A - A = 0$.

(e) $0 - A = -A$.

Proof. In each part of this proposition, you must show that two matrices are equal. Definition 4.8 is therefore used to show that all elements of the corresponding matrices are equal.

(a) Observe that $A + B$ and $B + A$ are both matrices, although their elements are computed in different ways. To show that these two matrices are *equal*, it is necessary to verify that each element of $A + B$ is equal to the corresponding element of $B + A$. So, let i and j be integers with $1 \leq i \leq m$ and $1 \leq j \leq n$. You must show that $(A+B)_{ij} = (B+A)_{ij}$, which is accomplished by transforming $(A+B)_{ij}$ into $(B + A)_{ij}$, as follows:

$$\begin{aligned}(A + B)_{ij} &= A_{ij} + B_{ij} && \text{(definition of matrix addition)}\\ &= B_{ij} + A_{ij} && \text{(by the commutative property of adding real numbers)}\\ &= (B + A)_{ij} && \text{(definition of matrix addition).}\end{aligned}$$

(b-e) The proofs of these properties are similar to the proof in part (a) and are left to the exercises.

This completes the proof.

QED

Rules pertaining to the multiplication of a matrix by a real number are similar to those of multiplying numbers, as shown in the next proposition.

Proposition 4.4 *If* $A, B, 0 \in \mathbb{R}^{m \times n}$ *and* s *and* t *are real numbers, then the following properties hold:*

(a) $s(A + B) = sA + sB$.

(b) $s(A - B) = sA - sB$.

(c) $(s + t)A = sA + tA$.

(d) $(s - t)A = sA - tA$.

(e) $s(tA) = (st)A$.

(f) $0A = 0$ *(the number* 0 *times the matrix* A *= the zero matrix).*

(g) $s0 = 0$ *(the number* s *times the zero matrix = the zero matrix).*

Proof. In each part of this proposition, you must show that two matrices are equal. Definition 4.8 is therefore used to show that all elements of the corresponding matrices are equal.

(a) To show that $s(A+B) = sA+sB$, let i and j be integers with $1 \leq i \leq m$ and $1 \leq j \leq n$. You must show that $[s(A + B)]_{ij} = (sA + sB)_{ij}$, but

$$\begin{aligned}
[s(A + B)]_{ij} &= s[(A + B)_{ij}] \\
&= s(A_{ij} + B_{ij}) \\
&= sA_{ij} + sB_{ij} \\
&= (sA)_{ij} + (sB)_{ij} \\
&= (sA + sB)_{ij}.
\end{aligned}$$

So, by Definition 4.8, $s(A + B) = sA + sB$.

(b-g) The proofs of these properties are similar to the proof in part (a) and are left to the exercises.

This completes the proof.

QED

Unlike adding and subtracting, the operation of *multiplying* two matrices differs in several important ways from those same operations on numbers. For example, if a and b are numbers, then you know that $ab = ba$. However, if A and B are *matrices*, then it is *not* necessarily *true* that $AB = BA$. In fact, you may not even be able to perform these computations because the dimensions of A and B are not appropriate. For example, if A is a (2×3) matrix and B is a (3×4) matrix, then AB is a (2×4) matrix. However, you *cannot* compute BA because the number of columns of B (4, in this case) is not equal to the number of rows of A (2, in this case). Even when you *can* compute both AB and BA, the resulting matrices might not be equal, as shown in the next example.

Example 4.13 – Two Matrices A and B for Which $AB \neq BA$

If

$$A = \begin{bmatrix} 2 & 0 \\ 0 & 3 \end{bmatrix} \text{ and } B = \begin{bmatrix} 0 & 1 \\ 1 & 0 \end{bmatrix},$$

then $AB \neq BA$ because

$$AB = \begin{bmatrix} 0 & 2 \\ 3 & 0 \end{bmatrix} \text{ and } BA = \begin{bmatrix} 0 & 3 \\ 2 & 0 \end{bmatrix}.$$

Another difference between multiplying two real numbers and two matrices is that, for real numbers a and b, if $ab = 0$, then either $a = 0$ or $b = 0$. The next example shows that this is *not* the case for matrices A and B.

Example 4.14 – Two Matrices A and B for Which $AB = 0$ and Yet $A \neq 0$ and $B \neq 0$

If

$$A = \begin{bmatrix} 1 & 2 \\ -1 & -2 \end{bmatrix} \text{ and } B = \begin{bmatrix} 2 & 4 \\ -1 & -2 \end{bmatrix},$$

then $A \neq 0$ and $B \neq 0$ and yet

$$AB = \begin{bmatrix} 1 & 2 \\ -1 & -2 \end{bmatrix} \begin{bmatrix} 2 & 4 \\ -1 & -2 \end{bmatrix} = \begin{bmatrix} 0 & 0 \\ 0 & 0 \end{bmatrix}.$$

In contrast, certain properties of matrix muliplication *are* the same as those of multiplying real numbers, as summarized in the next proposition.

Proposition 4.5 *If s is a real number and $A, B, C, 0$, and I (the identity matrix) are matrices for which you can perform the following operations on the basis of their dimensions, then the following properties hold:*

(a) $A(B + C) = AB + AC$.

(b) $A(B - C) = AB - AC$.

(c) $(A + B)C = AC + BC$.

(d) $(A - B)C = AC - BC$.

(e) $(AB)C = A(BC)$.

(f) $s(AB) = (sA)B = A(sB)$.

(g) $IA = A$ *and* $BI = B$.

(h) $0A = 0$ *and* $B0 = 0$.

Proof. In each part of this proposition, you must show that two matrices are equal. Definition 4.8 is therefore used to show that all corresponding elements of the matrices are equal.

(a) It is first necessary to be sure that A, B, and C have dimensions for which you can compute $A(B + C)$, AB, and AC. This is possible provided that $A \in \mathbb{R}^{m \times p}$ and that $B, C \in \mathbb{R}^{p \times n}$. So now let i and j be integers with $1 \leq i \leq m$ and $1 \leq j \leq n$. You must show that

$[A(B+C)]_{ij} = (AB+AC)_{ij}$, but

$$\begin{aligned}
[A(B+C)]_{ij} &= A_{i.} \cdot (B+C)_{.j} \\
&= A_{i.} \cdot (B_{.j} + C_{.j}) \\
&= (A_{i.} \cdot B_{.j}) + (A_{i.} \cdot C_{.j}) \\
&= (AB)_{ij} + (AC)_{ij} \\
&= (AB+AC)_{ij}.
\end{aligned}$$

(b-h) The proofs of these properties are similar to the proof in part (a) and are left to the exercises.

This completes the proof.

QED

Part (g) of Proposition 4.5 is particularly important because it states that, for matrices, the identity matrix plays the same role as the number 1 does for real numbers – when you multiply any number by 1 (or any matrix by the identity matrix), you get back what you started with. Similarly, part (h) of Proposition 4.5 indicates that for matrices, the zero matrix behaves like the number 0 does for real numbers – zero times anything is zero.

In this section, you have learned about matrices and how to perform various unary and binary operations on matrices. In all cases, those operations result in a matrix. You have also seen some of the properties these operations satisfy. In Section 4.3, you will see how matrices and vectors are used in problem solving.

4.3 The Algebra of Solving Linear Equations

In Chapter 1, the problem of solving one linear equation in one unknown is generalized sequentially to the problem of solving a system of n linear equations in n unknowns. In this section, you will see how matrices and vectors are used to develop a solution procedure for that problem.

4.3.1 The Problem of Solving a System of Linear Equations

Recall Example 1.1 in Section 1.1.1, in which you want to find a value for the variable x that satisfies

$$106x = 15,900.$$

This problem is unified with another similar one to create the following problem of solving a single linear equation.

The Problem of Solving a Linear Equation

Given values for the real numbers a and b, find a value for the variable x so that

$$ax = b. \tag{4.10}$$

In Section 1.4.2, this problem is generalized to the following one.

The Problem of Solving Two Linear Equations in Two Unknowns

Given values for the real numbers $a, b, c, d, e,$ and f, find values for the variables x and y so that

$$\begin{aligned} ax + by &= e \\ cx + dy &= f. \end{aligned} \tag{4.11}$$

A further generalization arises when you consider the following problem of solving n linear equations in n unknowns.

The Problem of Solving n Linear Equations in n Unknowns (Scalar Notation)

Given a positive integer n and values for the real numbers a_{ij}, for each $i, j = 1, \ldots, n$ and for the real numbers $b_1, \ldots, b_n$, find values for the variables $x_1, \ldots, x_n$ so that

$$\begin{aligned} a_{11}x_1 + a_{12}x_2 + \cdots + a_{1n}x_n &= b_1 \\ a_{21}x_1 + a_{22}x_2 + \cdots + a_{2n}x_n &= b_2 \\ \vdots \qquad\qquad & \;\;\vdots\;\; \vdots \\ a_{n1}x_1 + a_{n2}x_2 + \cdots + a_{nn}x_n &= b_n. \end{aligned} \tag{4.12}$$

The first step in developing a solution procedure is to use vectors and matrices to write (4.12) in a simpler form that is easier to work with. For example, think of the n unknown variables collectively as an n-vector $\mathbf{x} = (x_1, \ldots, x_n)$. Likewise, view the values of $b_1, \ldots, b_n$ on the right-hand side of (4.12) as the components of an n-vector, say, $\mathbf{b} = (b_1, \ldots, b_n)$. The remaining values of a_{ij} in (4.12) constitute a rectangular table of numbers that you can represent by the following $(n \times n)$ matrix:

$$A = \begin{bmatrix} a_{11} & a_{12} & \ldots & a_{1n} \\ a_{21} & a_{22} & \ldots & a_{2n} \\ \vdots & \vdots & \vdots & \vdots \\ a_{n1} & a_{n2} & \ldots & a_{nn} \end{bmatrix}.$$

With the n-vectors **x** and **b** and the $(n \times n)$ matrix A, you can restate (4.12) in terms of matrix multiplication by thinking of **x** and **b** as *column* vectors, that is, as $(n \times 1)$ matrices. The result is the following more concise description of the problem of solving n linear equations.

The Problem of Solving n Linear Equations in n Unknowns (Matrix-Vector Notation)

Given a positive integer n, values for the elements of an $(n \times n)$ matrix A, and an n-vector $\mathbf{b} = (b_1, \ldots, b_n)^t$, find values for the n-vector of variables $\mathbf{x} = (x_1, \ldots, x_n)^t$ so that

$$A\mathbf{x} = \mathbf{b}. \tag{4.13}$$

The remainder of this section is devoted to developing a systematic approach for solving (4.13). In that regard, recall from Chapter 1 that you want to use the *known* data to obtain the *unknown* solution. In this case, the data are the matrix A and the n-vector **b**. The unknown information is the values of the components of the n-vector **x**.

4.3.2 The Basic Approach to Solving Linear Equations

Recall from Section 1.1.1 that you would ideally like to obtain a *closed-form* solution to (4.13), that is, a solution consisting of a simple expression or formula in terms of the data. Unfortunately, no such solution is known in general. However, one *special case* of (4.13) for which a closed-form solution exists is the problem in (4.10) of solving one linear equation in one unknown, whose solution is as follows.

Solution to One Linear Equation in One Unknown

$$x = \frac{b}{a} \text{ (provided that } a \neq 0\text{).} \tag{4.14}$$

There are several ways to generalize the result in (4.14) so that you can solve (4.13). For instance, from your previous experience with solving *two* linear equations, recall the following approach. Multiply the first equation in (4.11) by d, the second equation by b, and then subtract the second equation from the first to obtain the following:

$$(ad - bc)x = de - bf. \tag{4.15}$$

You can now obtain a value for x because (4.15) consists of one equation in one unknown, which you already know how to solve. The result, in this case, is

$$x = \frac{de - bf}{ad - bc} \text{ (assuming that } ad - bc \neq 0\text{).} \tag{4.16}$$

A similar process results in the value for y, which is given in the following solution for (4.11).

Solution to Two Linear Equations in Two Unknowns

Assuming that $ad - bc \neq 0$, the solution to (4.11) is as follows:

$$x = \frac{de - bf}{ad - bc} \text{ and } y = \frac{af - ce}{ad - bc}.$$

It is possible to generalize this approach in a systematic way so that you can solve n linear equations in n unknowns. Here, however, the objective is to show how matrices and vectors are used in developing a solution procedure for this problem. So consider again the single linear equation

$$ax = b,$$

whose solution is

$$x = \frac{b}{a} \text{ (assuming that } a \neq 0). \tag{4.17}$$

Another approach to generalizing (4.17) so that you can solve the problem in (4.13) is to replace, in (4.17), the real number x with the n-vector $\mathbf{x}$, the real number a with the matrix A, and the real number b with the n-vector $\mathbf{b}$, to obtain

$$\mathbf{x} = \frac{\mathbf{b}}{A}.$$

The trouble is that this expression contains a syntax error because you cannot divide the n-vector $\mathbf{b}$ by the matrix A. Another approach is needed.

There are several alternatives, one of which is to rewrite (4.17) so that generalization *is* possible. Doing so requires eliminating the operation of division. For example, you can rewrite (4.17) as follows:

$$x = a^{-1}b. \tag{4.18}$$

In (4.18), the operation of *multiplying* a^{-1} by b appears instead of division. Now, generalizing (4.18) to matrices and vectors results in the following:

$$\mathbf{x} = A^{-1}\mathbf{b}. \tag{4.19}$$

If you assume that A^{-1} is an $(n \times n)$ matrix, then no syntax error arises because you *can* perform the matrix multiplication in (4.19). The remaining question is, What, precisely, is A^{-1}? The answer is provided in what follows.

4.3.3 The Inverse of a Matrix

To determine what A^{-1} in (4.19) is, consider the corresponding value of a^{-1} in (4.18). In (4.18), a^{-1} is a number that has certain properties in relation to a. In particular,

$$aa^{-1} = a^{-1}a = 1. \tag{4.20}$$

Analogously, in (4.19), A^{-1} is an $(n \times n)$ matrix that should have certain properties in relation to A. To determine what those properties are, try to generalize (4.20). If you replace the symbol a everywhere with A, you obtain

$$AA^{-1} = A^{-1}A = 1,$$

which results in a syntax error because it is not possible to use the last equality sign to compare the *matrix* $A^{-1}A$ with the *number* 1.

One way to resolve this error is to replace the number 1 with an appropriate $(n \times n)$ matrix. To determine what that matrix is, observe that the number 1 has the property that when you multiply 1 by any other number, the result is that number. Correspondingly, you need a matrix which, when multiplied with any other matrix, results in that other matrix. From Proposition 4.5(g) in Section 4.2.5, that desired matrix is the $(n \times n)$ identity matrix, I. Thus, the generalization of (4.20) requires you to find a matrix A^{-1} with the property that

$$AA^{-1} = A^{-1}A = I.$$

This type of matrix is formalized in the following definition.

Definition 4.9 *Supppose that $A \in \mathbb{R}^{n \times n}$. A matrix $A^{-1} \in \mathbb{R}^{n \times n}$ is an* **inverse** *of A if and only if*

$$AA^{-1} = A^{-1}A = I.$$

The inverse of an $(n \times n)$ matrix A is important because, if you *can* find A^{-1}, then you can solve the system of linear equations in (4.13). That solution is

$$\mathbf{x} = A^{-1}\mathbf{b}, \tag{4.21}$$

because then,

$$A\mathbf{x} = A(A^{-1}\mathbf{b}) = (AA^{-1})\mathbf{b} = I\mathbf{b} = \mathbf{b}.$$

Example 4.15 – Using the Inverse to Solve a System of Linear Equations

Consider the following system of linear equations:

$$\begin{aligned} 2x_1 - x_2 &= 5 \\ x_1 + x_2 &= 4 \end{aligned}$$

in which

Continued

$$A = \begin{bmatrix} 2 & -1 \\ 1 & 1 \end{bmatrix}, \quad \mathbf{x} = \begin{bmatrix} x_1 \\ x_2 \end{bmatrix}, \text{ and } \mathbf{b} = \begin{bmatrix} 5 \\ 4 \end{bmatrix}.$$

The matrix

$$A^{-1} = \begin{bmatrix} \frac{1}{3} & \frac{1}{3} \\ -\frac{1}{3} & \frac{2}{3} \end{bmatrix}$$

is the inverse of A because

$$AA^{-1} = \begin{bmatrix} 2 & -1 \\ 1 & 1 \end{bmatrix} \begin{bmatrix} \frac{1}{3} & \frac{1}{3} \\ -\frac{1}{3} & \frac{2}{3} \end{bmatrix} = \begin{bmatrix} 1 & 0 \\ 0 & 1 \end{bmatrix} = I$$

and

$$A^{-1}A = \begin{bmatrix} \frac{1}{3} & \frac{1}{3} \\ -\frac{1}{3} & \frac{2}{3} \end{bmatrix} \begin{bmatrix} 2 & -1 \\ 1 & 1 \end{bmatrix} = \begin{bmatrix} 1 & 0 \\ 0 & 1 \end{bmatrix} = I.$$

As a result, from (4.21), the solution to $A\mathbf{x} = \mathbf{b}$ is

$$\mathbf{x} = A^{-1}\mathbf{b} = \begin{bmatrix} \frac{1}{3} & \frac{1}{3} \\ -\frac{1}{3} & \frac{2}{3} \end{bmatrix} \begin{bmatrix} 5 \\ 4 \end{bmatrix} = \begin{bmatrix} 3 \\ 1 \end{bmatrix} = \begin{bmatrix} x_1 \\ x_2 \end{bmatrix}.$$

That is, $x_1 = 3$ and $x_2 = 1$ solves the original system of equations.

The issue now is how to find the inverse of a matrix A because, from (4.21), once you have A^{-1}, the solution to the system of linear equations in (4.13) is

$$\mathbf{x} = A^{-1}\mathbf{b}.$$

4.3.4 Closed-Form Inverses of Certain Matrices

You would ideally like to find a closed-form expression for A^{-1} in terms of the elements of A. No such expression is known for a general $(n \times n)$ matrix, so a numerical method is used to compute A^{-1}. However, you *can* find A^{-1} in closed form for certain special matrices, as you will now see.

The Inverse of a (2×2) Matrix

For the (2×2) matrix

$$A = \begin{bmatrix} a & b \\ c & d \end{bmatrix}$$

in which $ad - bc \neq 0$,

$$A^{-1} = \frac{1}{ad-bc}\begin{bmatrix} d & -b \\ -c & a \end{bmatrix} = \begin{bmatrix} \frac{d}{ad-bc} & \frac{-b}{ad-bc} \\ \frac{-c}{ad-bc} & \frac{a}{ad-bc} \end{bmatrix}$$

because

$$\begin{aligned} AA^{-1} &= \frac{1}{ad-bc}\begin{bmatrix} a & b \\ c & d \end{bmatrix}\begin{bmatrix} d & -b \\ -c & a \end{bmatrix} \\ &= \frac{1}{ad-bc}\begin{bmatrix} ad-bc & 0 \\ 0 & ad-bc \end{bmatrix} \\ &= \begin{bmatrix} 1 & 0 \\ 0 & 1 \end{bmatrix} \\ &= I \end{aligned}$$

and

$$\begin{aligned} A^{-1}A &= \frac{1}{ad-bc}\begin{bmatrix} d & -b \\ -c & a \end{bmatrix}\begin{bmatrix} a & b \\ c & d \end{bmatrix} \\ &= \frac{1}{ad-bc}\begin{bmatrix} ad-bc & 0 \\ 0 & ad-bc \end{bmatrix} \\ &= \begin{bmatrix} 1 & 0 \\ 0 & 1 \end{bmatrix} \\ &= I. \end{aligned}$$

As a result of having found the inverse of a (2×2) matrix, you can apply (4.21) to obtain the following solution to a (2×2) system of linear equations.

Solution to Two Linear Equations in Two Unknowns

By using (4.21), the solution to (4.11), assuming that $ad - bc \neq 0$, is

$$\begin{aligned} \begin{bmatrix} x \\ y \end{bmatrix} &= A^{-1}\begin{bmatrix} e \\ f \end{bmatrix} \\ &= \frac{1}{ad-bc}\begin{bmatrix} d & -b \\ -c & a \end{bmatrix}\begin{bmatrix} e \\ f \end{bmatrix} \\ &= \begin{bmatrix} \frac{de-bf}{ad-bc} \\ \frac{af-ce}{ad-bc} \end{bmatrix}. \end{aligned}$$

The Inverse of a Diagonal Matrix

Recall from Section 4.2.2 that a diagonal matrix is a square matrix that has the following form:

$$A = \begin{bmatrix} A_{11} & 0 & \dots & 0 \\ 0 & A_{22} & \dots & 0 \\ \vdots & \vdots & \ddots & \vdots \\ 0 & 0 & \dots & A_{nn} \end{bmatrix}.$$

In the event that none of the diagonal elements $A_{11}, A_{22}, \dots, A_{nn}$ is zero,

$$A^{-1} = \begin{bmatrix} \frac{1}{A_{11}} & 0 & \dots & 0 \\ 0 & \frac{1}{A_{22}} & \dots & 0 \\ \vdots & \vdots & \ddots & \vdots \\ 0 & 0 & \dots & \frac{1}{A_{nn}} \end{bmatrix}$$

because

$$AA^{-1} = \begin{bmatrix} A_{11} & 0 & \dots & 0 \\ 0 & A_{22} & \dots & 0 \\ \vdots & \vdots & \ddots & \vdots \\ 0 & 0 & \dots & A_{nn} \end{bmatrix} \begin{bmatrix} \frac{1}{A_{11}} & 0 & \dots & 0 \\ 0 & \frac{1}{A_{22}} & \dots & 0 \\ \vdots & \vdots & \ddots & \vdots \\ 0 & 0 & \dots & \frac{1}{A_{nn}} \end{bmatrix}$$

$$= \begin{bmatrix} A_{11}\frac{1}{A_{11}} & 0 & \dots & 0 \\ 0 & A_{22}\frac{1}{A_{22}} & \dots & 0 \\ \vdots & \vdots & \ddots & \vdots \\ 0 & 0 & \dots & A_{nn}\frac{1}{A_{nn}} \end{bmatrix}$$

$$= \begin{bmatrix} 1 & 0 & \dots & 0 \\ 0 & 1 & \dots & 0 \\ \vdots & \vdots & \ddots & 0 \\ 0 & 0 & \dots & 1 \end{bmatrix}.$$

A similar computation shows that $A^{-1}A = I$.

Because the $(n \times n)$ identity matrix I is a special case of a diagonal matrix in which the diagonal elements are each 1, you now know by the foregoing result that $I^{-1} = I$.

The Inverse of a Permutation Matrix

Recall from Section 4.2.2 that a permutation matrix is a square matrix in which each row and column has exactly one 1 and all other elements are 0. As you are asked to verify in the exercises, the inverse of a permutation matrix P is the transpose of P, that is,

$$P^{-1} = P^t.$$

Example 4.16 – The Inverse of a Permutation Matrix

If $P = \begin{bmatrix} 0 & 0 & 1 \\ 1 & 0 & 0 \\ 0 & 1 & 0 \end{bmatrix}$, then $P^{-1} = P^t = \begin{bmatrix} 0 & 1 & 0 \\ 0 & 0 & 1 \\ 1 & 0 & 0 \end{bmatrix}$.

The Inverse of an Eta Matrix

Recall from Section 4.2.2 that an eta matrix, E, is the identity matrix except possibly for one column, say, column k. Thus, E has the following form:

$$E = \begin{bmatrix} 1 & 0 & \ldots & E_{1k} & \ldots & 0 \\ 0 & 1 & \ldots & E_{2k} & \ldots & 0 \\ \vdots & \vdots & \ddots & \vdots & \vdots & \vdots \\ 0 & 0 & \ldots & E_{kk} & \ldots & 0 \\ \vdots & \vdots & \vdots & \vdots & \ddots & \vdots \\ 0 & 0 & \ldots & E_{nk} & \ldots & 1 \end{bmatrix}.$$

As you are asked to show in the exercises, if the diagonal element $E_{kk} \neq 0$, then the inverse of E, namely, E^{-1}, is also an eta matrix in which column k differs from that of the identity matrix. Specifically,

$$E^{-1} = \begin{bmatrix} 1 & 0 & \ldots & -\dfrac{E_{1k}}{E_{kk}} & \ldots & 0 \\ 0 & 1 & \ldots & -\dfrac{E_{2k}}{E_{kk}} & \ldots & 0 \\ \vdots & \vdots & \ddots & \vdots & \vdots & \vdots \\ 0 & 0 & \ldots & \dfrac{1}{E_{kk}} & \ldots & 0 \\ \vdots & \vdots & \vdots & \vdots & \ddots & \vdots \\ 0 & 0 & \ldots & -\dfrac{E_{nk}}{E_{kk}} & \ldots & 1 \end{bmatrix}.$$

Example 4.17 – The Inverse of an Eta Matrix

Observe that column $k = 3$ of the eta matrix

$$E = \begin{bmatrix} 1 & 0 & 4 \\ 0 & 1 & -6 \\ 0 & 0 & 2 \end{bmatrix}$$

differs from column 3 of the identity matrix. Because the diagonal element $E_{kk} = E_{33} = 2$ is *not* 0, you can compute E^{-1} as the following eta matrix:

$$E^{-1} = \begin{bmatrix} 1 & 0 & -\frac{4}{2} \\ 0 & 1 & -\frac{-6}{2} \\ 0 & 0 & \frac{1}{2} \end{bmatrix} = \begin{bmatrix} 1 & 0 & -2 \\ 0 & 1 & 3 \\ 0 & 0 & \frac{1}{2} \end{bmatrix}.$$

4.3.5 A Numerical Method for Finding the Inverse of an $(n \times n)$ Matrix: Gaussian Elimination

You have just seen how to find closed-form expressions for the inverses of certain matrices. For a general $(n \times n)$ matrix A, a numerical method for finding the inverse is used. According to Definition 4.9 in Section 4.3.3, the method must produce an $(n \times n)$ matrix A^{-1} for which

$$A^{-1}A = I \text{ and } AA^{-1} = I.$$

To begin with, an $(n \times n)$ matrix A^{-1} that satisfies only the following property is produced:

$$A^{-1}A = I. \tag{4.22}$$

The basic idea behind the method, called **Gaussian elimination**, is to find n matrices, $E^1, \ldots, E^n$, for which

$$(E^n E^{n-1} \cdots E^1)A = I. \tag{4.23}$$

Comparing (4.23) with (4.22), you might guess that

$$A^{-1} = E^n E^{n-1} \cdots E^1.$$

To be more specific, the method starts by finding an $(n \times n)$ *eta* matrix, E^1, such that the first column of $E^1 A$ is the same as the first column of I (the remaining columns of $E^1 A$ being irrelevant). Having found E^1, the next step is to find an eta matrix E^2 so that the first *two* columns of $E^2(E^1 A)$ are the same as the first two columns of I. Proceeding systematically in this manner, when the eta matrix E^n is computed, you will have that all n columns of $E^n \cdots E^1 A$ are the same as the n columns of I, so (4.23) is satisfied.

The next question is how to find these n eta matrices. Equivalently stated, suppose you have already found $k-1$ eta matrices, $E^1, \ldots, E^{k-1}$ so that the first $k-1$ columns of the matrix $C = E^{k-1} \ldots E^1 A$ are the same as the first $k-1$ columns of I, that is,

$$E^{k-1} \cdots E^1 A = C = \begin{bmatrix} 1 & 0 & \ldots & 0 & C_{1k} & \ldots & C_{1n} \\ 0 & 1 & \ldots & 0 & C_{2k} & \ldots & C_{2n} \\ \vdots & \vdots & \ddots & \vdots & \vdots & \ddots & \vdots \\ 0 & 0 & \ldots & 1 & C_{k-1,k} & \ldots & C_{k-1,n} \\ 0 & 0 & \ldots & 0 & C_{kk} & \ldots & C_{kn} \\ \vdots & \vdots & \vdots & \vdots & \vdots & \ddots & \vdots \\ 0 & 0 & \ldots & 0 & C_{nk} & \ldots & C_{nn} \end{bmatrix}.$$

You can find the next eta matrix, E^k, *provided that* $C_{kk} \neq 0$. In particular, E^k is the identity matrix except in column k, and that column is computed by the following formula (which you are asked to verify in the exercises):

$$E_{ik}^k = \begin{cases} -\dfrac{C_{ik}}{C_{kk}}, & \text{if } i \neq k \\ \\ \dfrac{1}{C_{kk}}, & \text{if } i = k. \end{cases} \tag{4.24}$$

Example 4.18 – An Example of Finding an Eta Matrix in Gaussian Elimination

Suppose you want to find the inverse of the following (3×3) matrix:

$$A = \begin{bmatrix} 2 & -2 & 0 \\ 4 & -4 & 1 \\ -6 & 5 & -3 \end{bmatrix}.$$

Accordingly, you must find $n = 3$ eta matrices, E^1, E^2, E^3 so that $E^3 E^2 E^1 A = I$. To find E^1 – which is I except in column 1 – set $k = 1$, and start with $C = A$. Noting that $C_{kk} = C_{11} = A_{11} = 2 \neq 0$, you can use (4.24) to find column $k = 1$ of E^1, so

$$E^1 = \begin{bmatrix} \frac{1}{2} & 0 & 0 \\ \\ -\frac{4}{2} & 1 & 0 \\ \\ -\frac{-6}{2} & 0 & 1 \end{bmatrix} = \begin{bmatrix} \frac{1}{2} & 0 & 0 \\ -2 & 1 & 0 \\ 3 & 0 & 1 \end{bmatrix}.$$

Continued

You can verify that

$$E^1 A = \begin{bmatrix} 1 & -1 & 0 \\ 0 & 0 & 1 \\ 0 & -1 & 3 \end{bmatrix},$$

from which you can see that the first column of $E^1 A$ is the same as the first column of I.

The next step in Example 4.18 is to set $k = 2$ and find the eta matrix E^2 so that the first two columns of $E^2(E^1 A)$ are the same as the first two columns of I. Letting $C = E^1 A$ (which is calculated at the end of Example 4.18), note that $C_{kk} = C_{22} = 0$ and so you *cannot* use the formula in (4.24) to find E^2.

The approach to overcoming this obstacle is to interchange row $k = 2$ of C with some other row r of C in such a way that, *after* the interchange, the new value of $C_{kk} \neq 0$. However, when interchanging rows k and r, you do *not* want to alter the fact that the first $k - 1$ columns of C are the same as the first $k - 1$ columns of I. You will preserve this desirable property provided that the value of r (the row to be interchanged with row k) is *greater than* k. In other words, if $C_{kk} = 0$, then look in column k of C for a row $r > k$ with $C_{rk} \neq 0$. If you find such a row r, then interchange row k with row r of C and proceed as before.

As you saw in Example 4.12 in Section 4.2.4, one way to interchange row k with row r of C is to multiply C on the *left* by a permutation matrix in which you interchange rows k and r of I. This process is illustrated for the matrix $C = E^1 A$ in Example 4.18.

Example 4.19 – Interchanging Two Rows In Gaussian Elimination

From the final matrix in Example 4.18,

$$C = E^1 A = \begin{bmatrix} 1 & -1 & 0 \\ 0 & 0 & 1 \\ 0 & -1 & 3 \end{bmatrix}.$$

Now $k = 2$ and because $C_{kk} = C_{22} = 0$, you must try to interchange row $k = 2$ with some row $r > 2$. In this case, the only possibility is to use row $r = 3$. Because $C_{32} = -1 \neq 0$, you *can* interchange rows 2 and 3 of C to obtain the following new matrix C. This is accomplished by multiplying $E^1 A$ on the left by the permutation matrix – say, P^2 – in which rows 2 and 3 of the identity matrix are interchanged:

Continued

$$C = P^2E^1A = \begin{bmatrix} 1 & 0 & 0 \\ 0 & 0 & 1 \\ 0 & 1 & 0 \end{bmatrix} \begin{bmatrix} 1 & -1 & 0 \\ 0 & 0 & 1 \\ 0 & -1 & 3 \end{bmatrix} = \begin{bmatrix} 1 & -1 & 0 \\ 0 & -1 & 3 \\ 0 & 0 & 1 \end{bmatrix}.$$

Now you can see that $C_{kk} = C_{22} = -1 \neq 0$, so you can compute the eta matrix E^2. Specifically, by using (4.24) to find column $k = 2$ of E^2, you obtain

$$E^2 = \begin{bmatrix} 1 & -\frac{-1}{-1} & 0 \\ 0 & \frac{1}{-1} & 0 \\ 0 & -\frac{0}{-1} & 1 \end{bmatrix} = \begin{bmatrix} 1 & -1 & 0 \\ 0 & -1 & 0 \\ 0 & 0 & 1 \end{bmatrix}.$$

You can verify that

$$E^2P^2E^1A = \begin{bmatrix} 1 & 0 & 3 \\ 0 & 1 & 3 \\ 0 & 0 & 1 \end{bmatrix},$$

from which you can see that the first *two* columns of $E^2P^2E^1A$ are the same as the first two columns of I.

The final step in Example 4.19 is to set $k = 3$ and find the last eta matrix, E^3, so that $E^3(P^2E^2E^1A)$ is I. The computations for E^3 are shown in the next example.

Example 4.20 – Finding the Final Eta Matrix in Example 4.18

To find E^3 – which is I, except in column 3 – set $k = 3$ and compute $C = P^2E^2E^1A$ which, as given at the end of Example 4.19, is

$$C = \begin{bmatrix} 1 & 0 & 3 \\ 0 & 1 & 3 \\ 0 & 0 & 1 \end{bmatrix}.$$

Noting that $C_{33} = 1 \neq 0$, you can use (4.24) to find column $k = 3$ of E^3:

$$E^3 = \begin{bmatrix} 1 & 0 & -\frac{3}{1} \\ 0 & 1 & -\frac{3}{1} \\ 1 & 0 & \frac{1}{1} \end{bmatrix} = \begin{bmatrix} 1 & 0 & -3 \\ 0 & 1 & -3 \\ 0 & 0 & 1 \end{bmatrix}.$$

You can verify that

$$E^3P^2E^2E^1A = I,$$

from which you can see that all three columns of $E^3P^2E^2E^1A$ are those of I.

By putting together the pieces from Examples 4.18, 4.19, and 4.20, the inverse of

$$A = \begin{bmatrix} 2 & -2 & 0 \\ 4 & -4 & 1 \\ -6 & 5 & -3 \end{bmatrix}$$

is

$$\begin{aligned} A^{-1} &= E^3 E^2 P^2 E^1 \\ &= \begin{bmatrix} 1 & 0 & -3 \\ 0 & 1 & -3 \\ 0 & 0 & 1 \end{bmatrix} \begin{bmatrix} 1 & -1 & 0 \\ 0 & -1 & 0 \\ 0 & 0 & 1 \end{bmatrix} \begin{bmatrix} 1 & 0 & 0 \\ 0 & 0 & 1 \\ 0 & 1 & 0 \end{bmatrix} \begin{bmatrix} \frac{1}{2} & 0 & 0 \\ -2 & 1 & 0 \\ 3 & 0 & 1 \end{bmatrix} \\ &= \begin{bmatrix} \frac{7}{2} & -3 & -1 \\ 3 & -3 & -1 \\ -2 & 1 & 0 \end{bmatrix}, \end{aligned}$$

for which you can verify that $AA^{-1} = A^{-1}A = I$.

A Summary of Gaussian Elimination

To find the inverse of an $(n \times n)$ matrix A using Gaussian elimination, try to find n eta matrices, $E^1, \ldots, E^n$, for which

$$(E^n \cdots E^1)A = I,$$

because then,

$$A^{-1} = E^n \cdots E^1.$$

Each new eta matrix, E^k, is chosen so that, when multiplied with all preceding eta matrices and A, the result is a matrix whose first k columns are the same as the first k columns of I.

In some cases, you cannot find E^k because of a division by 0. In this event, it is necessary to use a permutation matrix, P^k, to interchange two rows of the current matrix prior to finding E^k. Thus, to find the inverse of an $(n \times n)$ matrix A using Gaussian elimination, you try to find n eta matrices, $E^1, \ldots, E^n$, and n permutation matrices (some of which can be I) such that

$$(E^n P^n \cdots E^1 P^1)A = I,$$

because then,

$$A^{-1} = E^n P^n \cdots E^1 P^1.$$

Specifically, suppose you have found the eta matrices $E^1, \ldots, E^{k-1}$ and the permutation matrices $P^1, \ldots, P^{k-1}$ for which the first $k-1$ columns of the matrix

$$C = E^{k-1} P^{k-1} \cdots E^1 P^1 A$$

are the same as the first $k - 1$ columns of I. In the event that $C_{kk} \neq 0$, you can set $P^k = I$, otherwise, you will need to find a row $r > k$ to interchange with row k of C for which $C_{rk} \neq 0$. If such a row r is not found, then the algorithm stops, otherwise, create the permutation matrix P^k by interchanging rows k and r of I.

After finding P^k, you can compute

$$C = P^k(E^{k-1}P^{k-1} \cdots E^1P^1A)$$

in which $C_{kk} \neq 0$ and then create the eta matrix E^k in which column k is defined as follows:

$$E^k_{ik} = \begin{cases} -\dfrac{C_{ik}}{C_{kk}}, \text{ if } i \neq k \\ \\ \dfrac{1}{C_{kk}}, \text{ if } i = k. \end{cases}$$

Of course, all other columns of E^k are those of the identity matrix.

In this section, you have seen how to solve a system of n linear equations in n unknowns by using the inverse of a matrix. In some cases, you can find the inverse in closed form. In general, however, Gaussian elimination is a numerical method for attempting to find the inverse.

From the examples you have seen so far, you should realize that an $(n \times n)$ matrix A might not *have* an inverse. For example, the (2×2) matrix

$$\begin{bmatrix} a & b \\ c & d \end{bmatrix}$$

has no inverse when $ad - bc = 0$. As another example, in performing Gaussian elimination, suppose that you cannot find the next eta matrix E^k because $C_{kk} = 0$. In this case, you attempt to find a row $r > k$ with $C_{rk} \neq 0$ to interchange with row k of C – but, what if there is no such row r? The algorithm stops and this method fails to produce the inverse. Although you might hope that some *other* method can produce the inverse, this is not possible, as you will learn in Section 4.4. This and other theoretical results related to solving linear equations are presented in Section 4.4.

4.4 The Theory of Solving Linear Equations

In Section 4.3, you learned how to use the inverse of a matrix to solve a system of linear equations. One important observation from the specific examples presented there is that, *it is not always possible to solve the equations.* Even the simplest such system:

$$ax = b,$$

has no solution when $a = 0$ and $b \neq 0$. In this section, you will learn more about *when* you can solve a system of linear equations. Throughout this section, all matrices are $(n \times n)$, I is the $(n \times n)$ identity matrix, and 0 is the $(n \times n)$ zero matrix. You should also review the properties of matrix and vector multiplication presented in Propositions 4.1 through 4.5.

4.4.1 Properties of the Inverse

Recall Definition 4.9 in Section 4.3.3 for the inverse of an $(n \times n)$ matrix A, which is an $(n \times n)$ matrix A^{-1} such that

$$A^{-1}A = AA^{-1} = I.$$

As just discussed, some matrices have inverses and others do not, which motivates the following definition.

Definition 4.10 *An $(n \times n)$ matrix A is* **nonsingular** *(or* **invertible***) if and only if there is an $(n \times n)$ matrix B such that*

$$AB = BA = I.$$

An $(n \times n)$ matrix A is **singular** *if A is not nonsingular.*

The next proposition shows that, for a nonsingular matrix A, the matrix B in Definition 4.10 is unique.

Proposition 4.6 *If the $(n \times n)$ matrix A is nonsingular, then there is a unique matrix B such that $AB = BA = I$.*

Proof. Because A is nonsingular, from Definition 4.10, *there is* an $(n \times n)$ matrix B such that

$$AB = BA = I. \tag{4.25}$$

To show that B is unique, use the direct uniqueness method (see Section 1.6.12) to assume that there is also an $(n \times n)$ matrix C such that

$$AC = CA = I. \tag{4.26}$$

It must be shown that $B = C$, or equivalently, that $B - C = 0$.

Subtracting (4.26) from (4.25) results in the following:

$$AB - AC = I - I,$$

or, equivalently,

$$A(B - C) = 0. \tag{4.27}$$

On multiplying both sides of (4.27) on the *left* by B and using (4.25) together with the properties of matrix multiplication in Proposition 4.5, you obtain the following:

$$B[A(B - C)] = (BA)(B - C) = I(B - C) = B - C$$

and

$$B[A(B - C)] = B0 = 0.$$

In other words, $B - C = 0$, so $B = C$, thus showing that B is unique and completing the proof. QED

As a result of Proposition 4.6, for a nonsingular matrix A, there is one and only one matrix B for which $AB = BA = I$. Observe that the matrix B is in fact the inverse of A. In other words, for a nonsingular matrix A, there is a unique inverse matrix, A^{-1}. The following proposition presents some additional properties of the inverse.

Proposition 4.7 *If A and B are nonsingular $(n \times n)$ matrices, then:*

(a) A^{-1} is nonsingular and $(A^{-1})^{-1} = A$.

(b) AB is nonsingular and $(AB)^{-1} = B^{-1}A^{-1}$.

Proof. The hypothesis that A and B are nonsingular matrices means, by definition, that there are matrices A^{-1} and B^{-1} such that

$$AA^{-1} = A^{-1}A = I \text{ and } BB^{-1} = B^{-1}B = I.$$

These inverses are used in the subsequent proofs.

(a) To show that A^{-1} is nonsingular, it is necessary to create the inverse of A^{-1}, which is a matrix C such that $A^{-1}C = CA^{-1} = I$. However, $C = A$ is that unique matrix because $A^{-1}A = AA^{-1} = I$, by the fact that A^{-1} is the inverse of A.

(b) To show that AB is nonsingular, it is necessary to create an inverse of AB, which is a matrix C such that $(AB)C = C(AB) = I$. However, $C = B^{-1}A^{-1}$ is that unique matrix because

$$\begin{aligned}(AB)C &= (AB)(B^{-1}A^{-1}) \\ &= [A(BB^{-1})]A^{-1} \\ &= [A(I)]A^{-1} \\ &= AA^{-1} \\ &= I.\end{aligned}$$

A similar computation shows that $C(AB) = I$, so, $C = B^{-1}A^{-1}$.

This completes the proof. QED

Recall from Chapter 1 that one of the uses of mathematics is to make certain operations and computations more efficient. For example, if you want to show that a particular matrix B is the inverse of a matrix A, according to the definition, you must prove two statements: $AB = I$ and $BA = I$. The next proposition provides a condition under which you can reach the same conclusion by having to prove only that $BA = I$, which is half the work.

Proposition 4.8 *Suppose that A is an $(n \times n)$ matrix. If B is a nonsingular $(n \times n)$ matrix for which $BA = I$, then $AB = I$ (so, in fact, $B = A^{-1}$).*

Proof. Because B is nonsingular, there is an $(n \times n)$ matrix B^{-1} such that $BB^{-1} = B^{-1}B = I$. Multiplying both sides of $BA = I$ in the hypothesis on the left by B^{-1}, it follows that

$$B^{-1}(BA) = (B^{-1}B)A = IA = A$$

and

$$B^{-1}(BA) = B^{-1}I = B^{-1}.$$

In other words, $A = B^{-1}$. Now by Proposition 4.7(a), $A = B^{-1}$ is nonsingular and

$$A^{-1} = (B^{-1})^{-1} = B.$$

So, $AB = AA^{-1} = I$ and the proof is complete. QED

4.4.2 Linearly Independent Vectors

As you learned in Section 4.2, you can think of an $(n \times n)$ matrix A as a collection of vectors in one of two ways: (1) as the rows of A, stacked on top of each other, or (2) as the columns of A, written next to each other. For the moment, consider A as a collection of its columns. The goal of this section is to show that if A is nonsingular, then the columns of A satisfy a certain property. Vice versa, if the columns of A satsify the certain property, then the matrix A is nonsingular. Moreover, this property provides the basis for methods such as Gaussian elimination that compute the inverse of a matrix and hence for solving linear equations. The objective now is to understand this property and to develop a symbolic definition.

Developing the Definition

To serve as a visual guide in this development, consider two vectors, $\mathbf{x} = (x_1, x_2)$ and $\mathbf{y} = (y_1, y_2)$, in two dimensions. Geometrically, the desirable property is that of "opening up properly." For example, the vectors $\mathbf{x}$ and $\mathbf{y}$ in Figure 4.11(a) and in Figure 4.11(b) open up properly. In contrast, the vectors $\mathbf{x}$ and $\mathbf{y}$ in Figure 4.12(a) and in Figure 4.12(b) do *not* open up properly. This is because the vectors in Figure 4.12(a) lie on top of each other and the vectors in Figure 4.12(b) point in exactly opposite directions. Translating this visual image to symbolic form is a challenging task described in what follows.

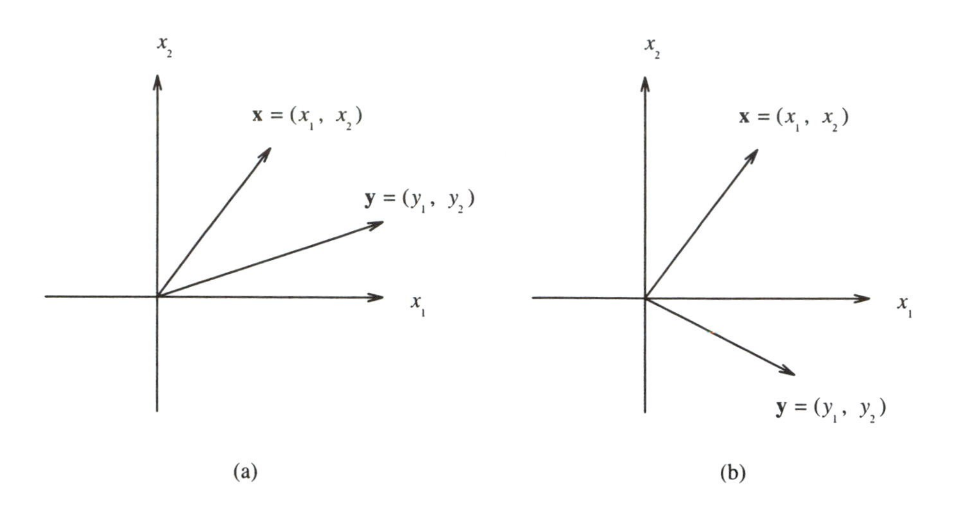

Figure 4.11: Examples of Vectors that Open Up Properly

Developing a Definition in Two Dimensions. One approach is to develop a definition for the property satisfied by the vectors in Figure 4.11. Alternatively, you can first develop a definition for the property of "not opening up properly," satisfied by the vectors in Figure 4.12, and then write the negation of that definition to capture the property of the vectors in Figure 4.11.

Following the latter approach, try to identify similarities between the pair of vectors in Figure 4.12(a) and the pair in Figure 4.12(b). For example, in Figure 4.12(a), $\mathbf{x}$ points in the same direction as $\mathbf{y}$ but has a different length. Recall from Section 4.1.3 that you can change the length of $\mathbf{y}$ (without affecting the direction) by multiplying $\mathbf{y}$ by a nonnegative real number, say, $t \geq 0$. Thus, by an appropriate choice of the real number $t \geq 0$, you can make the vectors $\mathbf{x}$ and $t\mathbf{y}$ the same. By using the quantifier *there is*, you can translate this observation regarding the vectors $\mathbf{x}$ and $\mathbf{y}$ in Figure 4.12(a) to the following symbolic form:

There is a real number $t \geq 0$ such that $\mathbf{x} = t\mathbf{y}$. (4.28)

Looking now at Figure 4.12(b), you will notice that $\mathbf{x}$ points in the *opposite* direction to $\mathbf{y}$. In this case, by multiplying $\mathbf{y}$ by an appropriate choice of the *nonpositive* real number $t \leq 0$, you can again make the vectors $\mathbf{x}$ and $t\mathbf{y}$ the same. By using the quantifier *there is*, this observation translates to the following symbolic form for the vectors $\mathbf{x}$ and $\mathbf{y}$ in Figure 4.12(b):

There is a real number $t \leq 0$ such that $\mathbf{x} = t\mathbf{y}$. (4.29)

Can you create a *single* statement that includes both (4.28) and (4.29) as special cases? One such unification is as follows:

There is a real number t such that $\mathbf{x} = t\mathbf{y}$. (4.30)

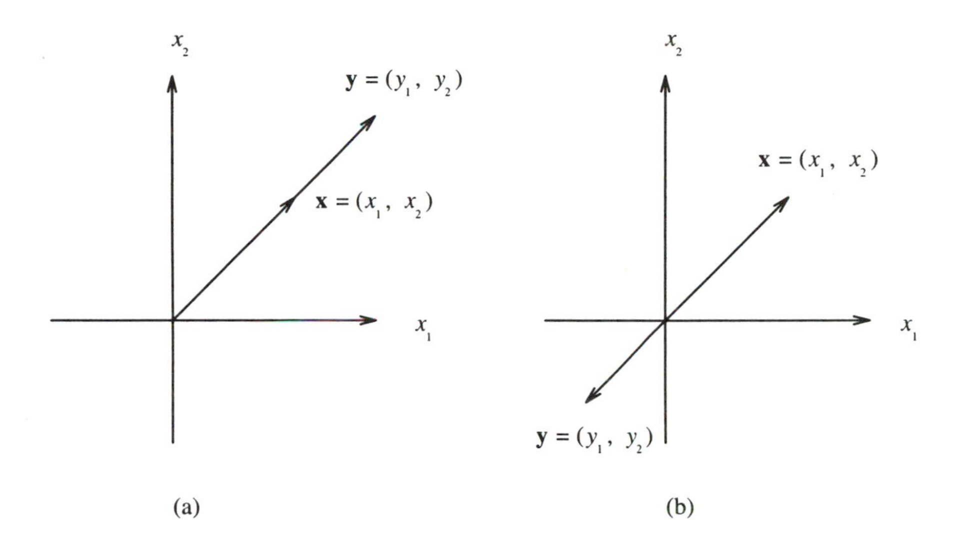

Figure 4.12: Examples of Vectors that Do Not Open Up Properly

Generalizing the Definition to Three Dimensions. The next step is to generalize (4.30) when the vectors lie in *three* dimensions. Indeed, (4.30) is still valid when $\mathbf{x}$ and $\mathbf{y}$ are three-dimensional vectors that lie on top of, or opposite to, each other. However, what if you are working with a *third* vector, $\mathbf{z}$, in three dimensions? Three such vectors that do not open up properly are shown in Figure 4.13. On the basis of Figure 4.13, you might write the following statement for this property:

There are real numbers t and u such that $\mathbf{x} = t\mathbf{y} + u\mathbf{z}$. (4.31)

Is the property in (4.31) correct for *all* groups of three vectors in three dimensions that do not open up properly? Only by extensive trial with other special cases of $\mathbf{x}$, $\mathbf{y}$, and $\mathbf{z}$ will you discover that (4.31) is *not* necessarily correct in all cases. For example, the vectors in Figure 4.14 do *not* satisfy (4.31) because $t\mathbf{y} + u\mathbf{z}$ always points in the same direction as $\mathbf{y}$ and $\mathbf{z}$. Rather, the vectors in Figure 4.14 satisfy the following property:

There are real numbers t and u such that $\mathbf{y} = t\mathbf{x} + u\mathbf{z}$. (4.32)

The vectors in Figure 4.14 also satisfy the property that

there are real numbers t and u such that $\mathbf{z} = t\mathbf{x} + u\mathbf{y}$. (4.33)

In fact, at least *one* of (4.31), (4.32), or (4.33) is *always true* for vectors $\mathbf{x}$, $\mathbf{y}$, and $\mathbf{z}$ in three dimensions that do not open up properly. The challenge is to write a *single* statement that covers all three of the special cases in (4.31), (4.32), and (4.33). Doing so requires cleverness by first putting all vectors on the same side of the equality sign and then realizing that there is a real number associated with *each* vector, that is,

There are real numbers a, b, and c such that $a\mathbf{x} + b\mathbf{y} + c\mathbf{z} = \mathbf{0}$. (4.34)

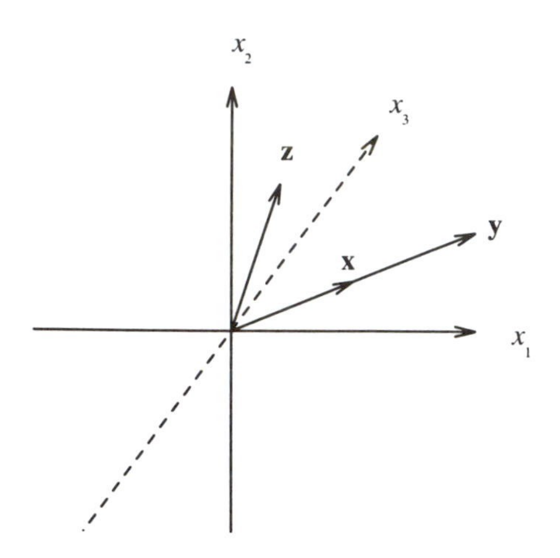

Figure 4.13: Three Vectors in $\mathbb{R}^3$ that Do Not Open Up Properly

However, the real numbers in the special cases are not just *any* real numbers – they have some special properties: in (4.31), $a = 1$; in (4.32), $b = 1$; and in (4.33), $c = 1$. One way to capture this fact is to require in (4.34) that *at least one* of the real numbers be 1. An alternative, but equivalent, way to say this (as you are asked to verify in the exercises) is the following:

$$\text{There are real numbers } a, b, c, \text{ not all } 0, \text{ with } a\mathbf{x} + b\mathbf{y} + c\mathbf{z} = \mathbf{0}. \quad (4.35)$$

Generalizing the Definition to n Dimensions. The next step is to generalize (4.35) to the case where the vectors lie in n dimensions. So suppose you have a group of k vectors in n dimensions that do not open up properly. Once again, subscript and superscript notation is helpful, so, let each of $\mathbf{x}^1, \ldots, \mathbf{x}^k$ be an n-vector. By introducing the term "linearly dependent" for the concept of "not opening up properly" and by generalizing (4.35) in the natural way, you obtain the following definition.

Definition 4.11 *The n-vectors $\mathbf{x}^1, \ldots, \mathbf{x}^k$ are* **linearly dependent** *if and only if there are real numbers $t_1, \ldots t_k$, not all zero, such that*

$$t_1\mathbf{x}^1 + \cdots + t_k\mathbf{x}^k = \mathbf{0}.$$

Alternatively, letting X be the $(n \times k)$ matrix in which column i of X is the n-vector $\mathbf{x}^i$ (that is, $X = [\mathbf{x}^1, \ldots, \mathbf{x}^k]$), the n-vectors $\mathbf{x}^1, \ldots, \mathbf{x}^k$ are **linearly dependent** *if and only if there is a k-vector $\mathbf{t} = (t_1, \ldots, t_k)$ with $\mathbf{t} \neq \mathbf{0}$ such that*

$$X\mathbf{t} = \mathbf{0}.$$

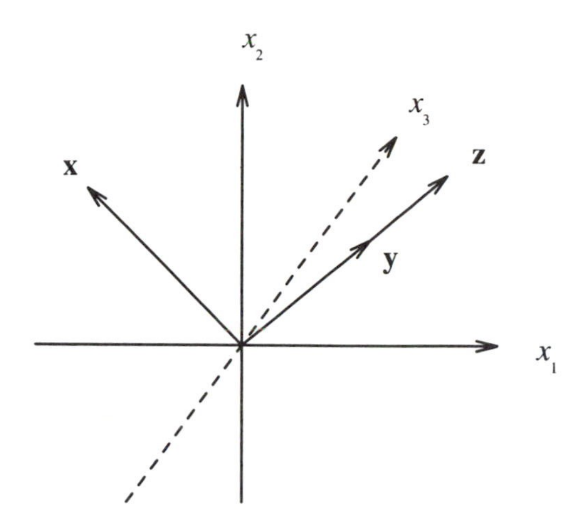

Figure 4.14: Three Other Vectors in $\mathbb{R}^3$ that Do Not Open Up Properly

As you learned in Section 2.2.2, only after carefully *validating* this property should you consider Definition 4.11 to be correct. That validation process, however, is not illustrated here.

Definition 4.11 applies to vectors in dimensions that you cannot visualize, which is one of the advantages of this symbolic definition. Another advantage is that you can state the negation of the definition. In particular, the negation of Definition 4.11 captures the concept of a group of n-vectors "opening up properly." By applying the rules for negating statements containing quantifiers to Definition 4.11, you obtain the following.

Definition 4.12 *The n-vectors $\mathbf{x}^1, \ldots, \mathbf{x}^k$ are* **linearly independent** *if and only if for all real numbers $t_1, \ldots, t_k$ with*

$$t_1\mathbf{x}^1 + \cdots + t_k\mathbf{x}^k = \mathbf{0},$$

it follows that

$$t_1 = \cdots = t_k = 0.$$

Alternatively, letting X be the $(n \times k)$ matrix in which column i of X is the n-vector $\mathbf{x}^i$ (that is, $X = [\mathbf{x}^1, \ldots, \mathbf{x}^k]$), the n-vectors $\mathbf{x}^1, \ldots, \mathbf{x}^k$ are **linearly independent** *if and only if for all k-vectors $\mathbf{t} = (t_1, \ldots, t_k)$ with*

$$X\mathbf{t} = \mathbf{0},$$

it follows that

$$\mathbf{t} = \mathbf{0}.$$

The vectors in Figure 4.11 are linearly independent. In contrast, the vectors in Figure 4.12 are linearly dependent.

Linear Independence and Solving Linear Equations

As mentioned previously in this section, there is a relation between the columns of an $(n \times n)$ matrix being linearly independent and the ability to solve a system of linear equations, which is presented in the next theorem.

Theorem 4.1 *The following statements are equivalent for an $(n \times n)$ matrix A:*

(a) A is nonsingular.

(b) For any n-vector $\mathbf{b}$, there is a unique n-vector $\mathbf{x}$ such that $A\mathbf{x} = \mathbf{b}$.

(c) The n columns of A are linearly independent vectors.

Proof. To prove the equivalence of these statements, it is shown that (a) imples (b), (b) implies (c), and (c) implies (a).

(a) implies (b)

Suppose that A is nonsingular. By definition, this means that there is an $(n \times n)$ matrix, A^{-1}, such that

$$AA^{-1} = A^{-1}A = I.$$

Now choose an n-vector $\mathbf{b}$, for which you must show that there is a unique n-vector $\mathbf{x}$ such that $A\mathbf{x} = \mathbf{b}$.

According to the uniqueness method, you must first produce an n-vector $\mathbf{x}$ for which $A\mathbf{x} = \mathbf{b}$. That n-vector is

$$\mathbf{x} = A^{-1}\mathbf{b}$$

because

$$A\mathbf{x} = A(A^{-1}\mathbf{b}) = (AA^{-1})\mathbf{b} = I\mathbf{b} = \mathbf{b}.$$

It remains to show that the value of $\mathbf{x} = A^{-1}\mathbf{b}$ is the *only* solution to

$$A\mathbf{x} = \mathbf{b}. \tag{4.36}$$

So, assume that $\mathbf{y}$ is also an n-vector that satisfies

$$A\mathbf{y} = \mathbf{b}. \tag{4.37}$$

You must show that $\mathbf{x} = \mathbf{y}$, or equivalently, that $\mathbf{x} - \mathbf{y} = \mathbf{0}$.

By subtracting (4.37) from (4.36), you obtain

$$A(\mathbf{x} - \mathbf{y}) = \mathbf{0}. \tag{4.38}$$

The result follows on multiplying (4.38) through on the left by A^{-1} because

$$A^{-1}[A(\mathbf{x}-\mathbf{y})] = (A^{-1}A)(\mathbf{x}-\mathbf{y}) = I(\mathbf{x}-\mathbf{y}) = \mathbf{x}-\mathbf{y}$$

and

$$A^{-1}[A(\mathbf{x}-\mathbf{y})] = A^{-1}\mathbf{0} = \mathbf{0}.$$

So, $\mathbf{x} = A^{-1}\mathbf{b}$ is the unique solution to the equation $A\mathbf{x} = \mathbf{b}$.

(b) implies (c)

Suppose that for any n-vector $\mathbf{b}$, there is a unique n-vector $\mathbf{x}$ such that $A\mathbf{x} = \mathbf{b}$. To show that the columns $A_{.1}, \ldots, A_{.n}$ are linearly independent, according to Definition 4.12, you can assume that $\mathbf{t} = (t_1, \ldots, t_n)$ is an n-vector such that

$$A\mathbf{t} = \mathbf{0}. \tag{4.39}$$

You must show that

$$\mathbf{t} = \mathbf{0}. \tag{4.40}$$

To reach the conclusion in (4.40), recall that, by the hypothesis, there is a unique solution to (4.39). You can see that the vector $\mathbf{0}$ is, in fact, a solution to (4.39). Thus, because $\mathbf{t}$ is also a solution to (4.39), by the uniqueness, it must be that $\mathbf{t} = \mathbf{0}$.

(c) implies (a)

Assume that the columns $A_{.1}, \ldots, A_{.n}$ are linearly independent vectors. You must show that the matrix A is nonsingular which, by definition, means that you must produce an $(n \times n)$ matrix, A^{-1}, such that

$$AA^{-1} = A^{-1}A = I.$$

Gaussian elimination, as described in Section 4.3, is used to do so. To see how, recall that Gaussian elimination attempts to create eta matrices $E^1, \ldots, E^n$ and permutation matrices $P^1, \ldots, P^n$ such that

$$(E^n P^n \cdots E^1 P^1)A = I. \tag{4.41}$$

The remainder of this proof involves showing that $A^{-1} = E^n P^n \cdots E^1 P^1$.

The assumption that the columns of A are linearly independent insures that Gaussian elimination *successfully* produces all the matrices in (4.41). This is proved by contradiction. Specifically, suppose that you are *not* able to use Gaussian elimination to find all of the matrices in (4.41). A contradiction is reached by showing that the columns of A are linearly *dependent*.

So suppose you have used Gaussian elimination to find $E^1, \ldots, E^{k-1}$ and $P^1, \ldots, P^{k-1}$, for which the first $k-1$ columns of the matrix

$$C = E^{k-1}P^{k-1} \cdots E^1 P^1 A \tag{4.42}$$

are the same as the first $k-1$ columns of I. What can prevent you from finding E^k?

The answer is that $C_{kk} = 0$, in which case, you try to find a row $r > k$ for which $C_{rk} \neq 0$. In the event that you cannot find such a row r, Gaussian elimination fails. In other words, if Gaussian elimination fails, it is because, for all rows $r \geq k, C_{rk} = 0$.

In this case, it is possible to show that the columns of C are linearly dependent which, by Definition 4.11, means that you must create real numbers $t_1, \ldots, t_n$, not all zero, for which

$$t_1 C_{.1} + \cdots + t_n C_{.n} = \mathbf{0}. \tag{4.43}$$

For each $i = 1, \ldots, n$, the desired value for the real number t_i is

$$t_i = \begin{cases} -C_{ik}, & \text{if } i < k \\ 1, & \text{if } i = k \\ 0, & \text{if } i > k. \end{cases} \tag{4.44}$$

Not all of the real numbers in (4.44) are 0 because $t_k = 1$. Also, these real numbers satisfy (4.43) because, from (4.44), from the fact that the first $k-1$ columns of C are the same as those of I, and from the fact that $C_{rk} = 0$ for $r \geq k$, it follows that

$$\begin{aligned} t_1 C_{.1} + \cdots + t_n C_{.n} &= \sum_{i=1}^{k-1} t_i C_{.i} + t_k C_{.k} + \sum_{i=k+1}^{n} t_i C_{.i} \\ &= \sum_{i=1}^{k-1} t_i I_{.i} + (1) C_{.k} + \sum_{i=k+1}^{n} (0) C_{.i} \\ &= \sum_{i=1}^{k-1} (-C_{ik}) I_{.i} + C_{.k} \\ &= \mathbf{0}. \end{aligned} \tag{4.45}$$

You now know that the columns of C are linearly dependent. The next step is to show that the values of $t_1, \ldots, t_n$ in (4.44) also make the columns of A linearly dependent. To see this, let $\mathbf{t}$ be the n-vector in which $\mathbf{t} = (t_1, \ldots, t_n)$. From (4.45),

$$C\mathbf{t} = \mathbf{0}. \tag{4.46}$$

Replacing C in (4.46) with the expression from (4.42), you have that

$$(E^{k-1} P^{k-1} E^{k-2} P^{k-2} \cdots E^1 P^1 A)\mathbf{t} = \mathbf{0}. \tag{4.47}$$

The eta matrix E^{k-1} is nonsingular because the diagonal element in row $k-1$ and column $k-1$ of E^{k-1} is not zero, so you can multiply (4.47) through on the left by $(E^{k-1})^{-1}$ to obtain

$$(P^{k-1} E^{k-2} P^{k-2} \cdots E^1 P^1 A)\mathbf{t} = \mathbf{0}. \tag{4.48}$$

Likewise, the permutation matrix P^{k-1} in (4.48) is nonsingular (because *all* permutation matrices are nonsingular), so, multiplying (4.48) through on the left by $(P^{k-1})^{-1}$, you obtain

$$(E^{k-2} P^{k-2} \cdots E^1 P^1 A)\mathbf{t} = \mathbf{0}. \tag{4.49}$$

Sequentially multiplying (4.49) by the inverses of the eta and permutation matrices, you eventually reach the conclusion that

$$A\mathbf{t} = \mathbf{0}.$$

This shows that the columns of A are linearly *dependent*, and this contradiction proves that Gaussian elimination successfully constructs a matrix B for which $BA = I$.

To complete the proof, it must also be shown that $AB = I$. But

$$B = E^n P^n \cdots E^1 P^1$$

is nonsingular because each E^i and P^i is nonsingular and the product of nonsingular matrices is nonsingular (see Proposition 4.7(b) in Section 4.4.1). But then, from Proposition 4.8 in Section 4.4.1, it follows that $AB = I$ and so $B = A^{-1}$, thus completing the proof.

QED

One of the interesting features of Theorem 4.1 is that you can understand more clearly what it means to say that an $(n \times n)$ matrix A is *not* nonsingular, that is, that A is *singular*. If you simply negate Definition 4.10 in Section 4.4.1 for a nonsingular matrix, you obtain the following statement:

> An $(n \times n)$ matrix A is *singular* if and only if there is no $(n \times n)$ matrix A^{-1} such that $AA^{-1} = A^{-1}A = I$.

In other words, all you can say is that there is no inverse matrix. In contrast, as a result of Theorem 4.1, if A is singular, you know that part (a) of the theorem is *not true*. Consequently, you also know that part (c) of that theorem is *not true*. By negating part (c) of Theorem 4.1, you have that

> An $(n \times n)$ matrix A is *singular* if and only if the columns of A are linearly dependent, that is, *there are* real numbers $t_1, \ldots, t_n$, not all 0, such that $t_1 A_{.1} + \cdots + t_n A_{.n} = \mathbf{0}$.

In this section, you have learned some of the theory behind solving linear equations and the role of nonsingular matrices and linearly independent vectors. In Section 4.5, you will see how abstraction is applied to develop an axiomatic system that includes n-vectors and much more.

4.5 Vector Spaces

In Section 4.1, you learned about n-vectors and their various operations. A similar development is given for matrices in Section 4.2. You may have recognized some of the following similarities between n-vectors and their operations and matrices and their operations (in which $\mathbf{x}$, $\mathbf{y}$, $\mathbf{z}$, and $\mathbf{0}$ are n-vectors and A, B, C, and 0 are matrices):

1. $\mathbf{x} + \mathbf{y} = \mathbf{y} + \mathbf{x}$ and $A + B = B + A$.
2. $\mathbf{x} + \mathbf{0} = \mathbf{0} + \mathbf{x} = \mathbf{x}$ and $A + 0 = 0 + A = A$.
3. $(\mathbf{x} + \mathbf{y}) + \mathbf{z} = \mathbf{x} + (\mathbf{y} + \mathbf{z})$ and $(A + B) + C = (A + B) + C$.

Because of these (and other) similarities, you should consider *unifying* n-vectors and matrices. In this section, you will see how this goal is accomplished through the process of abstraction and developing an axiomatic system that contains both n-vectors and matrices as special cases. You should review the material in Section 2.3 on developing abstract and axiomatic systems as well as unary operators in Section 2.3.1 and binary operators in Section 1.5.2.

4.5.1 Developing an Abstract System

As described in Section 2.3.1, the first step is to create an abstract system by thinking of *objects* rather than specific items. So, create an appropriate set of objects together with one or more operations that you can perform on those objects, as described next.

To work with n-vectors, you not only need n-vectors, but also real numbers. Correspondingly, you should create the following two sets when you use abstraction:

F = a set of objects having the properties of real numbers and
V = a set of objects having the properties of n-vectors.

Throughout this section, the elements of F are called **scalars**, which are general objects rather than the specific real numbers used in Section 4.1. An element $s \in F$ might be a real number, a complex number, or some other object that satisfies properties similar to those of real numbers. The elements of V are called *vectors*, which are general objects rather than the specific n-vectors in Section 4.1. In particular, an element $\mathbf{v} \in V$ can be an n-vector, a matrix, a function, a set, or some other object. Herein lies the advantage of abstraction – the ability to study and understand many different mathematical concepts while having to work with only one structure.

The next step is to be able to perform operations with the elements in F and V. Because these elements are *objects*, you cannot directly use the binary operations of addition, subtraction, and multiplication that apply to real numbers and n-vectors. Nevertheless, you want to perform *similar* operations on the objects in F and V. As seen in Section 4.1, for n-vectors and real numbers, these binary operations fall into the following four groups:

1. Operations that combine two real numbers to produce a real number (such as addition, subtraction, multiplication, and division).
2. Operations that combine two n-vectors to produce an n-vector (such as addition and subtraction).

3. Operations that combine a real number and an n-vector to produce an n-vector (such as multiplying an n-vector by a real number).

4. Operations that combine two n-vectors to produce a real number (such as the dot product).

Correspondingly, you should create four types of binary operations for the sets F and V.

Binary Operations on the Scalars in F

Consider the binary operation of adding two real numbers. You might therefore create a corresponding binary operation, denoted, say, by the symbol +, for combining two *objects* $s, t \in F$. That is, you should think of

$$s + t$$

as an element of F that corresponds to "adding" the two scalars s and t. Because the symbol + is also used to add two *real numbers*, be sure to interpret this symbol correctly on the basis of the context in which this symbol is used.

The operation of subtracting two real numbers a and b is another binary operation, but you can also perform this operation by adding the *negative* of b to a, that is,

$$a - b = a + (-b).$$

Thus, rather than creating a new binary operation for subtracting two elements in F, you can use a *unary* operation called the **negative of a scalar** that, together with addition, constitutes the binary operation of subtraction. Specifically, for an element $t \in F$, you should think of

$$-t$$

as an element of F that corresponds to the negative of t. Then, for $s, t \in F$, the combined operations of

$$s + (-t)$$

is an element of F that corresponds to subtracting t from s.

Turning to the multiplication of two real numbers, you can create a corresponding binary operation, denoted by $*$, for combining two objects $s, t \in F$. That is, you should think of

$$s * t$$

as an element of F that corresponds to "multiplying" the two scalars s and t. Sometimes, the symbol $*$ is omitted, in which case the two elements of F are simply written next to each other, as in st.

Binary Operations on the Vectors in V

Now consider the binary operation of adding two n-vectors to create an n-vector. You might therefore create a corresponding binary operation – denoted, say, by the symbol $\oplus$ – for combining two objects **u** and **v** in V. That is, you should think of

$$\mathbf{u} \oplus \mathbf{v}$$

as an element of V that corresponds to "adding" the vectors **u** and **v**.

The specific way in which you compute $\oplus$ depends on the specific objects in V. For example, if V contains n-vectors, then $\oplus$ is the operation of adding two n-vectors, as you learned in Section 4.1.3. In contrast, if V contains matrices, then $\oplus$ is the operation of adding two matrices, as you learned in Section 4.2.4. If V contains other objects, such as *functions*, for example, then you need to specify precisely what $\oplus$ means, that is, how $\oplus$ combines two functions to produce a function.

Consider now the operation of subtracting two n-vectors **x** and **y**. As with subtracting real numbers, this is a binary operation, but you can also perform this operation by adding the *negative* of **y** to **x**, that is,

$$\mathbf{x} - \mathbf{y} = \mathbf{x} + (-\mathbf{y}).$$

Thus, rather than creating a new binary operation for subtracting two elements in V, you can use a *unary* operation called the **negative of a vector**, denoted by $-$. Specifically, for an element **v** in V, you should think of

$$-\mathbf{v}$$

as an element of V that corresponds to the negative of **v**. Then, for **u** and **v** in V, the combined operations of

$$\mathbf{u} \oplus (-\mathbf{v})$$

is an element of V that corresponds to subtracting **v** from **u**.

Operations Between Scalars in F and Vectors in V

The next operation you learned in Section 4.1.3 is that of multiplying a real number by an n-vector. You might therefore create a corresponding binary operation, denoted by $\odot$, for combining a scalar $s \in F$ with a vector $\mathbf{v} \in V$. That is, you should think of

$$s \odot \mathbf{v}$$

as an element of V that corresponds to "multiplying" the scalar s by the vector **v**.

The specific way in which you compute $s \odot \mathbf{v}$ depends on the specific objects in F and V. For example, if F contains real numbers and V contains n-vectors, then $s \odot \mathbf{v}$ is the operation of multiplying the real number s by the n-vector **v**, as you learned in Section 4.1.3. In contrast, if V contains matrices, then $s \odot \mathbf{v}$ is the operation of multiplying the real number s by the *matrix* $\mathbf{v}$, as you learned in Section 4.2.4. If V contains other objects, such as *functions*, for example, then you need to specify precisely what $\odot$ means, that is, how $\odot$ combines a scalar and a function to produce a function.

Sometimes, the symbol $\odot$ is omitted, in which case, the scalar is simply written next to the vector. Because this same notation is also used to combine two *scalars*, be sure to interpret this notation correctly on the basis of the context in which the notation is used. For example, if $s, t \in F$ and $\mathbf{v} \in V$, then the following interpretations apply:

$st = s * t$ (this combines two scalars to produce a scalar),

$s\mathbf{v} = s \odot \mathbf{v}$ (this combines a scalar and a vector to produce a vector),

$(st)\mathbf{v} = (s * t) \odot \mathbf{v}$ (first combines the two scalars s and t to produce the scalar $s*t$; then combines the resulting scalar $s * t$ with the vector $\mathbf{v}$ to produce the vector $(s * t) \odot \mathbf{v}$).

Operations On Two Vectors in V that Produce a Scalar in F

The final type of operation you learned in Section 4.1.3 is the dot product for combining two n-vectors to produce a *real number*. You might therefore create a corresponding binary operation for combining two vectors $\mathbf{u}, \mathbf{v} \in V$ to create a *scalar*, denoted by $\langle \mathbf{u}, \mathbf{v} \rangle$. That is, you should think of

$$\langle \mathbf{u}, \mathbf{v} \rangle$$

as an element of F that corresponds to the "dot product" of the vectors $\mathbf{u}$ and $\mathbf{v}$.

You already know how to compute the dot product of two n-vectors to produce a real number (see Section 4.1.3). However, consider the special case in which V contains *matrices*. In Section 4.2, there is no operation that combines two matrices to produce a real number. Because the dot product appears to apply only to n-vectors but not to matrices, the dot-product operation is *not* included in the abstract system being developed here. However, in the exercises you are asked to develop a *separate* abstract and axiomatic system to study the properties of the dot product of two vectors.

By putting together all of the pieces, you now have the abstract system $(F, +, *, V, \oplus, \odot)$ consisting of the following items:

1. A set F of scalars together with the binary operations of addition $(+)$ and multiplication $(*)$. The unary operation of computing the negative of a scalar also allows you to *subtract* two scalars.

2. A set V of vectors together with the binary operation of addition $(\oplus)$. The unary operation of computing the negative of a vector also allows you to *subtract* two vectors.

3. A binary operation ($\odot$) for combining a scalar and a vector to produce a vector.

4.5.2 Developing an Axiomatic System

The next step is to create an axiomatic system for the abstract system $(F, +, *, V, \oplus, \odot)$. This requires identifying those elements in F and V that have special properties as well as the properties you want the operations $+, *, \oplus$, and $\odot$ to satisfy.

Identifying Special Elements

When you think about F and V, use the special case of the real numbers and n-vectors as a guide and ask yourself which elements have special properties with regard to the binary operations on those sets.

Identifying Special Scalars. When you think of the set F of scalars as the *real numbers*, both 0 and 1 are special. The number 0 is special with regard to addition because

$$\text{for all real numbers } a,\ a + 0 = 0 + a = a. \tag{4.50}$$

In a similar way, the number 1 is special with regard to multiplication because

$$\text{for all real numbers } a,\ (a)1 = 1(a) = a. \tag{4.51}$$

In the context of the abstract system, you would therefore want the set F of scalars to contain two special elements – denoted, respectively, by 0 and 1 – that satisfy properties similar to those in (4.50) and (4.51). Specifically, you want the following to hold:

$$\text{There is a } 0 \in F \text{ such that for all } s \in F,\ s + 0 = 0 + s = s. \tag{4.52}$$

Likewise, for multiplication, you also want the following condition:

$$\text{There is a } 1 \in F \text{ such that for all } s \in F,\ s * 1 = 1 * s = s. \tag{4.53}$$

Identifying Special Vectors. When you think of the set V of vectors as the set of n-vectors, the zero vector plays a special role with regard to addition because

$$\text{for all } n\text{-vectors } \mathbf{x},\ \mathbf{x} + \mathbf{0} = \mathbf{0} + \mathbf{x} = \mathbf{x}. \tag{4.54}$$

In the context of the abstract system, you would therefore want the set V of vectors to contain a special element – denoted by $\mathbf{0}$ – that satisfies a property similar to the one in (4.54). Specifically, you want the following condition to hold:

$$\text{There is a } \mathbf{0} \in V \text{ such that for all } \mathbf{v} \in V,\ \mathbf{v} \oplus \mathbf{0} = \mathbf{0} \oplus \mathbf{v} = \mathbf{v}. \tag{4.55}$$

Identifying the Axioms

The axioms are chosen to reflect the important properties you wish to study regarding the abstract system. After working with various special cases – such as real numbers and n-vectors – mathematicians have identified a list of axioms that are classified broadly in the following three groups:

1. Axioms pertaining to the operations of $+$, $*$, and taking the negative of an element in F.

2. Axioms pertaining to the operation $\oplus$ and taking the negative of an element in V.

3. Axioms pertaining to the operation of $\odot$ that combines a scalar and a vector to produce a vector.

The axioms in group (1) pertaining to the scalars in F are abstracted from the properties that real numbers satisfy with regard to addition, mutiplication, and taking the negative. The axioms in group (2) pertaining to the vectors in V are abstracted from the properties satisfied by the special case of n-vectors presented in Proposition 4.1 in Section 4.1.4, as are the axioms in group (3) pertaining to the multiplication of scalars and vectors. The complete list gives rise to the axiomatic system of a *vector space*, as stated in the following definition.

Definition 4.13 *A* **vector space** *is an abstract system consisting of* $(F, +, *, V, \oplus, \odot)$ *in which* $+$ *and* $*$ *are closed binary operations on* F, $\oplus$ *is a closed binary operation on* V, *and* $\odot$ *combines an element in* F *with an element in* V *to produce an element in* V *satisfying all of the following axioms:*

Axioms for Scalars

1. $\exists$ *an element* $0 \in F$ *such that* $\forall s \in F,\ s + 0 = 0 + s = s$.

2. $\forall s \in F,\ \exists$ *an element* $-s \in F$ *such that* $s + (-s) = (-s) + s = 0$.

3. $\forall r, s, t \in F,\ (r + s) + t = r + (s + t)$.

4. $\forall s, t \in F,\ s + t = t + s$.

5. $\exists$ *an element* $1 \in F$ *such that* $\forall s \in F,\ s * 1 = 1 * s = s$.

6. $\forall s \in F$ *with* $s \neq 0,\ \exists$ *an element* $s^{-1} \in S$ *such that* $s * s^{-1} = s^{-1} * s = 1$.

Continued

7. $\forall r, s, t \in F,\ (r * s) * t = r * (s * t)$.
8. $\forall s, t \in F,\ s * t = t * s$.
9. $\forall r, s, t \in F,\ r*(s+t) = (r*s)+(r*t)$ and $(s+t)*r = (s*r)+(t*r)$.

Axioms for Vectors

10. $\exists$ an element $\mathbf{0} \in V$ such that $\forall\, \mathbf{v} \in V,\ \mathbf{v} \oplus \mathbf{0} = \mathbf{0} \oplus \mathbf{v} = \mathbf{v}$.
11. $\forall\, \mathbf{v} \in V,\ \exists$ an element $-\mathbf{v} \in V$ such that $\mathbf{v} \oplus (-\mathbf{v}) = (-\mathbf{v}) \oplus \mathbf{v} = \mathbf{0}$.
12. $\forall\, \mathbf{u}, \mathbf{v}, \mathbf{w} \in V,\ (\mathbf{u} \oplus \mathbf{v}) \oplus \mathbf{w} = \mathbf{u} \oplus (\mathbf{v} \oplus \mathbf{w})$.
13. $\forall\, \mathbf{u}, \mathbf{v} \in V,\ \mathbf{u} \oplus \mathbf{v} = \mathbf{v} \oplus \mathbf{u}$.

Axioms for Scalars and Vectors

14. $\forall s \in F$ and $\forall\, \mathbf{u},\ \mathbf{v} \in V,\ s \odot (\mathbf{u} \oplus \mathbf{v}) = (s \odot \mathbf{u}) \oplus (s \odot \mathbf{v})$.
15. $\forall s, t \in F$ and $\forall\, \mathbf{v} \in V,\ (s + t) \odot \mathbf{v} = (s \odot \mathbf{v}) \oplus (t \odot \mathbf{v})$.
16. $\forall s, t \in F$ and $\forall\, \mathbf{v} \in V,\ (s * t) \odot \mathbf{v} = s \odot (t \odot \mathbf{v})$.
17. $\forall\, \mathbf{v} \in V,\ 1 \odot \mathbf{v} = \mathbf{v}$.

Examples of Vectors Spaces

Several examples of vector spaces are presented in the remainder of this section. In these specific cases, F is the set of real numbers for which, from previous experience, you know that Axioms 1 through 9 in Definition 4.13 are satisfied. Therefore, only the remaining axioms are discussed. Also, for each example, the specific meaning of the operations $\oplus$ (that combines two vectors) and $\odot$ (that combines a scalar and a vector) are explained.

The Vector Space of n-Vectors. The sets $F = \mathbb{R}$ and $V = \mathbb{R}^n$ constitute a vector space when $\oplus$ is vector addition and $\odot$ is the multiplication of a real number and an n-vector, as described in Section 4.1.3. That is, for the real number t and the n-vectors $\mathbf{x} = (x_1, \ldots, x_n)$ and $\mathbf{y} = (y_1, \ldots, y_n)$, the operations

$$\mathbf{x} \oplus \mathbf{y} = (x_1 + y_1, \ldots, x_n + y_n)$$

and

$$t \odot \mathbf{x} = (tx_1, \ldots, tx_n)$$

result in $F = \mathbb{R}$ and $V = \mathbb{R}^n$ being a vector space. Axioms 10 through 17 in Definition 4.13 are established in Proposition 4.1 in Section 4.1.4 and are not repeated here.

The Vector Space of $(m \times n)$ Matrices. The sets $F = \mathbb{R}$ and $V = \mathbb{R}^{m \times n}$ consisting of all $(m \times n)$ matrices constitute a vector space when $\oplus$ is matrix addition and $\odot$ is the multiplication of a real number and a matrix, as described in Section 4.2.4. That is, for the real number t and the $(m \times n)$ matrices A and B, the operations

$$(A \oplus B)_{ij} = A_{ij} + B_{ij}, \text{ for all } i = 1, \ldots, m \text{ and for all } j = 1, \ldots, n$$

and

$$(t \odot A)_{ij} = tA_{ij}, \text{ for all } i = 1, \ldots, m \text{ and for all } j = 1, \ldots, n$$

result in $F = \mathbb{R}$ and $V = \mathbb{R}^{m \times n}$ being a vector space. Axioms 10 through 17 in Definition 4.13 are established in Propositions 4.3, 4.4, and 4.5 in Section 4.2.5 and are not repeated here.

The Vector Space of Real-Valued Functions. Another example of a vector space arises when $F = \mathbb{R}$ and V is the set of all real-valued functions of one variable, that is,

$$V = \{f : \mathbb{R} \longrightarrow \mathbb{R}\}.$$

In this case, $\oplus$ combines two functions f and g in V to create the function $f \oplus g : \mathbb{R} \to \mathbb{R}$, defined as follows:

$$(f \oplus g)(x) = f(x) + g(x), \text{ for all } x \in \mathbb{R}.$$

The operation $\odot$ combines a real number a and a function $f \in V$ to create the function $a \odot f : \mathbb{R} \to \mathbb{R}$, defined as follows:

$$(a \odot f)(x) = af(x), \text{ for all } x \in \mathbb{R}.$$

Axioms 10 through 17 in Definition 4.13 are now verified.

For Axiom 10:

You must show the existence of a zero vector – that is, a zero function, in this case. The function $0 : \mathbb{R} \to \mathbb{R}$ defined by

$$0(x) = 0, \text{ for all } x \in \mathbb{R}$$

is such a function because for any function $f \in V$, $f \oplus 0 = 0$ since for each $x \in \mathbb{R}$,

$$(f \oplus 0)(x) = f(x) + 0(x) = f(x) + 0 = f(x).$$

Similarly, $0 \oplus f = 0$.

For Axiom 11:

Let $f \in V$, for which you must create a function that is the negative of f. In this case, define the function $-f$ as follows:

$$(-f)(x) = -f(x), \text{ for all } x \in \mathbb{R}.$$

You can see that $f \oplus (-f) = 0$ because, for each $x \in \mathbb{R}$,

$$[f \oplus (-f)](x) = f(x) + [-f(x)] = f(x) - f(x) = 0.$$

Likewise, $(-f) \oplus f = 0$.

For Axiom 12:

For $f, g, h \in V$, it follows that for each $x \in \mathbb{R}$,

$$\begin{aligned}[(f \oplus g) \oplus h](x) &= (f \oplus g)(x) + h(x)\\ &= [f(x) + g(x)] + h(x)\\ &= f(x) + [g(x) + h(x)]\\ &= f(x) + (g \oplus h)(x)\\ &= [f \oplus (g \oplus h)](x).\end{aligned}$$

(Give reasons for each of the foregoing steps.)

For Axiom 13:

For $f, g \in V$, it follows that for each $x \in \mathbb{R}$,

$$(f \oplus g)(x) = f(x) + g(x) = g(x) + f(x) = (g \oplus f)(x).$$

(Give reasons for each of the foregoing steps.)
For the remaining axioms, let $a, b, x \in \mathbb{R}$ and $f, g \in V$.

For Axiom 14:

$$\begin{aligned}[a \odot (f \oplus g)](x) &= a[(f \oplus g)(x)]\\ &= a[f(x) + g(x)]\\ &= af(x) + ag(x)\\ &= (a \odot f)(x) + (a \odot g)(x)\\ &= [(a \odot f) \oplus (a \odot g)](x).\end{aligned}$$

(Give reasons for each of the foregoing steps.)

For Axiom 15:

$$\begin{aligned}[(a + b) \odot f](x) &= (a + b)f(x)\\ &= af(x) + bf(x)\\ &= (a \odot f)(x) + (b \odot f)(x)\\ &= [(a \odot f) \oplus (b \odot f)](x).\end{aligned}$$

(Give reasons for each of the foregoing steps.)

For Axiom 16:

$$\begin{aligned}[(a * b) \odot f](x) &= (ab)f(x)\\ &= a[bf(x)]\\ &= a[(b \odot f)(x)]\\ &= [a \odot (b \odot f)](x).\end{aligned}$$

(Give reasons for each of the foregoing steps.)

For Axiom 17:

$$(1 \odot f)(x) = 1f(x) = f(x).$$

Because all the axioms in Definition 4.13 hold, the set $F = \mathbb{R}$ and the set V of real-valued functions of one variable form a vector space under these definitions of $\oplus$ and $\odot$.

4.5.3 Deriving Results for the Axiomatic System

As you learned in Chapter 2, an axiomatic system allows you to study many problems and structures simultaneously while having to derive only one set of results. That is, any result you obtain for an axiomatic system applies not only to all the special cases you already know but also to any *new* special cases you come across in the future. In this section, some results pertaining to a general vector space are presented.

Recall that the fewest possible number of axioms are chosen to be included in the axiomatic system. Other important properties are then derived from the basic axioms, such as the properties in the following proposition.

Proposition 4.9 *If* $(F, +, *, V, \oplus, \odot)$ *is a vector space and* $\mathbf{u} \in V$ *and* $t \in F$, *then the following hold:*

(a) The vector $\mathbf{0}$ *is the only zero vector.*

(b) The vector $-\mathbf{u}$ *is the only negative of the vector* $\mathbf{u}$.

(c) $0 \odot \mathbf{u} = \mathbf{0}$.

(d) $t \odot \mathbf{0} = \mathbf{0}$.

(e) $(-1) \odot \mathbf{u} = -\mathbf{u}$.

(f) If $t \odot \mathbf{u} = \mathbf{0}$, *then* $t = 0$ *or* $\mathbf{u} = \mathbf{0}$.

Proof. In each part, you must prove that two vectors are equal. This is accomplished in various ways by using the axioms in Definition 4.13.

(a) To show that $\mathbf{0}$ is the only zero vector, suppose that $\mathbf{z}$ is also a vector such that for all $\mathbf{v} \in V$,

$$\mathbf{z} \oplus \mathbf{v} = \mathbf{v} \oplus \mathbf{z} = \mathbf{v}. \tag{4.56}$$

According to the uniqueness method, you must show that $\mathbf{z} = \mathbf{0}$. Because (4.56) holds *for all* $\mathbf{v} \in V$, in particular, it holds for $\mathbf{v} = \mathbf{0}$. So, by replacing $\mathbf{v}$ in (4.56) with $\mathbf{0}$, you obtain that

$$\mathbf{z} \oplus \mathbf{0} = \mathbf{0} \oplus \mathbf{z} = \mathbf{0}. \tag{4.57}$$

From Axiom 10, you know that $\mathbf{z} \oplus \mathbf{0} = \mathbf{z}$ so, from (4.57) it follows that $\mathbf{z} = \mathbf{0}$. This means that $\mathbf{0}$ is the only zero vector.

(b) To show that $-\mathbf{u}$ is the only negative of the vector $\mathbf{u}$, suppose that $\mathbf{v}$ also satisfies $\mathbf{u} \oplus \mathbf{v} = \mathbf{v} \oplus \mathbf{u} = \mathbf{0}$. According to the uniqueness method, you must show that $\mathbf{v} = -\mathbf{u}$. But now,

$$\mathbf{u} \oplus \mathbf{v} = \mathbf{0} = \mathbf{u} \oplus (-\mathbf{u}),$$

so, adding $-\mathbf{u}$ to the left of both sides yields

$$(-\mathbf{u}) \oplus (\mathbf{u} \oplus \mathbf{v}) = (-\mathbf{u}) \oplus [\mathbf{u} \oplus (-\mathbf{u})],$$

or, equivalently, by Axiom 12,

$$[(-\mathbf{u}) \oplus \mathbf{u}] \oplus \mathbf{v} = [(-\mathbf{u}) \oplus \mathbf{u}] \oplus (-\mathbf{u}).$$

But by Axiom 11, $(-\mathbf{u}) \oplus \mathbf{u} = \mathbf{0}$ so

$$\mathbf{0} \oplus \mathbf{v} = \mathbf{0} \oplus (-\mathbf{u}),$$

from which it follows by Axiom 10 that $\mathbf{v} = -\mathbf{u}$.

(c) To show that $0 \odot \mathbf{u}$ is the zero vector, let $-(0 \odot \mathbf{u})$ be the negative of $0 \odot \mathbf{u}$ (which exists by Axiom 11). Then,

$$\begin{aligned}
0 \odot \mathbf{u} &= (0 \odot \mathbf{u}) \oplus \mathbf{0} && \text{(Axiom 10)} \\
&= (0 \odot \mathbf{u}) \oplus \{(0 \odot \mathbf{u}) \oplus [-(0 \odot \mathbf{u})]\} && \text{(Axiom 11)} \\
&= [(0 \odot \mathbf{u}) \oplus (0 \odot \mathbf{u})] \oplus [-(0 \odot \mathbf{u})] && \text{(Axiom 12)} \\
&= [(0 + 0) \odot \mathbf{u}] \oplus [-(0 \odot \mathbf{u})] && \text{(Axiom 14)} \\
&= (0 \odot \mathbf{u}) \oplus [-(0 \odot \mathbf{u})] && \text{(Axiom 1)} \\
&= \mathbf{0}. && \text{(Axiom 11)}
\end{aligned}$$

(d) First note that for any scalar t, $t * 0 = 0$. This is because

$$\begin{aligned}
(t * 0) + (t * 0) &= t * (0 + 0) && \text{(Axiom 9)} \\
&= t * 0. && \text{(Axiom 1)}
\end{aligned} \tag{4.58}$$

By adding the scalar $-(t * 0)$ (which exists by Axiom 2) to both sides of (4.58), you obtain

$$[(t * 0) + (t * 0)] + [-(t * 0)] = (t * 0) + [-(t * 0)],$$

or equivalently, from Axiom 3, that

$$(t * 0) + \{(t * 0) + [-(t * 0)]\} = (t * 0) + [-(t * 0)]. \tag{4.59}$$

From Axiom 2, $(t * 0) + [-(t * 0)] = 0$, so (4.59) becomes

$$(t * 0) + 0 = 0,$$

from which it follows by Axiom 1 that

$$t * 0 = 0. \tag{4.60}$$

The result for part (d) of this proposition now follows by noting that

$$\begin{aligned}
t \odot \mathbf{0} &= t \odot (0 \odot \mathbf{0}) && \text{[part (c)]} \\
&= (t * 0) \odot \mathbf{0} && \text{(Axiom 16)} \\
&= 0 \odot \mathbf{0} && \text{[(4.60)]} \\
&= \mathbf{0}. && \text{[part (c)]}
\end{aligned}$$

(e) To show that $(-1) \odot \mathbf{u} = -\mathbf{u}$, according to part (b), it suffices to show that $\mathbf{u} \oplus [(-1) \odot \mathbf{u}] = \mathbf{0}$. But

$$\begin{aligned} \mathbf{u} \oplus [(-1) \odot \mathbf{u}] &= (1 \odot \mathbf{u}) \oplus [(-1) \odot \mathbf{u}] && \text{(Axiom 17)} \\ &= [1 + (-1)] \odot \mathbf{u} && \text{(Axiom 15)} \\ &= 0 \odot \mathbf{u} && \text{(Axiom 2)} \\ &= \mathbf{0}. && \text{[part (c)]} \end{aligned}$$

(f) According to the either/or method (see Section 1.6.11), suppose that

$$t \odot \mathbf{u} = \mathbf{0} \text{ and } t \neq 0. \tag{4.61}$$

You must show that $\mathbf{u} = \mathbf{0}$. However, because $t \neq 0$, by Axiom 6, there is a scalar t^{-1} such that $t * t^{-1} = t^{-1} * t = 1$. When you now multiply the first equality in (4.61) through by t^{-1}, you obtain

$$t^{-1} \odot (t \odot \mathbf{u}) = t^{-1} \odot \mathbf{0}, \tag{4.62}$$

which, by Axiom 16 and part (c) yields that

$$(t^{-1} * t) \odot \mathbf{u} = \mathbf{0}. \tag{4.63}$$

From the fact that $t^{-1} * t = 1$, you can rewrite (4.63) as follows:

$$1 \odot \mathbf{u} = \mathbf{0}. \tag{4.64}$$

Finally, from Axiom 17, you have the desired result that $\mathbf{u} = \mathbf{0}$.

This completes the proof.

QED

In this section, you have seen how the axiomatic system of a vector space includes n-vectors, $(m \times n)$ matices, and real-valued functions of one variable as special cases. Any results obtained for the axiomatic system apply not only to these special cases, but also to any new vector spaces you encounter in the future.

Chapter Summary

In this chapter, you have seen how the concepts of Chapters 1 and 2 are applied in the area of linear algebra for studying vectors and matrices. A review of the various topics follows.

Vectors

In Section 4.1, you learned that an n-vector – such as $\mathbf{x} = (x_1, \ldots, x_n)$ – is an ordered list of n real numbers that you can visualize either as a point in n dimensions, whose coordinates are the components of $\mathbf{x}$, or as an arrow, whose tail is at the origin and whose head is at the point $\mathbf{x}$. You also saw special n-vectors (such as $\mathbf{0}$ and the standard unit vectors) and how to perform various geometric and corresponding algebraic operations on n-vectors and real numbers. Most of these unary and binary operations result in an n-vector but some – such as multiplying two n-vectors and finding the norm of an n-vector – result in a real number. You also saw many of the properties that these various operations satisfy.

Matrices

In Section 4.2, you learned that an $(m \times n)$ matrix is a rectangular table of numbers organized in m rows and n columns. These matrices are a generalization of vectors because you can think of a matrix as a collection of row vectors, stacked on top of each other, or, alternatively, as a collection of column vectors, written next to each other. There are many special matrices such as the zero matrix, the identitiy matrix, permutation matrices, and eta matrices. You saw how to perform various operations on $(m \times n)$ matrices and real numbers and the properties these operations satisfy. Special care is needed when multiplying two matrices A and B because you can compute AB only if the number of columns of A is the same as the number of rows of B. Also, even if you *can* compute AB and BA, in general, $AB \neq BA$.

Solving Linear Equations

In Section 4.3, you saw that matrices and vectors are useful in solving a system of n linear equations in n unknowns. By letting A be the $(n \times n)$ matrix of known coefficients and $\mathbf{b}$ be the column vector of known right-hand-side values, you can represent the problem of solving a system of linear equations as that of wanting to find an n-vector $\mathbf{x}$ such that

$$A\mathbf{x} = \mathbf{b}.$$

One approach to solving this system is to find the inverse matrix, A^{-1}, because then,

$$\mathbf{x} = A^{-1}\mathbf{b}.$$

In some special cases, finding a closed-form expression for A^{-1} in terms of the elements in A is possible. For example, a closed-form expression exists for the inverses of (2×2), diagonal, permutation, and eta matrices. In general, however, a numerical method, such as Gaussian elimination, is needed to find A^{-1}.

One important observation from these results is that it is not always *possible* to solve a system of linear equations. In particular, an $(n \times n)$ matrix

A does not always *have* an inverse. In Section 4.4, you learned the theory of solving linear equations and that a nonsingular $(n \times n)$ matrix is a matrix that *does* have an inverse. You also discovered the geometric concept of linearly independent vectors, which are vectors that "open up properly." This geometric idea was then translated to a symbolic definition. The three concepts of a nonsingular matrix, linearly independent vectors, and the ability to solve a system of linear equations uniquely are shown to be equivalent in Theorem 4.1.

Vector Spaces

Because there are many similarities between n-vectors and matrices and their operations, it is possible to unify these concepts by developing an appropriate abstract system, as you saw in Section 4.5. This system involves two sets of objects – scalars (which are an abstraction of real numbers) and vectors (which are an abstraction of n-vectors and matrices) – together with several unary and binary operations on these objects.

The axiomatic system of a vector space enables you to study the properties that these operations satisfy. The axioms themselves are generalizations of the properties of n-vectors and their operations (except for the operation of multiplying two n-vectors). Several results for vector spaces are then proved on the basis of these axioms. The results thus obtained apply to all the special cases, including the vector space of n-vectors, of matrices, and of real-valued functions of one variable, as well as to any *new* vector spaces you encounter.

Exercises

Exercise 4.1 Perform each of the indicated operations with $\mathbf{x} = (-1, 1, 0)$ and $\mathbf{y} = (0, 2, 3)$.

(a) $\mathbf{x} + 2\mathbf{y}$.

(b) $||\mathbf{x}||$.

(c) $\mathbf{x} \cdot \mathbf{y}$.

(d) Is $\mathbf{x} \leq \mathbf{y}$? Why or why not? Explain.

Exercise 4.2 Perform each of the indicated operations with $\mathbf{x} = (1, 2, -1)$ and $\mathbf{y} = (3, -1, 0)$.

(a) $2\mathbf{x} - 3\mathbf{y}$.

(b) $||\mathbf{y} - \mathbf{x}||$.

(c) $\mathbf{x} \cdot 2\mathbf{y}$.

(d) Is $\mathbf{x} \le \mathbf{y}$? Is $\mathbf{y} \le \mathbf{x}$? Why or why not? Explain.

Exercise 4.3 Suppose that $\mathbf{x}$ is an n-vector drawn at the origin and that t is a real number whose value varies. What geometric figure results by considering $t\mathbf{x}$ for all possible values of t. That is, describe the following set:

$$\{t\mathbf{x} : t \text{ is a real number}\}.$$

What is this set when $\mathbf{x} = \mathbf{0}$?

Exercise 4.4 Suppose that $\mathbf{x}$ and $\mathbf{d}$ are given n-vectors and that t is a real number whose value varies. What geometric figure results by considering $\mathbf{x} + t\mathbf{d}$ for all possible values of t. That is, describe the following set:

$$\{\mathbf{x} + t\mathbf{d} : t \text{ is a real number}\}.$$

What is this set when $\mathbf{d} = \mathbf{0}$? What is this set when $\mathbf{x} = \mathbf{0}$ and $\mathbf{d} \neq \mathbf{0}$?

Exercise 4.5 Suppose that $\mathbf{x} = (x_1, x_2)$ is a given vector in two dimensions drawn at the origin.

(a) Translate the visual image of "the line through the origin that is perpendicular to $\mathbf{x}$" to symbolic form. (Hint: Think of the dot product.)

(b) Observe that the line in part (a) perpendicular to the vector $\mathbf{x}$ divides the plane into two *parts* (each called a *half space*). That is, the line and all points on one side of the line are a half space. Translate these two half-spaces to symbolic form using the given vector $\mathbf{x}$ in two dimensions.

(c) Generalize your results in parts (a) and (b) to a given n-vector $\mathbf{x}$. That is, write a symbolic representation of the *plane* that goes through the origin and is perpendicular to $\mathbf{x}$ together with the two associated half spaces created by this plane.

Exercise 4.6 Suppose that $\mathbf{x} = (x_1, x_2)$ and $\mathbf{y} = (y_1, y_2)$ are given vectors in two dimensions drawn at the origin with $\mathbf{y} \neq \mathbf{0}$. Translate the visual image of the *projection of* $\mathbf{x}$ *onto* $\mathbf{y}$, as shown by the vector $\mathbf{p}$ in the following figure, to symbolic form:

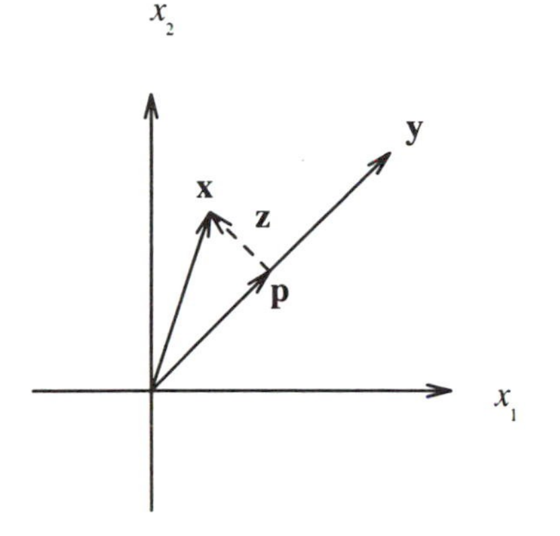

That is, express the unknown vector **p** in terms of the known vectors **x** and **y**. Generalize your result to given n-vectors **x** and **y**. (Hint: The vector **p** points in the same direction as **y** but has different length. To find the length of **p**, use the fact that **z** = **x** − **p** is perpendicular to **y**.)

Exercise 4.7 Suppose that t, x, y, and z are real numbers with $t \geq 0$ that satisfy the following condition:

$$\left|\frac{x}{y} - z\right| \leq t. \tag{4.65}$$

(a) Identify all syntax errors that arise in the following generalization of (4.65) when **y** is an n-vector:

$$|x - z\mathbf{y}| \leq t|\mathbf{y}|.$$

(b) Generalize (4.65) when both **y** and **z** are n-vectors. Be sure your resulting condition contains no syntax errors. (Hint: Consider using the norm of an n-vector.)

Exercise 4.8 Suppose that k is a positive integer and x is a real number.

(a) What difficulties arise in trying to generalize the expression x^k to the case where x is an n-vector, say, $\mathbf{x} = (x_1, \ldots, x_n)$?

(b) What difficulties arise in trying to generalize the expression x^k to the case where k is an n-vector, say, $\mathbf{k} = (k_1, \ldots, k_n)$?

Exercise 4.9 Suppose that **x**, **y**, and **z** are n-vectors and t and u are real numbers. Complete Proposition 4.1 in Section 4.1.4 by proving each of the following properties.

(a) $(\mathbf{x} + \mathbf{y}) + \mathbf{z} = \mathbf{x} + (\mathbf{y} + \mathbf{z})$.

(b) $\mathbf{x} + \mathbf{0} = \mathbf{0} + \mathbf{x} = \mathbf{x}$.

(c) $\mathbf{x} + (-\mathbf{x}) = \mathbf{0}$.

(d) $t(\mathbf{x} + \mathbf{y}) = (t\mathbf{x}) + (t\mathbf{y})$.

(e) $(t + u)\mathbf{x} = (t\mathbf{x}) + (u\mathbf{x})$.

(f) $1\mathbf{x} = \mathbf{x}$.

Exercise 4.10 Suppose that **x**, **y**, and **z** are n-vectors and t and u are real numbers. Complete Proposition 4.2 in Section 4.1.4 by proving each of the following properties.

(a) $||t\mathbf{x}|| = |t|\ ||\mathbf{x}||$.

(b) $\mathbf{x} \cdot \mathbf{x} = ||\mathbf{x}||^2$, that is, $||\mathbf{x}|| = \sqrt{\mathbf{x} \cdot \mathbf{x}}$.

(c) $\mathbf{x} \cdot (\mathbf{y} + \mathbf{z}) = (\mathbf{x} \cdot \mathbf{y}) + (\mathbf{x} \cdot \mathbf{z})$.

(d) $t(\mathbf{x} \cdot \mathbf{y}) = (t\mathbf{x}) \cdot \mathbf{y} = \mathbf{x} \cdot (t\mathbf{y})$.

Exercise 4.11 Perform each of the indicated operations using the following matrices or explain why you cannot do so:

$$A = \begin{bmatrix} 2 & 1 \\ 0 & -1 \end{bmatrix}, \ B = \begin{bmatrix} 1 & 3 \\ -1 & 2 \end{bmatrix}, \ C = \begin{bmatrix} 1 & 0 & 0 \\ 0 & 2 & -1 \end{bmatrix}.$$

(a) C^t.

(b) $A + 2B$.

(c) AB.

(d) AC.

(e) CA.

Exercise 4.12 Perform each of the indicated operations using the following matrices or explain why you cannot do so:

$$A = \begin{bmatrix} 1 & 0 \\ 0 & 2 \\ 0 & -1 \end{bmatrix}, \ B = \begin{bmatrix} -1 & 0 \\ 1 & 2 \end{bmatrix}, \ C = \begin{bmatrix} -1 & 1 \\ 2 & 0 \\ 1 & -3 \end{bmatrix}.$$

(a) $2A - B$.

(b) AB.

(c) BA.

(d) C^tC.

Exercise 4.13 Suppose that A is an $(m \times n)$ matrix and $\mathbf{x}$ is a vector. To multiply A on the *right* by $\mathbf{x}$, think of the vector $\mathbf{x}$ as a *matrix*.

(a) What must be the dimensions of the matrix $\mathbf{x}$ so that you can compute $A\mathbf{x}$? What are the dimensions of the resulting matrix $A\mathbf{x}$?

(b) For the values of

$$A = \begin{bmatrix} 1 & -1 & 0 \\ 2 & 0 & -2 \\ -3 & 1 & 3 \end{bmatrix} \text{ and } \mathbf{x} = \begin{bmatrix} -1 \\ 1 \\ 2 \end{bmatrix}$$

compute the value of $A\mathbf{x}$ by using the *rows* of A according to the following formula:

$$(A\mathbf{x})_i = A_{i.} \cdot \mathbf{x}, \text{ for } i = 1, 2, 3.$$

(c) For the values of A and $\mathbf{x}$ in part (b), compute $A\mathbf{x}$ by using the *columns* of A according to the following formula:

$$A\mathbf{x} = A_{.1}x_1 + A_{.2}x_2 + A_{.3}x_3.$$

Exercise 4.14 Suppose that A is an $(m \times n)$ matrix and $\mathbf{x}$ is a vector. To multiply A on the *left* by $\mathbf{x}$, think of the vector $\mathbf{x}$ as a *matrix*.

(a) What must be the dimensions of the matrix $\mathbf{x}$ so that you can compute $\mathbf{x}A$? What are the dimensions of the resulting matrix $\mathbf{x}A$?

(b) For the values of A and $\mathbf{x}$ in Exercise 4.13, compute the value of $\mathbf{x}A$ by using the *rows* of A according to the following formula:

$$\mathbf{x}A = x_1A_{1.} + x_2A_{2.} + x_3A_{3.}.$$

(c) For the values of A and $\mathbf{x}$ in Exercise 4.13, compute $\mathbf{x}A$ by using the *columns* of A according to the following formula:

$$(\mathbf{x}A)_j = \mathbf{x} \cdot A_{.j}, \text{ for } j = 1, 2, 3.$$

Exercise 4.15 Suppose that $A \in \mathbb{R}^{m \times p}$ and $B \in \mathbb{R}^{p \times n}$ and that you want to find row i and column j of AB.

(a) One approach is to compute AB and then identify row i and column j. Explain why a better alternative is to prove that $(AB)_{i.} = A_{i.}B$ and $(AB)_{.j} = AB_{.j}$.

(b) Prove the statements in part (a).

Exercise 4.16 Suppose that $A \in \mathbb{R}^{m\times n}, B \in \mathbb{R}^{n\times p}, C \in \mathbb{R}^{p\times q}$. From the properties of matrix multiplication, you know that

$$A(BC) = (AB)C.$$

(a) In terms of m, n, p, and q, how many multiplications are needed to compute $A(BC)$?

(b) In terms of m, n, p, and q, how many multiplications are needed to compute $(AB)C$?

(c) For the special case in which $m = 1, n = 100, p = 99$, and $q = 500$, which of these two alternatives is more efficient?

Exercise 4.17 Suppose that $A, B, C, 0 \in \mathbb{R}^{m\times n}$. Complete Proposition 4.3 in Section 4.2.5 by proving each of the following properties:

(a) $(A + B) + C = A + (B + C)$.

(b) $A + 0 = 0 + A = A$.

(c) $A - A = 0$.

(d) $0 - A = -A$.

Exercise 4.18 Suppose that $A, B, 0 \in \mathbb{R}^{m\times n}$ and s and t are real numbers. Complete Proposition 4.4 in Section 4.2.5 by proving each of the following properties:

(a) $s(A - B) = sA - sB$.

(b) $(s + t)A = sA + tA$.

(c) $(s - t)A = sA - tA$.

(d) $s(tA) = (st)A$.

(e) $0A = 0$ (the real number zero times A equals the zero matrix).

(f) $s0 = 0$ (s times the zero matrix equals the zero matrix).

Exercise 4.19 Suppose that A, B, and C are matrices for which you can perform the following operations on the basis of their dimensions. Prove each of the following properties in Proposition 4.5 in Section 4.2.5.

(a) $A(B - C) = AB - AC$.

(b) $(A + B)C = AC + BC$.

(c) $(A - B)C = AC - BC$.

Exercise 4.20 Suppose that s is a real number and that $A, B, C, 0$, and I (the identity matrix) are matrices for which you can perform the following operations on the basis of their dimensions. Prove each of the following properties in Proposition 4.5 in Section 4.2.5.

(a) $(AB)C = A(BC)$.

(b) $s(AB) = (sA)B = A(sB)$.

(c) $IA = A$ and $BI = B$.

(d) $0A = 0$ and $B0 = 0$.

Exercise 4.21 Show that if P is the $(m \times m)$ permutation matrix in which rows i and j of the identity matrix I are interchanged and $A \in \mathbb{R}^{m \times n}$, then PA is the matrix in which rows i and j of A are interchanged.

Exercise 4.22 Show that if P is the $(n \times n)$ permutation matrix in which columns i and j of the identity matrix I are interchanged and $A \in \mathbb{R}^{m \times n}$, then AP is the matrix in which columns i and j of A are interchanged.

Exercise 4.23

(a) Show that if P is a permutation matrix, then P is nonsingular and $P^{-1} = P^t$.

(b) Show that if E is an eta matrix in which column k differs from column k of the identity matrix I and $E_{kk} \neq 0$, then E is nonsingular and E^{-1} is the eta matrix in which column k is defined as follows:

$$E_{ik}^{-1} = \begin{cases} -\dfrac{E_{ik}}{E_{kk}}, & \text{if } i \neq k \\[2ex] \dfrac{1}{E_{kk}}, & \text{if } i = k. \end{cases}$$

Exercise 4.24 In Gaussian elimination, suppose you have computed the following matrix in which the first $k-1$ columns are the same as the first $k-1$ columns of the identity matrix I:

$$C = \begin{bmatrix} 1 & 0 & \ldots & 0 & C_{1k} & \ldots & C_{1n} \\ 0 & 1 & \ldots & 0 & C_{2k} & \ldots & C_{2n} \\ \vdots & \vdots & \ddots & \vdots & \vdots & \ddots & \vdots \\ 0 & 0 & \ldots & 1 & C_{k-1,k} & \ldots & C_{k-1,n} \\ 0 & 0 & \ldots & 0 & C_{kk} & \ldots & C_{kn} \\ \vdots & \vdots & \vdots & \vdots & \vdots & \ddots & \vdots \\ 0 & 0 & \ldots & 0 & C_{nk} & \ldots & C_{nn} \end{bmatrix}$$

Show that if $C_{kk} \neq 0$ and E^k is the eta matrix defined as follows:

$$E_{ik}^k = \begin{cases} -\dfrac{C_{ik}}{C_{kk}}, & \text{if } i \neq k \\[2ex] \dfrac{1}{C_{kk}}, & \text{if } i = k, \end{cases}$$

then the first k columns of $E^k C$ are equal to the first k columns of the identity matrix I.

Exercise 4.25 Generalize the following concepts from an $(n \times n)$ matrix A to an $(m \times n)$ matrix A or explain why you cannot do so.

(a) The columns of A are linearly independent vectors.

(b) A is nonsingular.

Exercise 4.26 Explain how to apply abstraction to the *elements* of a matrix. Identify four different special cases of the elements of such a matrix.

Exercise 4.27 Prove that the following two statements are equivalent for the n-vectors $\mathbf{x}$, $\mathbf{y}$, and $\mathbf{z}$:

(i) There are real numbers u, v, and w, at least one of which is 1, such that $u\mathbf{x} + v\mathbf{y} + w\mathbf{z} = \mathbf{0}$.

(ii) There are real numbers a, b, and c, not all 0, such that $a\mathbf{x} + b\mathbf{y} + c\mathbf{z} = \mathbf{0}$.

Exercise 4.28 Prove that the complex numbers $\mathbb{C} = \{a + bi : a, b \in \mathbb{R}\}$ satisfy Axioms 1 through 9 in Definition 4.13 for a vector space under the following operations on $a + bi,\ c + di \in \mathbb{C}$:

$$(a + bi) + (c + di) = (a + c) + (b + d)i$$

and

$$(a + bi) * (c + di) = (ac - bd) + (ad + bc)i.$$

Exercise 4.29 Using the notation $(F, +, *, V, \oplus, \odot)$ for a vector space as in Section 4.5, generalize Definition 4.12 to create a definition of the vectors $\mathbf{v}^1, \ldots, \mathbf{v}^k \in V$ being *linearly independent*.

Exercise 4.30 Apply each part of Proposition 4.9 in Section 4.5.3 to the special case of the vector space in which F is the set of real numbers and V is the set of real-valued functions of one variable. That is, explain what each part of Proposition 4.9 means in the context of this specific vector space and its operations.

Exercise 4.31 Recall that the dot product of the n-vectors $\mathbf{x} = (x_1, \ldots, x_n)$ and $\mathbf{y} = (y_1, \ldots, y_n)$ is an operation that results in the following real number:

$$\mathbf{x} \cdot \mathbf{y} = x_1 y_1 + \cdots + x_n y_n.$$

You can verify for yourself that this operation satisfies the following properties for all n-vectors $\mathbf{x}$, $\mathbf{y}$, and $\mathbf{z}$ and for all real numbers t:

(i) $\mathbf{x} \cdot \mathbf{y} = \mathbf{y} \cdot \mathbf{x}$.

(ii) $(\mathbf{x} + \mathbf{y}) \cdot \mathbf{z} = (\mathbf{x} \cdot \mathbf{z}) + (\mathbf{y} \cdot \mathbf{z})$.

(iii) $(t\mathbf{x}) \cdot \mathbf{y} = t(\mathbf{x} \cdot \mathbf{y})$.

(iv) $\mathbf{x} \cdot \mathbf{x} \geq 0$ and $\mathbf{x} \cdot \mathbf{x} = 0$ if and only if $\mathbf{x} = \mathbf{0}$.

Generalize these results to the abstract concepts of vectors in a set V and scalars in a set F. Specifically, using the notation $(F, +, *, V, \oplus, \odot)$ for a vector space as in Section 4.5 together with the notation $\langle \mathbf{x}, \mathbf{y} \rangle$ for combining two vectors $\mathbf{x}, \mathbf{y} \in V$ to produce a scalar in F, create an axiomatic system that generalizes the foregoing axioms to an arbitrary vector space and not just to n-vectors and real numbers.

Chapter 5

SELECTED TOPICS IN ABSTRACT ALGEBRA

In this chapter, you will see how the ideas from Chapters 1 and 2 are applied to selected topics from **abstract algebra** – the study of numbers and systems that behave like numbers. Throughout this chapter, the following sets of numbers are used:

$$\begin{aligned} \mathbb{N} &= \{1, 2, 3, \ldots\}, \\ \mathbb{Z} &= \{\ldots, -2, -1, 0, 1, 2, \ldots\}, \\ \mathbb{Q} &= \left\{\tfrac{p}{q} : p \text{ and } q \text{ are integers with } q \neq 0\right\}, \\ \mathbb{R} &= \{\text{real numbers}\}. \end{aligned}$$

You should also review the material on binary operations in Section 1.5.2, on equivalence relations in Section 2.3.3, on sets in Section 3.1, and on functions in Section 3.2.

5.1 Integer Arithmetic

The objective in this section is to study the binary operations of addition, subtraction, multiplication, and division of integers. For the subsequent purposes of abstraction and developing an axiomatic system, keep the following two questions in mind:

1. Which integers have special properties with regard to each of these operations?
2. What properties do these operations satisfy?

5.1.1 Binary Operations on Integers

From previous experience with integers, you know the basic operations of addition, subtraction, multiplication, and division. The purpose of this section is to study these operations in detail.

Addition

The binary operation of addition (+) combines two integers a and b to produce a new integer, $a + b$. That is, + is a *closed* binary operation on the integers. The number 0 is special because when you add any number to 0, you get that number as a result:

(Property of 0) For all $a \in \mathbb{Z}$, $a + 0 = 0 + a = a$.

Two of the basic properties of addition are:

(Associative Law) For all $a, b, c \in \mathbb{Z}$, $(a + b) + c = a + (b + c)$.
(Commutative Law) For all $a, b \in \mathbb{Z}$, $a + b = b + a$.

Subtraction

Rather than considering subtraction as another separate binary operation on the integers, an alternative approach is to think of $a - b$ as adding the *negative* of b to a, that is,

$$a - b = a + (-b).$$

Thus, to study subtraction, you might just as well study the unary operation of taking the *negative* of an integer. The negative of an integer b is the "opposite" of b. You can identify the following properties that integers have with regard to this unary operation and especially to the number 0:

For all $a \in \mathbb{Z}$, there is an integer $-a$ such that
$a + (-a) = (-a) + a = 0$.
For all $a \in \mathbb{Z}$, $-(-a) = a$.
For all $a \in \mathbb{Z}$, the integer $-a$ is unique, that is,
$-a$ is the only integer b for which $a + b = b + a = 0$.

Multiplication

Consider now the closed binary operation of multiplication ($\times$) that combines the two integers a and b to produce the new integer $a \times b$. Again, the objective is to identify special numbers and properties with regard to multiplication. In this case, the numbers 1 and 0 are special because of the following properties:

(Property of 1) For all $a \in \mathbb{Z}$, $a \times 1 = 1 \times a = a$.
(Property of 0) For all $a \in \mathbb{Z}$, $a \times 0 = 0 \times a = 0$.

As with addition, the following associative and commutative laws hold for multiplication:

(Associative Law) For all $a, b, c \in \mathbb{Z}$, $(a \times b) \times c = a \times (b \times c)$.
(Commutative Law) For all $a, b \in \mathbb{Z}$, $a \times b = b \times a$.

From here on, the symbol $\times$ is omitted when writing multiplication involving integer variables, for example, $a \times b$ is written as ab. Also, when multiplication occurs together with addition or subtraction, the multiplication is performed *first*. Thus, for example, $a + bc$ means $a + (bc)$.

Division of Integers

Turning now to division, first notice that division is *not* a closed binary operation on the integers. This is because the result of dividing one integer by another may not be an integer. For example, 2 divided by 4 is not an integer. On the basis of this observation, there are two approaches to studying division.

Working with the Rationals. When one integer is divided by another nonzero integer, you may not get an integer but you *do* get a *rational* number. Thus, one way to study division is to work with the set $\mathbb{Q}$ of rational numbers. Proceeding along this line, recall that subtraction is a form of addition. In an analogous manner, division is a form of multiplication. Specifically, think of dividing the rational number a by the rational number $b \neq 0$ as multiplying a by the *reciprocal* of b, that is,

$$\frac{a}{b} = ab^{-1}.$$

Thus, to study division, you might just as well study the unary operation of taking the *reciprocal* of a rational number. The reciprocal of a rational number b is the "opposite" of b. You can identify the following properties that rational numbers have with regard to this unary operation and especially to the number 1:

For all $a \in \mathbb{Q}$ with $a \neq 0$, there is an element $a^{-1} \in \mathbb{Q}$ such that $aa^{-1} = a^{-1}a = 1$.
For all $a \in \mathbb{Q}$ with $a \neq 0$, $(a^{-1})^{-1} = a$.
For all $a \in \mathbb{Q}$ with $a \neq 0$, the number a^{-1} is unique, that is, a^{-1} is the only rational number b for which $ab = ba = 1$.

More is said about this approach to division in Section 5.2.

Working with Integers. Another approach to studying division of integers is to work exclusively with *integers*. For example, when you divide 30 by 7, you get the integer 4, with an integer remainder of 2. The objective now is to *prove* that you can always perform this type of division with integers in a unique way. To do so, it is necessary to use the following *axiom*, which is assumed to be *true*, pertaining to every nonempty set of positive integers ($\mathbb{N}$).

The Least-Integer Principle

Every nonempty set of positive integers has a least element. That is, for every nonempty set $M \subseteq \mathbb{N}$, there is an element $x \in M$ such that, for all $y \in M$, $x \leq y$.

Now it is possible to prove that you can always divide two integers and obtain a remainder, as described in the following theorem and proof.

Theorem 5.1 (The Division Algorithm) *If a and b are integers with $a > 0$, then there are unique integers q and r such that*

$$b = aq + r, \text{ where } 0 \leq r < a.$$

Proof. According to the uniqueness methods (see Section 1.6.12), it is first necessary to construct the integers q and r and then to show that they are unique.

The values of q and r are obtained from the least element of the following nonempty set of nonnegative integers:

$$M = \{\text{integers } w \geq 0 : \text{ there is an integer } k \text{ such that } w = b - ak\}.$$

To see that $M \neq \emptyset$, note that if $b > 0$, then, by setting $k = 0$, it follows that $w = b \in M$. Alternatively, if $b \leq 0$, then, by setting $k = b$, it follows that $w = b - ak = b - ab = b(1 - a) \geq 0$ because $b \leq 0$ and $1 - a \leq 0$ (recall that $a \geq 1$ from the hypothesis).

The desired values of q and r are obtained from the fact that the nonempty set M of nonnegative integers has a least element. In particular, if $0 \in M$, then 0 is the least element of M. Otherwise, M has a least element by the Least-Integer Principle. In either case, let r be the least element of M. Because $r \in M$, by the defining properties of M, (1) $r \geq 0$ and (2) there is an integer q such that $r = b - aq$, or equivalently, $b = aq + r$.

It remains only to show that the remainder $r < a$, so assume that $r \geq a$. A contradiction is obtained by showing that r is *not* the least element of M. In fact, $w = b - a(q + 1)$ is a smaller element of M than r. You can see that $w \in M$ because

$$w = b - a(q + 1) = b - aq - a = r - a \geq 0.$$

Also, $w < r$ as a result of the hypothesis that $a > 0$ because

$$w = b - a(q + 1) = b - aq - a < b - aq = r.$$

Having constructed values of q and r for which

$$b = aq + r, \text{ and} \tag{5.1}$$

$$0 \leq r < a, \tag{5.2}$$

you must now show that these values are *unique*. So suppose that m and n are also integers for which

$$b = am + n, \text{ and} \tag{5.3}$$

$$0 \leq n < a. \tag{5.4}$$

You must show that $q = m$ and $r = n$. Subtracting (5.3) from (5.1) and rearranging results in

$$a(q - m) = n - r. \tag{5.5}$$

But from (5.2) and (5.4), it follows that

$$-a < n - r < a. \tag{5.6}$$

Combining (5.5) and (5.6) says that

$$-a < a(q - m) < a,$$

from which it follows that $q - m = 0$, that is, $m = q$. Substituting $m = q$ in (5.5) now yields that $n - r = 0$, that is, $n = r$. Thus, there are unique integers q and r for which $b = aq + r$ and $0 \leq r < a$, completing the proof.

QED

As a result of Theorem 5.1, it now makes sense to talk about *the* remainder on dividing b by a. From Theorem 5.1, that remainder is r.

5.1.2 Divisibility of Integers

As you know, the result of dividing one integer by another does not, in general, result in an integer. In some cases, however, it *does*. For example, 12 divided by 3 is the integer 4. This favorable situation is studied now in greater detail. Properties of such integers are also examined.

The Definition of Divisibility

On the basis of the foregoing discussion, the objective is to create a definition of what it means for an integer a to divide an integer b *evenly*? Use the following examples as a guide in developing the definition:

1. 3 divides 12 evenly.
2. 5 divides 30 evenly.
3. 2 *does not* divide 7 evenly.
4. 4 *does not* divide 9 evenly.

As explained in Section 2.2.2, to create this definition, identify and describe similarities shared by the integers in (1) and (2), keeping in mind the vocabulary of mathematics (such as the quantifiers *there is* and *for all*). After doing so, *verify* the definition to be sure that the property *includes* the integers in (1) and (2) and *excludes* the integers in (3) and (4).

For example, one common property shared by the integers in (1) and (2) is that the remainder obtained by performing the division is 0, which is *not* the case with the integers in (3) and (4). In terms of Theorem 5.1, if a divides b evenly, there is a unique integer q such that $b = aq + 0$. This observation is stated concisely in the following definition.

Definition 5.1 *An integer a* **divides** *an integer b, written $a \mid b$, if and only if there is an integer k such that $b = ak$.*

Observe that the word "evenly" has been dropped and is understood implicitly. The symbol | is a *binary relation* that is used to compare two integers (just like the binary relations $<$ and $\leq$). Thus, the statement

$$a \mid b$$

means that a *does* divide b, or equivalently, that the result of checking whether a divides b is *true*. Analogously, the statement

$$a \nmid b$$

means that a does *not* divide b, or equivalently, that the result of checking whether a divides b is *false*. These ideas are illustrated in the following example.

Example 5.1 – Divisible and Nondivisible Integers

The following comparisons are all *true*:

$$\begin{array}{rl} 3 \mid 12 & \text{(because } 12 = 3 \times 4), \\ 5 \mid 30 & \text{(because } 30 = 5 \times 6), \\ 2 \nmid 7 & \text{(because there is no integer } k \text{ such that } 7 = 2k), \\ 4 \nmid 9 & \text{(because there is no integer } k \text{ such that } 9 = 4k). \end{array}$$

Properties of Divisible Integers

Now you will see various properties of divisible integers. Several straightforward ones are presented in the following proposition and are used in the subsequent development.

Proposition 5.1 *Suppose that a, b, and c are integers.*

(a) If $a \mid b$, then $a \mid (-b)$.

(b) If $a \mid b$ and $b \mid c$, then $a \mid c$.

(c) If $a \mid b$ and $a \mid c$, then $a \mid (b + c)$.

(d) If $a \mid b$ and $b \mid a$ then $a = \pm b$.

Proof. The proofs of these statements are left to the exercises. QED

Another property of divisible integers relates to the concept of a *common divisor*, as defined next.

Definition 5.2 *An integer d is a* **common divisor** *of the integers a and b if and only if $d \mid a$ and $d \mid b$.*

For example, 2 is a common divisor of 12 and -20 because 2 divides 12 and 2 divides -20. Other common divisors of 12 and -20 are $\pm 1, \pm 2$, and ± 4. The numbers 4 and -4 are important because they are, respectively, the largest and smallest common divisors of 12 and -20. Collectively, these two values are called the *greatest common divisors* of 12 and -20, a general definition of which follows.

Definition 5.3 *An integer d is a* **greatest common divisor** *of the integers a and b if and only if (1) d is a common divisor of a and b and (2) for any common divisor c of a and b, $c \mid d$.*

How do you know that *there is* a greatest common divisor for two integers? The answer is provided in the following theorem.

Theorem 5.2 *If a and b are integers, at least one of which is not 0, then there is a unique greatest common divisor d of a and b such that $d > 0$.*

Proof. (The Euclidean Algorithm) According to the uniqueness methods (see Section 1.6.12), it is first necessary to show that *there is* a positive greatest common divisor – say, d – of a and b and subsequently to show that d is unique.

The construction of d is based on the following three possibilities: $a > 0$, $a < 0$, or $a = 0$. Consequently, a proof by cases (see Section 1.6.11) is appropriate.

Case 1: $a > 0$.

In this case, repeated applications of the division algorithm are used to produce the value of d. To begin with, from the division algorithm applied to b and a, there are unique integers q_1 and r_1 such that

$$b = aq_1 + r_1, \text{ with } 0 \leq r_1 < a.$$

If $r_1 = 0$, then you can verify that $d = a > 0$ is the greatest common divisor of a and b. If $r_1 > 0$, then you can apply the divison algorithm to a and r_1 to obtain unique integers q_2 and r_2 such that

$$a = r_1 q_2 + r_2, \text{ with } 0 \leq r_2 < r_1.$$

Each time this process is repeated, the new remainder (r_2, in this case) is strictly less than the previous remainder (r_1, in this case). Thus, this process must eventually yield a remainder of 0, say, through the following sequence of iterations:

$$\begin{aligned} b &= aq_1 + r_1 & (5.7)\\ a &= r_1q_2 + r_2 & (5.8)\\ r_1 &= r_2q_3 + r_3 & (5.9)\\ &\vdots & \\ r_{k-3} &= r_{k-2}q_{k-1} + r_{k-1} & (5.10)\\ r_{k-2} &= r_{k-1}q_k + r_k & (5.11) \end{aligned}$$

In (5.11), $r_k = 0$, so now it must be shown that r_{k-1} is a greatest common divisor of a and b.

The first step is to show that r_{k-1} is a common divisor of a and b. From (5.11), $r_{k-2} = r_{k-1}q_k$, so r_{k-1} divides r_{k-2}. Now you can see that r_{k-1} divides the right-hand side of (5.10), so r_{k-1} divides r_{k-3}. Progressively working *backward* through the foregoing equations leads to the fact that r_{k-1} divides a, from (5.8), and finally, that r_{k-1} divides b, from (5.7).

It remains to show that if c divides both a and b, then c divides d. To do so, work *forward* through the foregoing equations. For example, from (5.7),

$$r_1 = b - aq_1,$$

and because c divides both a and b, it follows that c divides r_1. Now, you can see from (5.8) that c divides $r_2 = a - r_1q_2$. In a similar manner, from (5.10), you can conclude that c divides r_{k-1}.

It has thus been shown that r_{k-1}, the last *nonzero* remainder obtained from this sequence of applications of the division algorithm, is in fact d, that is, r_{k-1} is a positive greatest common divisor of a and b.

Case 2: $a < 0$.

In this case, you obtain the greatest common divisor d of $-a$ and b using the method in Case 1, starting with $-a$ and b instead of a and b. The resulting value of d is also the greatest common divisor of a and b because a and $-a$ have the same divisors [see Proposition 5.1(a)].

Case 3: $a = 0$.

In this case, $d = |b| > 0$ because this value of d divides both $a = 0$ and b. Also, if c divides both $a = 0$ and b, then there is an integer m such that $b = cm$, so

$$d = |b| = |mc| = \begin{cases} |m|c, & \text{if } c \geq 0 \\ -|m|c, & \text{if } c < 0. \end{cases}$$

In either case, $c \mid d$ and so $d = |b|$ is a greatest common divisor of $a = 0$ and b.

It remains to show that the value of $d > 0$ created in any of these cases is unique. So, suppose that $e > 0$ is also a positive greatest common divisor of a and b. From Definition 5.3, it follows that $d \mid e$ and $e \mid d$. The only way this can happen is if $d = \pm e$ [see Proposition 5.1(d)], but because d and e are both positive, it must be that $d = e$.

QED

In view of Theorem 5.2, there is a unique positive greatest common divisor of the integers a and b when not both are 0 which, hereafter, is denoted by (a, b). The Euclidean algorithm for finding the value of (a, b) is illustrated in the following example.

Example 5.2 – The Euclidean Algorithm for Finding the Greatest Common Divisor of Two Integers

To find the value of $(a, b) = (108, 1095)$, the Euclidean algorithm results in the following sequence of computations:

$b =$	$a \times$	$q +$	r
1095 =	108 ×	10 +	15
108 =	15 ×	7 +	3
15 =	3 ×	5 +	0

The last nonzero remainder is 3, so (108, 1095) = 3.

You can use the integers obtained from the Euclidean algorithm to express (a, b) in terms of a and b. For instance, in Example 5.2, 3 is the greatest common divisor of 108 and 1095, so you can express 3 in terms of 108 and 1095, as follows:

$$3 = 108(71) - 1095(7).$$

A systematic method for expressing (a, b) in terms of a and b is given in the proof of the following theorem.

Theorem 5.3 *If a and b are integers, at least one of which is not zero, then there are integers m and n such that $(a, b) = am + bn$.*

Proof. It is necessary to construct the values of the integers m and n. This construction is done on the basis of the same three cases as in the proof of Theorem 5.2.

Case 1: $a > 0$.

In this case, the proof is by induction on k, the number of iterations of the Euclidean algorithm used in computing (a, b). If $k = 1$, then $b = aq + 0$. In this case, $(a, b) = a$ and so $m = 1$ and $n = 0$ satsify $(a, b) = a = a(1) + b(0) = am + bn$.

Suppose the result is *true* whenever $k - 1$ iterations of the Euclidean algorithm are used to find the greatest common divisor of two integers. Now consider the following k iterations of the Euclidean algorithm:

$$
\begin{aligned}
b &= aq_1 + r_1 && (5.12)\\
a &= r_1q_2 + r_2 && (5.13)\\
r_1 &= r_2q_3 + r_3 && (5.14)\\
&\vdots\\
r_{k-3} &= r_{k-2}q_{k-1} + r_{k-1} && (5.15)\\
r_{k-2} &= r_{k-1}q_k + r_k && (5.16)
\end{aligned}
$$

In (5.16), $r_k = 0$, so $r_{k-1} = (a, b)$. Also, note that if you were to compute (r_1, a), you would perform the $k - 1$ iterations beginning with (5.13), to obtain the same greatest common divisor as that of (a, b). Thus, by the induction hypothesis, there are integers n_1 and m_1 such that

$$r_{k-1} = (a, b) = r_1n_1 + am_1. \qquad (5.17)$$

The integers m_1 and n_1 in (5.17) are used to construct the values of m and n. Specifically, from (5.12),

$$r_1 = b - aq_1,$$

which, when substituted in (5.17) yields

$$
\begin{aligned}
(a, b) &= r_1n_1 + am_1\\
&= (b - aq_1)n_1 + am_1\\
&= bn_1 - aq_1n_1 + am_1\\
&= a(m_1 - q_1n_1) + bn_1. && (5.18)
\end{aligned}
$$

From (5.18), the desired values of m and n are $m = m_1 - q_1n_1$ and $n = n_1$.

Case 2: $a < 0$.

In this case, apply the Euclidean algorithm to find $(-a, b)$ and obtain the values of m and n as in Case 1.

Case 3: $a = 0$.

In this case, you have from Case 3 in Theorem 5.2 that

$$(a, b) = |b| = \begin{cases} b, & \text{if } b \geq 0 \\ -b, & \text{if } b < 0 \end{cases} = \begin{cases} a(0) + b(1), & \text{if } b \geq 0 \\ a(0) + b(-1), & \text{if } b < 0. \end{cases}$$

Thus, in this case, $m = 0$ and $n = \pm 1$.

QED

The next example illustrates how Theorem 5.3 is used to express (a, b) in terms of a and b.

Example 5.3 – Expressing (a, b) in Terms of a and b

From Example 5.2, (108, 1095) = 3. To express 3 in terms of 108 and 1095, use the middle row in the table of computations in Example 5.2 to write

$$3 = 108 - 15(7). \tag{5.19}$$

Now, from the first row in the table of computations in Example 5.2, you know that

$$15 = 1095 - 108(10). \tag{5.20}$$

Replacing the value of 15 in (5.19) with the expression in (5.20) yields

$$\begin{aligned} 3 &= 108 - 15(7) \\ &= 108 - [1095 - 108(10)](7) \\ &= 108 - 1095(7) + 108(70) \\ &= 108(71) - 1095(7). \end{aligned}$$

The numbers $m = 71$ and $n = 7$ satisfy the properties in Theorem 5.3.

In this section, you have seen how some integers divide other integers evenly. You have also learned various properties of these types of integers. In general, however, the remainder of dividing two integers is *not* 0. Still, you can obtain some interesting and useful results, as you will now see.

5.1.3 Integers Modulo n

If you fix the divisor to a value of $n > 0$, then the remainder on dividing an integer a by n is some value r between 0 and $n - 1$. For example, if $n = 5$, then the remainder on dividing $a = 12$ by $n = 5$ is $r = 2$. There are many other integers whose remainder on division by $n = 5$ is also 2 (for example, 17). In this sense, with n fixed at a value of 5, both 12 and 17 are the *same*, in that their remainder on division by 5 is 2. Another way to express the notion that 12 and 17 have the same remainder when divided by 5 is to notice that the remainder of 2 cancels when 12 is subtracted from 17, thus, 17 – 12 is divisible by 5. This observation is formalized in the following definition.

Definition 5.4 *Let n be a positive integer. Two integers a and b are* **congruent modulo** n*, written $a \equiv b \pmod{n}$, if and only if $n \mid (a - b)$.*

With n fixed, the symbol $\equiv$ is a *binary relation* that is used to compare two integers. Thus, the statement

$$a \equiv b \quad (mod\ n)$$

means that n *does* divide $a - b$, or equivalently, that the result of checking whether n divides $a - b$ is *true*. Analogously, the statement

$$a \not\equiv b \pmod{n}$$

means that n does *not* divide $a - b$, or equivalently, that the result of checking whether n divides $a - b$ is *false*. These ideas are illustrated in the following example.

Example 5.4 – Congruence Modulo n

The following comparisons are all *true*:

$$\begin{aligned} 17 &\equiv 2 \pmod{5} \quad [\text{because } 5 \mid (17-2)], \\ 9 &\not\equiv 2 \pmod{5} \quad [\text{because } 5 \nmid (9-2)], \\ 14 &\equiv 8 \pmod{3} \quad [\text{because } 3 \mid (14-8)], \\ 14 &\not\equiv 7 \pmod{3} \quad [\text{because } 3 \nmid (14-7)]. \end{aligned}$$

The next theorem establishes that congruence modulo n is an equivalence relation (see Section 2.3.3).

Theorem 5.4 *For each positive integer n, congruence modulo n is an equivalence relation on the set of integers.*

Proof. According to Definition 2.19 for an equivalence relation, it must be shown that congruence modulo n is reflexive, symmetric, and transitive. So, let a, b, and c be integers and recall the properties in Proposition 5.1.

Reflexive: You can see that $a \equiv a \pmod{n}$ because $n \mid (a - a)$, that is, $n \mid 0$ and so $\equiv$ is reflexive.

Symmetric: Suppose that $a \equiv b \pmod{n}$, so $n \mid (a - b)$. Consequently, $n \mid (b - a)$, so $b \equiv a \pmod{n}$ and thus $\equiv$ is symmetric.

Transitive: Suppose that $a \equiv b \pmod{n}$ and that $b \equiv c \pmod{n}$, so $n \mid (a - b)$ and $n \mid (b - c)$. But then, $n \mid [(a - b) + (b - c)]$, that is, $n \mid (a - c)$. Consequently, $a \equiv c \pmod{n}$ and so $\equiv$ is transitive.

QED

Recall that an equivalence relation on a set gives rise to a collection of *equivalence classes* (see Section 2.3.3). In the special case of congruence modulo n, the equivalence class generated by the integer a is the following:

$$[a] = \{b \in \mathbb{Z} : b \equiv a \pmod{n}\}.$$

For example, for $n = 3$, there are the following equivalence classes:

$$\begin{array}{lcl} [0] &=& \{\ldots,-6,-3,0,3,6,\ldots\}, \\ [1] &=& \{\ldots,-5,-2,1,4,7,\ldots\}, \\ [2] &=& \{\ldots,-4,-1,2,5,8,\ldots\}, \\ [3] &=& \{\ldots,-6,-3,0,3,6,\ldots\}, \\ [4] &=& \{\ldots,-5,-2,1,4,7,\ldots\}, \\ [5] &=& \{\ldots,-4,-1,2,5,8,\ldots\}, \\ \vdots & \vdots & \quad\quad \vdots \end{array}$$

Observe that $[0] = [3]$, $[1] = [4]$, and $[2] = [5]$. Thus, although there are only three *distinct* sets – namely, $[0]$, $[1]$, and $[2]$ – there are many different ways to *represent* those sets. In fact, you can prove that for any integer k, $[0] = [3k]$, $[1] = [3k+1]$, and $[2] = [3k+2]$. In general, for $n > 0$ and any integer a, $[a] = [a \pmod n]$.

As proved in Theorem 2.1 in Section 2.3.3, these equivalence classes *partition* the set of integers, that is, every integer belongs to exactly one of these equivalence classes. In other words, every integer is equivalent to either 0, 1, or 2 (mod 3). This observation is generalized in the following theorem.

Theorem 5.5 *Suppose that n is a positive integer. Every integer is congruent (mod n) to exaclty one of the integers $0, 1, \ldots, n-1$ and hence the n distinct equivalence classes $[0], [1], \ldots, [n-1]$ partition $\mathbb{Z}$.*

Proof. Let a be an integer. When a is divided by n, from the divison algorithm, there are integers q and r with $0 \le r < n$ such that

$$a = nq + r. \tag{5.21}$$

Thus, $a - r = nq$ and so $n \mid (a - r)$, which means that $a \equiv r \pmod n$. This proves that each integer is congruent (mod n) to one of the integers $0, 1, \ldots, n-1$.

It remains to show the uniqueness. So suppose that a is congruent to s (mod n), where $0 \le s < n$. It is shown that $s = r$. By definition of congruence, $n \mid (a - s)$. Thus, there is an integer t such that $a - s = nt$, or equivalently,

$$a = nt + s. \tag{5.22}$$

By the uniqueness of the division algorithm, from (5.21) and (5.22), it follows that $r = s$. This shows that the value to which a is congruent (mod n) is unique, thus completing the proof.

QED

In this section you have learned how to perform modulo arithmetic and the properties satisfied by the associated equivalence relation. Now you will learn about special integers that can be divided only by 1 and themselves.

5.1.4 Prime Numbers

In some cases, you can divide a positive integer by *many* positive integers. For example, you can divide 100 by 1, 2, 5, 10, 20, 25, 50, and 100. In contrast, you can divide some positive integers only by 1 and themselves. For example, the only positive integers that divide 11 are 1 and 11. These types of integers are special and are defined formally as follows.

> **Definition 5.5** *An integer $p > 1$ is* **prime** *if and only if the only positive integers that divide p are 1 and p.*

One reason that prime numbers are important is that you can decompose any integer greater than 1 into a product of primes numbers. For example, you can decompose 66 into the product of the prime numbers 2, 3, and 11. The proof of this decomposition is the goal of this section, which requires two facts about the greatest common divisor and primes.

The first fact is that if a, b, and c are integers for which a divides the product bc and if the greatest common divisor of a and b is 1, then $a \mid c$. For example, if $a = 6, b = 5$, and $c = 18$, then $6 \mid (5 \cdot 18)$ and $(6,5) = 1$, so $6 \mid 18$. In contrast, for $a = 10, b = 4$, and $c = 25$, you have that $10 \mid (4 \cdot 25)$ but because $(10,4) = 2 \neq 1$, you *cannot* conclude that $4 \mid 25$. This fact is established in the following proposition.

> **Proposition 5.2** *If a, b, and c are integers for which $a \mid (bc)$ and $(a,b) = 1$, then $a \mid c$.*

Proof. To show that $a \mid c$, by definition, it must be shown that

there is an integer k such that $c = ka$. (5.23)

To produce this integer k, use the fact that $a \mid (bc)$ to obtain that

there is an integer q such that $bc = qa$. (5.24)

Also, applying Theorem 5.3 to the hypothesis that $(a,b) = 1$ means that

there are integers m and n such that $1 = am + bn$. (5.25)

Multiplying (5.25) through by c yields

$$c = cam + cbn. \tag{5.26}$$

Replacing cb in (5.26) with qa from (5.24) yields

$$c = cam + qan = (cm + qn)a. \tag{5.27}$$

From (5.27), the desired value of the integer k in (5.23) is $k = cm + qn$.

QED

The second fact needed to show that you can decompose a positive integer into a product of primes is established in the next proposition.

Proposition 5.3 *If p is a prime and $a_1, \ldots, a_n$ are integers for which $p \mid (a_1 a_2 \cdots a_n)$, then there is an integer i with $1 \leq i \leq n$ such that $p \mid a_i$.*

Proof. The proof is by induction on n. The case when $n = 1$ is obvious. So assume that the proposition is *true* whenever a prime number divides a product of $n - 1$ integers. To prove the result for $n \geq 2$, suppose that $p \mid (a_1 a_2 \cdots a_n)$. The proof proceeds by cases.

Case 1: $p \mid (a_1 a_2 \cdots a_{n-1})$.

In this case, from the induction hypothesis, p divides one of $a_1, \ldots, a_{n-1}$.

Case 2: $p \nmid (a_1 a_2 \cdots a_{n-1})$.

In this case, $(p, a_1 a_2 \cdots a_{n-1}) = 1$, as you are asked to show in the exercises. Applying Proposition 5.2 with $a = p$, $b = a_1 a_2 \cdots a_{n-1}$, and $c = a_n$ yields that $p \mid a_n$.

In either case, p divides one of the integers $a_1, \ldots, a_n$, thus completing the proof.

QED

Now it is possible to prove that primes are the building blocks of the integers, as stated in the following theorem.

Theorem 5.6 (The Fundamental Theorem of Arithmetic) *For any integer $a > 1$, there are primes $p_1, \ldots, p_n$ such that $a = p_1 p_2 \cdots p_n$ and this decomposition is unique except for the order in which the primes appear.*

Proof. It is first necessary to show the existence of these prime numbers and then to show that they are unique.

By contradiction, suppose that there in an integer $a > 1$ such that a *cannot* be written as the product of primes. Let b be the *first* such integer (this b exists by the Least-Integer Principle). Because b is *not* prime, you can write b as the product of two positive integers – say, b_1 and b_2 – that are each greater than 1 and less than b. Now, because b_1 and b_2 are less than b, by the induction hypothesis, you can write both b_1 and b_2 as a product of primes and hence also $b = b_1 b_2$. This contradicts the fact that b could not be expressed as a product of primes, and so, in fact, any integer $a > 1$ can be written as a finite product of primes.

It remains to prove the uniqueness. So suppose that there are two ways to decompse $a > 1$ as a product of primes, namely,

$$a = p_1 p_2 \cdots p_n = q_1 q_2 \cdots q_m. \tag{5.28}$$

First, it is shown that p_1 is equal to some q_j. This is true because $p_1 \mid a$ and so $p_1 \mid (q_1 q_2 \cdots q_m)$. By Proposition 5.3, it follows that $p_1 \mid q_j$ for some integer j with $1 \le j \le m$. But because q_j is prime and $p_1 \mid q_j$, it must be that $p_1 = 1$ or $p_1 = q_j$. But p_1 is also prime, so $p_1 \neq 1$ and hence it must be that $p_1 = q_j$. Canceling p_1 and q_j in (5.28) results in the following:

$$p_2 p_3 \cdots p_n = q_1 q_2 \cdots q_{j-1} q_{j+1} \cdots q_m. \tag{5.29}$$

Repeating this argument means that each of $p_2, \ldots, p_n$ is equal to one of the primes on the right side of (5.29). Neither list in (5.29) can run out before the other because otherwise, you would have that 1 is equal to a product of primes, which cannot happen. Thus, in fact, $n = m$ and the decomposition of a into a finite product of primes is unique, except for the order in which they are written.

QED

The following is an example of Theorem 5.6.

Example 5.5 – Factoring an Integer into Prime Numbers

The unique way to factor the integer 360 is

$$360 = 2 \times 2 \times 2 \times 3 \times 3 \times 5 = 2^3 \times 3^2 \times 5^1.$$

In Section 5.1 you have learned about the properties of addition, subtraction, multiplication, and division of integers. You have also seen the role played by the special integers 0 and 1. In Section 5.2, some of these results are unified into an axiomatic system through the process of abstraction.

5.2 Group Theory

One objective of this section is to develop an axiomatic system for studying addition and subtraction of integers, as discussed in Section 5.1. Recall that the operation of addition combines two integers to produce a new integer and that you can think of subtraction as the operation of adding the *negative* of an integer. Thus, to understand both addition and subtraction, you need only study the set $\mathbb{Z}$ of integers together with the binary operation + and the unary operation of negation.

In an analogous manner, the operation of multiplication combines two integers to produces a new integer. However, division requires special care because the result of dividing one integer by another is *not* necessarily an integer (for example, 4 divided by 3 is not an integer). In Section 5.1, you learned one way to study division – the division algorithm – in which the result of dividing the integer a by the integer b is the integer q, with an integer remainder of r. In this section, an alternative approach to dividing a by b is explored, namely, the use of the rational numbers:

$$\mathbb{Q} = \left\{ \frac{p}{q} : p \text{ and } q \text{ are integers with } q \neq 0 \right\}.$$

The advantage of working with $\mathbb{Q}$ is that the operation of dividing one rational number by another (nonzero) rational number *is* a rational number. That is, if p, q, r, and s are integers with $q \neq 0$, $r \neq 0$, and $s \neq 0$, and if

$$a = \frac{p}{q} \text{ and } b = \frac{r}{s},$$

then

$$\frac{a}{b} = \frac{p/q}{r/s} = \frac{p}{q}\left(\frac{s}{r}\right) = \frac{ps}{qr}. \tag{5.30}$$

From (5.30), you can see that the operation of dividing the rational number a by the rational number b is the same as *multiplying* a by the *inverse* of b, that is,

$$\frac{a}{b} = ab^{-1}.$$

Thus, to understand both multiplication and division, you need only study the set $\mathbb{Q}$ of rationals together with the binary operation of multiplication and the unary operation of taking the inverse.

The objective now is to develop an axiomatic system that includes, as special cases, the set $\mathbb{Z}$ of integers with the binary operation of addition and the set $\mathbb{Q}$ of rational numbers with the binary operation of multiplication.

5.2.1 Groups

As you learned in Chapter 1, unification requires identifying similarities and differences. You can then use abstraction to develop an appropriate axiomatic system. This process is illustrated in what follows.

Identifying Similarities and Differences

In Section 5.1, you learned several properties pertaining to addition of integers. Those properties are summarized in the second column of Table 5.1. Consider now the rationals $\mathbb{Q}$ and the operation of multiplication. Some of the corresponding properties are summarized in the last column of Table 5.1. Identify as many similarities and differences as you can from Table 5.1. Compare your list with the following.

Property	Adding Integers	Multiplying Rationals
Closedness	$\forall a, b \in \mathbb{Z},\ a + b \in \mathbb{Z}$.	$\forall a, b \in \mathbb{Q},\ ab \in \mathbb{Q}$.
The number 0	$\forall a \in \mathbb{Z}, a + 0 = 0 + a = a$. There is only one number having this property.	$\forall a \in \mathbb{Q},\ a(0) = 0(a) = 0$. There is only one number having this property.
The number 1		$\forall a \in \mathbb{Q},\ a(1) = 1(a) = a$. There is only one number having this property.
Inverses	$\forall a \in \mathbb{Z}$, $a + (-a) = (-a) + a = 0$. $\forall a \in \mathbb{Z},\ -a$ is unique.	$\forall a \in \mathbb{Q}$ with $a \neq 0$, $aa^{-1} = a^{-1}a = 1$. $\forall a \in \mathbb{Q}$ with $a \neq 0$, a^{-1} is unique.
Associativity	$\forall a, b, c \in \mathbb{Z}$, $(a + b) + c = a + (b + c)$.	$\forall a, b, c \in \mathbb{Q}$, $(ab)c = a(bc)$.
Commutativity	$\forall a, b \in \mathbb{Z},\ a + b = b + a$.	$\forall a, b \in \mathbb{Q},\ ab = ba$.

Table 5.1: A Comparison of Adding Integers and Multiplying Rationals

Similarities

1. Both binary operations are *closed* on their respective sets. That is, addition is closed on $\mathbb{Z}$ and multiplication is closed on $\mathbb{Q}$.

2. In both cases, there are numbers that play a special role: 0, in the case of addition and 0 and 1, in the case of multiplication.

3. Most numbers a have an "inverse" number: $-a$, in the case of addition and a^{-1}, in the case of multiplication when $a \neq 0$.

4. The associative law holds in both cases.

5. The commutative law holds in both cases.

Differences

1. In the case of addition, there is only *one* special number (0). In the case of multiplication, there are *two* special numbers (1 and 0).

2. With addition, *every* number a has an inverse, namely, $-a$. With multiplication, the number 0 has no inverse.

The objective now is to find a way to eliminate the foregoing differences so that unification is possible. Examining these differences carefully, you will notice that, with regard to multiplication, the number 0 causes some special problems because you cannot find the inverse of 0. One way to overcome this obstacle is to remove the number 0 from consideration when dealing with multiplication. To this end, let

$$\mathbb{Q}_0 = \mathbb{Q} - \{0\}.$$

When you consider multiplication on the set $\mathbb{Q}_0$, you have all of the foregoing similarities to addition and *none* of the differences because now:

1. Each operation has only one special number: 0, in the case of addition and 1, in the case of multiplication.

2. In both cases, *every* number a has an inverse: $-a$, in the case of addition and a^{-1}, in the case of multiplication.

Now that these differences have been eliminated, you can develop an axiomatic system that includes these two special cases.

Developing an Axiomatic System

Consider now the integers together with the operation of addition and the rationals (except for 0) together with the operation of multiplication. In both cases, there is a set of numbers and a numerical operation. In the spirit of abstraction, consider, therefore, a set G of *objects*, rather than specific numbers. Analogously, instead of a specific *numerical* operation, such as addition or multiplication, consider a *closed* binary operation, $\odot$, that combines two elements of G to create a new element of G. Thus, $(G, \odot)$ constitutes an *abstract system* in which the special case of $(\mathbb{Z}, +)$ represents the integers together with addition and that of $(\mathbb{Q}_0, \times)$ represents the rationals (except for 0) together with multiplication.

In creating an *axiomatic* system for $(G, \odot)$, it is necessary to identify those properties that you wish to isolate and study. Also, the *fewest* number of axioms should be included. In this case, from Table 5.1, three axioms are chosen as the basis for the system, as stated in the following definition (the axiom corresponding to the commutative law is not included here but *is* included in the axiomatic system presented in Section 5.2.2).

Definition 5.6 *Let G be a set and $\odot$ be a closed binary operation on G. The pair $(G, \odot)$ is a* **group** *if and only if the following axioms hold:*

1. (Existence of an Identity Element) *There is an element $e \in G$ such that for all $a \in G$, $a \odot e = e \odot a = a$. (The element e is called the* **identity element**.*)*

2. (Existence of an Inverse Element) *For each $a \in G$, there is an element $b \in G$ such that $a \odot b = b \odot a = e$. (The element b is called the* **inverse** *of a.)*

3. (Associative Law) *For all $a, b, c \in G$, $(a \odot b) \odot c = a \odot (b \odot c)$.*

Observe that for the special case of $(\mathbb{Z}, +)$, the identity element is 0 because for all $a \in \mathbb{Z}$, $a+0 = 0+a = a$, and for $(\mathbb{Q}_0, \times)$, the identity element is 1 because for all $a \in \mathbb{Q}_0$, $a(1) = 1(a) = a$. Also, for $(\mathbb{Z}, +)$, the inverse of the element $a \in \mathbb{Z}$ is $-a$ and for $(\mathbb{Q}_0, \times)$, the inverse of $a \in \mathbb{Q}_0$ is a^{-1}.

As mentioned previously, when developing an axiomatic system, the fewest number of axioms is included. It is for this reason that the following properties are *not* stated as axioms:

1. The identity element e is unique.

2. For each $a \in G$, there is a unique inverse element.

These two foregoing properties are not included as axioms because you can prove each one from the already chosen axioms in Definition 5.6, as you will now see.

Proposition 5.4 *If $(G, \odot)$ is a group, then there is a unique identity element.*

Proof. Because $(G, \odot)$ is a group, *there is* an identity element $e \in G$. To see that e is unique, suppose that $f \in G$ is also an identity element. It is shown that $e = f$.

Both e and f are identity elements, so,

$$\text{For all } a \in G, \ a \odot e = e \odot a = a, \tag{5.31}$$
$$\text{For all } a \in G, \ a \odot f = f \odot a = a. \tag{5.32}$$

Replacing a with f in (5.31) and a with e in (5.32) yields

$$e \odot f = f, \tag{5.33}$$
$$e \odot f = e. \tag{5.34}$$

From (5.33) and (5.34) it follows that $e = f$ and so the identity element is unique, thus completing the proof.

QED

As a result of Proposition 5.4, a group has a unique identity element, hereafter denoted by e. The next proposition establishes the uniqueness of the inverse of an element.

Proposition 5.5 *If $(G, \odot)$ is a group, then for any $a \in G$, there is a unique inverse of a, that is, a unique element $b \in G$ such that $a \odot b = b \odot a = e$.*

Proof. Let $a \in G$. Axiom (2) of a group insures that *there is* an element $b \in G$ such that

$$a \odot b = b \odot a = e. \tag{5.35}$$

According to the uniqueness method, suppose that $c \in G$ is also an inverse of a, that is,

$$a \odot c = c \odot a = e. \tag{5.36}$$

Now

$$\begin{aligned} b &= b \odot e && \text{(property of the identity element } e\text{)} \\ &= b \odot (a \odot c) && \text{[from (5.36)]} \\ &= (b \odot a) \odot c && \text{(associativity)} \\ &= e \odot c && \text{[from (5.35)]} \\ &= c && \text{(property of the identity element } e\text{).} \end{aligned}$$

It has now been shown that $b = c$ and so a has a unique inverse, thus completing the proof.

QED

The unique inverse of an element $a \in G$ is written hereafter as a^{-1}. In the context of a group, a^{-1} is the inverse of the element a with respect to the binary operation $\odot$ and should *not* be interpreted as 1 divided by a because division does not exist in a general group. Furthermore, if the operation $\odot$ is actually addition of numbers, then a^{-1} is the *additive* inverse of a and is usually denoted by $-a$ instead of a^{-1}.

Finite Groups

In both the special cases of $(\mathbb{Z}, +)$ and $(\mathbb{Q}_0, \times)$, the sets $\mathbb{Z}$ and $\mathbb{Q}_0$ each contain an *infinite* number of elements. You will now see an example of a group $(G, \odot)$ in which G has a *finite* number of elements. In fact, for each positive integer n, it is possible to create a group $(G, \odot)$ in which G has exactly n elements.

Let n be a fixed positive integer. Recall from Section 5.1.3 that congruence modulo n is an equivalence relation. As shown in Theorem 5.5, this relation partitions the integers into n equivalence classes, denoted by $[0], [1], \ldots, [n-1]$. Keep in mind that for each $a = 0, \ldots, n-1$, $[a]$ is a *set*, that is,

$$[a] = \{k \in \mathbb{Z} : k \equiv a \pmod{n}\}.$$

To construct a group, let G consist of the n sets $[0], \ldots, [n-1]$, hereafter denoted by $\mathbb{Z}_n$, that is,

$$\mathbb{Z}_n = \{[0], \ldots, [n-1]\}.$$

However, a group consists not only of a set, such as $\mathbb{Z}_n$, but also a binary operation that combines two elements of the set. How, then, can you combine two elements – say, $[a]$ and $[b]$ – of $\mathbb{Z}_n$? One approach is the following:

$$[a] \oplus [b] = [a + b].$$

For example, with $n = 4$,

$$[2] \oplus [3] = [2+3] = [5] = [5 \pmod 4)] = [1].$$

Here, you see that $[5] = [1]$. Thus, when you consider $[a] \oplus [b]$ for integers a and b, there are *other* integers – say, c and d – for which $[a] = [c]$ and $[b] = [d]$. The next proposition shows that, even in this case, $[a] \oplus [b] = [c] \oplus [d]$. That is, the operation $\oplus$ is *well defined* in the sense that *whatever* integers are used to represent the sets $[a]$ and $[b]$, the operation $\oplus$ produces the same result.

Proposition 5.6 *If a, b, c, and d are integers for which $[a] = [c]$ and $[b] = [d]$, then $[a] \oplus [b] = [c] \oplus [d]$.*

Proof. Because $[a] = [c]$ and $[b] = [d]$, there are integers p and q such that

$$a = c + pn, \tag{5.37}$$

$$b = d + qn. \tag{5.38}$$

Now, by the definition of $\oplus$,

$$[a] \oplus [b] = [a+b] \text{ and } [c] \oplus [d] = [c+d].$$

To show that $[a+b] = [c+d]$, you must show that there is an integer r such that

$$a + b = (c+d) + rn.$$

By adding (5.37) and (5.38), you can see that $r = p + q$ because

$$\begin{aligned} a + b &= (c+d) + (p+q)n \\ &= (c+d) + rn. \end{aligned}$$

Thus, $[a] \oplus [b] = [c] \oplus [d]$, as desired.

QED

The next proposition establishes that $(\mathbb{Z}_n, \oplus)$ is a group.

Proposition 5.7 *$(\mathbb{Z}_n, \oplus)$ is a group.*

Proof. Observe first that $\oplus$ is closed on $\mathbb{Z}_n$ because if $[a], [b] \in \mathbb{Z}_n$, then $[a] \oplus [b] = [a+b] \in \mathbb{Z}_n$. Now it is necessary to show that the three axioms in Definition 5.6 are satisfied for $(\mathbb{Z}_n, \oplus)$. For example, the identity element is $[0]$ because, for any element $[a] \in \mathbb{Z}_n$,

$$[0] \oplus [a] = [0+a] = [a]$$

and

$$[a] \oplus [0] = [a+0] = [a].$$

Also, the inverse of $[a] \in \mathbb{Z}_n$ is $[-a]$ because

$$[a] \oplus [-a] = [a + (-a)] = [0]$$

and

$$[-a] \oplus [a] = [(-a) + a] = [0].$$

Finally, for associativity, if $[a], [b], [c] \in \mathbb{Z}_n$, then

$$\begin{aligned}
([a] \oplus [b]) \oplus [c] &= [a + b] \oplus [c] && \text{(definition of } \oplus) \\
&= [(a + b) + c] && \text{(definition of } \oplus) \\
&= [a + (b + c)] && \text{(associativity of } +) \\
&= [a] \oplus [b + c] && \text{(definition of } \oplus) \\
&= [a] \oplus ([b] \oplus [c]) && \text{(definition of } \oplus).
\end{aligned}$$

Thus, the three axioms in Definition 5.6 hold and so $(\mathbb{Z}_n, \oplus)$ is a group.

QED

As a result of Proposition 5.7, you know that for any positive integer n, $(\mathbb{Z}_n, \oplus)$ is a group with n elements.

Another operation on two elements $[a]$ and $[b]$ in $\mathbb{Z}_n$ is

$$[a] \odot [b] = [ab].$$

Although $(\mathbb{Z}_n, \odot)$ satisfies Axioms 1 and 3 of Definition 5.6 (as you are asked to verify in the exercises), $(\mathbb{Z}_n, \odot)$ fails to be a group because not every element has an inverse. In particular, [0] has no inverse.

5.2.2 Additional Examples of Groups

The concept of a group arises in Section 5.2.1 by considering the special cases of $(\mathbb{Z}, +)$ and $(\mathbb{Q}_0, \times)$. Can you think of other examples of sets of objects together with a binary operation that might also be considered as special cases of a group? (Hint: Think of other mathematical objects for which there are identity and inverse elements.)

One possibility is *functions*, for which there is *sometimes* an identity function and an inverse function (see Section 3.2.3). A second possibility is *matrices*, for which there is an identity matrix and *sometimes* an inverse matrix (see Section 4.3.3). Both functions and matrices are examined in more detail now.

Functions

The objective is to determine if and when a collection of functions constitutes a group. Each function must have a domain and a codomain. To begin with, suppose that a single set S serves as both the domain and the codomain of each function, that is, $f : S \to S$. When forming a group, you need both a set of objects (functions, in this case) and a binary operation that combines two such objects. The set of objects to be considered for the group is

$$M(S) = \{f : S \longrightarrow S\}.$$

A binary operation is needed to combine two functions f and g in $M(S)$. One such operation is that of *composition*, in which $f \circ g : S \to S$ is defined by $(f \circ g)(x) = f(g(x))$, as presented in Section 3.2.2. The question now is the following: Is $(M(S), \circ)$ a group?

First note that composition is *closed* on $M(S)$ because, if $f, g \in M(S)$, then $f : S \to S$ and $g : S \to S$ and consequently, $f \circ g : S \to S$. Thus the function $f \circ g \in M(S)$ and so composition is closed on $M(S)$. It remains to determine whether the three axioms in Definition 5.6 hold.

Existence of an Identity Element. Is there an identity element, that is, a function $i \in M(S)$ such that for any function $f \in M(S)$, $f \circ i = i \circ f = f$? The answer is *yes* because, as you can verify, the function $i : S \to S$ defined by $i(x) = x$ satisfies this property.

Existence of an Inverse Element. The second property for $(M(S), \circ)$ to be a group is the existence of an inverse, that is, for any function $f \in M(S)$, there should exist a function $g \in M(S)$ such that $f \circ g = g \circ f = i$ (where i is the identity function). As discussed in Section 3.2.3, in general there may not exist such a function g, so $(M(S), \circ)$ is *not* a group. However, when f is *bijective* (see Definition 3.17 in Section 3.2.3), there *is* an inverse function, as proved in Theorem 3.1. Thus, rather than working with *all* functions in $M(S)$, consider only those functions that are bijective, so let

$$\overline{M(S)} = \{f \in M(S) : f \text{ is bijective}\}.$$

Perhaps $(\overline{M(S)}, \circ)$ forms a group. Once again, you can verify that composition is closed on $(\overline{M(S)}, \circ)$ because, if f and g are bijective functions, then so is $f \circ g$. Also, the function $i \in \overline{M(S)}$ defined by $i(x) = x$ is an identity element and, according to Theorem 3.1, each function $f \in \overline{M(S)}$ has an inverse. It remains to verify that $\circ$ is associative.

Associativity. To see that $\circ$ is associative, let $f, g, h \in \overline{M(S)}$. It follows that $(f \circ g) \circ h = f \circ (g \circ h)$ because, for all $x \in S$,

$$\begin{aligned}((f \circ g) \circ h)(x) &= (f \circ g)(h(x)) \\ &= f(g(h(x))) \\ &= (f \circ (g \circ h))(x).\end{aligned}$$

From these observations, $(\overline{M(S)}, \circ)$ is a group that is denoted by $\text{Sym}(S)$ and is called the **symmetric group of** S.

Matrices

You learned about matrices in Chapter 4. The objective now is to determine if and when a collection of matrices constitutes a group. Each matrix has a certain number of rows and columns. To begin with, suppose that all the matrices under consideration have n rows and n columns. When forming a group, you need both a set of objects (matrices, in this case) and a binary operation that combines two such objects. The set of objects to be considered for the group is

$$\mathbb{R}^{n \times n} = \{\text{all } (n \times n) \text{ matrices of real numbers}\}.$$

A binary operation is needed to combine two matrices A and B in $\mathbb{R}^{n \times n}$. One such operation is that of matrix multiplication, as presented in Section 4.2.4. The question now is the following: Is $(\mathbb{R}^{n \times n}, \times)$ a group?

First note that matrix multiplication is *closed* on $\mathbb{R}^{n \times n}$ because, if $A, B \in \mathbb{R}^{n \times n}$, then $A \times B$ is also an $(n \times n)$ matrix and so $A \times B \in \mathbb{R}^{n \times n}$. It remains to determine whether the three axioms in Definition 5.6 hold.

Existence of an Identity Element. Is there an identity element, that is, a matrix $I \in \mathbb{R}^{n \times n}$ such that for any matrix $A \in \mathbb{R}^{n \times n}$, $A \times I = I \times A = A$? The answer is *yes* because, as you can verify, the following $(n \times n)$ identity matrix I satisfies this property:

$$I = \begin{bmatrix} 1 & 0 & \cdots & 0 \\ 0 & 1 & \cdots & 0 \\ \vdots & \vdots & \ddots & \vdots \\ 0 & 0 & \cdots & 1 \end{bmatrix}. \tag{5.39}$$

Existence of an Inverse Element. The second property for $(\mathbb{R}^{n \times n}, \times)$ to be a group is the existence of an inverse, that is, for any matrix $A \in \mathbb{R}^{n \times n}$, there should exist a matrix $B \in \mathbb{R}^{n \times n}$ such that $A \times B = B \times A = I$ (where I is the identity matrix). As discussed in Section 4.3.3, in general there may not exist such a matrix B, so $(\mathbb{R}^{n \times n}, \times)$ is *not* a group. However, when A is *nonsingular* (see Definition 4.10 in Section 4.4.1), there *is* an inverse matrix. Therefore, rather than working with *all* matrices in $\mathbb{R}^{n \times n}$, consider only those matrices that are nonsingular, so let

$$\overline{\mathbb{R}^{n \times n}} = \{\text{nonsingular } (n \times n) \text{ matrices}\}.$$

Perhaps $(\overline{\mathbb{R}^{n \times n}}, \times)$ forms a group. Once again, you can verify that matrix multiplication is closed on $\overline{\mathbb{R}^{n \times n}}$ because, if A and B are nonsingular $(n \times n)$ matrices, then so is the matrix $A \times B$ (see Proposition 4.7(b) in Section 4.4.1). That is, if $A, B \in \overline{\mathbb{R}^{n \times n}}$, then $A \times B \in \overline{\mathbb{R}^{n \times n}}$, so matrix multiplication is closed on $\overline{\mathbb{R}^{n \times n}}$. Also, the matrix $I \in \overline{\mathbb{R}^{n \times n}}$ in (5.39) is an identity element and each matrix $A \in \overline{\mathbb{R}^{n \times n}}$ has an inverse by Definition 4.10. It remains to verify that matrix multiplication is associative.

Associativity. From Proposition 4.5(e) in Section 4.2.5, you know that if $A, B, C \in \overline{\mathbb{R}^{n\times n}}$, then $(A\times B)\times C = A\times(B\times C)$. Thus, matrix multiplication is associative.

As a result of these observations, $(\overline{\mathbb{R}^{n\times n}}, \times)$ is a group.

Abelian Groups

Recall from Table 5.1 in Section 5.2.1 that the groups $(\mathbb{Z}, +)$ and $(\mathbb{Q}_0, \times)$ both satisfy the *commutative* property. That is, if $a, b \in \mathbb{Z}$, then $a + b = b + a$. Likewise, if $a, b \in \mathbb{Q}_0$, then $ab = ba$. You can also prove that the group $(\mathbb{Z}_n, \oplus)$ satisfies the commutative property. In contrast, the group $(\overline{M(S)}, \circ)$ does *not* satisfy commutativity. For example, suppose that $S = \{1, 2, 3\}$ and that $f, g : S \to S$ are defined as follows:

$$f(1) = 2,\ f(2) = 3,\ f(3) = 1,$$
$$g(1) = 3,\ g(2) = 2,\ g(3) = 1.$$

Then

$$(g \circ f)(1) = g(f(1)) = g(2) = 2,$$

whereas

$$(f \circ g)(1) = f(g(1)) = f(3) = 1.$$

Also, the group $(\overline{\mathbb{R}^{n\times n}}, \times)$ does not satisfy commutativity. For example, suppose that A and B are $(n \times n)$ matrices defined as follows:

$$A = \begin{bmatrix} 1 & 2 \\ 3 & 4 \end{bmatrix} \text{ and } B = \begin{bmatrix} 0 & 1 \\ 1 & 0 \end{bmatrix}.$$

Then

$$A \times B = \begin{bmatrix} 1 & 2 \\ 3 & 4 \end{bmatrix} \times \begin{bmatrix} 0 & 1 \\ 1 & 0 \end{bmatrix} = \begin{bmatrix} 2 & 1 \\ 4 & 3 \end{bmatrix},$$

whereas

$$B \times A = \begin{bmatrix} 0 & 1 \\ 1 & 0 \end{bmatrix} \times \begin{bmatrix} 1 & 2 \\ 3 & 4 \end{bmatrix} = \begin{bmatrix} 3 & 4 \\ 1 & 2 \end{bmatrix}.$$

As you have just seen, neither functions nor matrices in general satisfy the commutative property whereas the groups $(\mathbb{Z}, +)$, $(\mathbb{Q}_0, \times)$, and $(\mathbb{Z}_n, \oplus)$ do satisfy this property. You can create a definition to identify those groups that *do* satisfy commutativity by adding the commutative property to the list of axioms of a group, as done in the following definition.

Definition 5.7 *Let G be a set and $\odot$ be a closed binary operation on G. The pair $(G, \odot)$ is an* **Abelian group** *if and only if $(G, \odot)$ is a group in which the commutative property holds, that is, if the following four axioms hold:*

1. (Existence of an Identity Element) *There is an identity element $e \in G$ such that for all $a \in G$, $a \odot e = e \odot a = a$.*

2. (Existence of an Inverse Element) *For each $a \in G$, there is an inverse element $b \in G$ such that $a \odot b = b \odot a = e$.*

3. (Associative Law) *For all $a, b, c \in G$, $(a \odot b) \odot c = a \odot (b \odot c)$.*

4. (Commutative Law) *For all $a, b \in G$, $a \odot b = b \odot a$.*

Observe that an Abelian group is a *special case* of a group because any Abelian group satisfies the three axioms of a group in Definition 5.6. In particular, $(\mathbb{Z}, +)$, $(\mathbb{Q}_0, \times)$, and $(\mathbb{Z}_n, \oplus)$ are special cases of groups. Furthermore, from the foregoing discussion, $(\overline{M(S)}, \circ)$ and $(\overline{\mathbb{R}^{n \times n}}, \times)$ are also groups, although these two are *not* Abelian groups.

From here on, unless otherwise indicated, the operation of a group G is denoted by $\odot$, e is the identity element, and the inverse of $a \in G$ is written a^{-1}. Be careful to interpret these notations appropriately for each special case. For example, for the group $(G, \odot) = (\mathbb{Z}, +)$, $a \odot b$ means $a + b$, e is the integer 0, and a^{-1} is $-a$. The next proposition presents some of the basic properties of a group.

Proposition 5.8 *In a group $(G, \odot)$, the following properties hold:*

(a) (Left Cancellation Law) *If $a, b, c \in G$ and $a \odot b = a \odot c$, then $b = c$.*

(b) (Right Cancellation Law) *If $a, b, c \in G$ and $b \odot a = c \odot a$, then $b = c$.*

(c) (Solving Equations) *If $a, b \in G$, then the unique solution to the equation $a \odot x = b$ for $x \in G$ is $x = a^{-1} \odot b$. Likewise, the unique solution to the equation $x \odot a = b$ for $x \in G$ is $x = b \odot a^{-1}$.*

(d) (Inverse of the Inverse) *If $a \in G$, then $(a^{-1})^{-1} = a$ (that is, the inverse of the element a^{-1} is a).*

(e) (Inverse of a Product) *If $a, b \in G$, then $(a \odot b)^{-1} = b^{-1} \odot a^{-1}$.*

Proof. Throughout this proof, let $a, b, c \in G$.

(a) Suppose that $a \odot b = a \odot c$. Applying a^{-1} to the *left* of both sides and using the axioms of a group, you obtain $a^{-1} \odot (a \odot b) = (a^{-1} \odot a) \odot b = e \odot b = b$ and similarly $a^{-1} \odot (a \odot c) = (a^{-1} \odot a) \odot c = e \odot c = c$, and so $b = c$.

(b) This proof is similar to that in part (a) and is left to the exercises.

(c) According to the uniqueness method (see Section 1.6.12), you must first show that $x = a^{-1} \odot b$ is in G and that this value of x is a solution to the equation $a \odot x = b$. Because a^{-1} and b are both in G and $\odot$ is closed on G, it follows that $a^{-1} \odot b$ is also in G. Also,

$$\begin{aligned} a \odot x &= a \odot (a^{-1} \odot b) && \text{(definition of } x\text{)} \\ &= (a \odot a^{-1}) \odot b && \text{(associativity)} \\ &= e \odot b && \text{(property of an inverse)} \\ &= b && \text{(property of the identity).} \end{aligned}$$

To show the uniqueness of this solution, suppose that $y \in G$ also satisfies $a \odot y = b$. Applying a^{-1} to the *left* of both sides yields that $y = a^{-1} \odot b = x$ and so the solution to this equation is unique.

The proof for the equation $x \odot a = b$ is similar and is left to the exercises.

(d) The inverse of a^{-1} is the unique element $b \in G$ such that $a^{-1} \odot b = b \odot a^{-1} = e$. However, $b = a$ satsifies this property, so $(a^{-1})^{-1} = a$.

(e) The inverse of $a \odot b$ is the unique element $x \in G$ such that $(a \odot b) \odot x = x \odot (a \odot b) = e$. However, $x = b^{-1} \odot a^{-1}$ is one element of G that satisfies this property because

$$\begin{aligned} (a \odot b) \odot x &= (a \odot b) \odot (b^{-1} \odot a^{-1}) && \text{(definition of } x\text{)} \\ &= [a \odot (b \odot b^{-1})] \odot a^{-1} && \text{(associativity)} \\ &= (a \odot e) \odot a^{-1} && \text{(property of an inverse)} \\ &= a \odot a^{-1} && \text{(property of the identity)} \\ &= e && \text{(property of an inverse).} \end{aligned}$$

Thus, $x = b^{-1} \odot a^{-1}$ is the inverse of $a \odot b$.

This completes the proof.

QED

As you learned in Chapter 2, one reason for developing an axiomatic system is that you can apply any results you obtain for that system to each of the special cases. For example, consider the group $(\mathbb{Z}_n, \oplus)$ with $n = 5$. According to Proposition 5.8(c), the unique solution to the equation

$$x \oplus [3] = [4]$$

is

$$\begin{aligned} x &= [4] \oplus [3]^{-1} \\ &= [4] \oplus [-3] \\ &= [4 - 3] \\ &= [1]. \end{aligned}$$

5.2.3 Subgroups

Suppose that $(G, \odot)$ is a group. In some cases, it is necessary to work with a subset H of G. When H itself forms a group with $\odot$, the result is called a *subgroup* of G, as stated in the following definition.

Definition 5.8 *Suppose that* $(G, \odot)$ *is a group and* H *is a subset of* G. *Then* $(H, \odot)$ *is a* **subgroup of** $(G, \odot)$ *if and only if* $(H, \odot)$ *forms a group.*

In other words, for a subset H of G, $(H, \odot)$ is a subgroup of $(G, \odot)$ if H together with the operation of $\odot$ from G satisfies the following properties:

1. The operator $\odot$ is *closed* on H, that is, if $a, b \in H$, then $a \odot b \in H$.
2. There is an identity element in H, that is, an element $f \in H$ such that for each $a \in H$, $a \odot f = f \odot a = a$.
3. For each $a \in H$ there is an inverse element $b \in H$ such that $a \odot b = b \odot a = f$.
4. For each $a, b, c \in H$, $(a \odot b) \odot c = a \odot (b \odot c)$.

One way to verify if $(H, \odot)$ is a subgroup of $(G, \odot)$ is to check the four foregoing conditions. However, as you learned in Chapter 1, one of the uses of mathematics is to solve problems more efficiently. For example, rather than having to check if H has its *own* identity element, you need only check if the identity element e of G is in H because, if so, then e is also the identity of H. This and additional savings in determining whether a subset H of G is a subgroup is presented in the next proposition.

Proposition 5.9 *Suppose that* $(G, \odot)$ *is a group whose identity element is* e *and that* H *is a subset of* G. *Then* $(H, \odot)$ *is a subgroup of* $(G, \odot)$ *if and only if the following conditions hold:*

(a) If $a, b \in H$, *then* $a \odot b \in H$.

(b) $e \in H$.

(c) If $a \in H$, *then* $a^{-1} \in H$.

Proof. Suppose that $(H, \odot)$ satisfies (a), (b), and (c). It must be shown that $(H, \odot)$ is a subgroup of $(G, \odot)$, that is, that $(H, \odot)$ satisfies the three axioms for being a group. First, however, you must be sure that $\odot$ is closed on H. Condition (a) insures that this is the case.

Next, you can see that e is the identity of H because $e \in H$ from condition (b) and also, for any $a \in H$, $a \in G$. Furthermore, because e is the identity of G, $a \odot e = e \odot a = a$.

To see that each element of H has an inverse in H, let $a \in H$, so $a \in G$ and hence a has an inverse in G, namely, $a^{-1} \in G$, which, by condition (c), is also in H. Now $a \odot a^{-1} = a^{-1} \odot a = e$, so a^{-1} is the inverse of a in H.

Finally, for associativity, let $a, b, c \in H$, so $a, b, c \in G$ and because $\odot$ is associative in G, it follows that $(a \odot b) \odot c = a \odot (b \odot c)$.

It remains to show that if $(H, \odot)$ is a subgroup of $(G, \odot)$ then conditions (a), (b), and (c) hold. The proofs of these facts are left to the exercises.

QED

Some examples of using Proposition 5.9 follow.

Example 5.6 – The Subgroup of Even Integers

Let H be the set of even integers. Then $(H, +)$ is a subgroup of the group $(\mathbb{Z}, +)$. This is because:

(a) If a and b are even integers, then so is $a + b$.

(b) 0, which is the identity of the integers, is an element of H.

(c) If a is an even integer, then its inverse, $-a$, is also an even integer.

Example 5.7 – The Subgroup Generated by an Element of a Group

Suppose that $(G, \odot)$ is a group with identity element e, and let $a \in G$. The following subset of G gives rise to a subgroup of G:

$$H = \{\ldots, a^{-2}, a^{-1}, e, a, a^2, \ldots\},$$

where

$$a^n = \underbrace{a \odot a \odot \cdots \odot a}_{n \text{ times}} \text{ and } a^{-n} = \underbrace{a^{-1} \odot a^{-1} \odot \cdots \odot a^{-1}}_{n \text{ times}}.$$

This is because:

(a) If a^m and a^n are both in H, then $a^m \odot a^n = a^{m+n} \in H$.

(b) e, which is the identity of G, is in H.

(c) $(a^n)^{-1} = a^{-n} \in H$.

In this section you have learned about the axiomatic systems that constitute groups, Abelian groups, and subgroups. With a group, you can study addition and subtraction of integers. However, you *cannot* use a group to study multiplication and division of integers unless you work with the *rational* numbers. If you ignore division, it is possible to develop an axiomatic system that allows you to study addition, subtraction, and multiplication of integers *simultaneously*, as shown in the next section.

5.3 Rings and Fields

In Section 5.2, you saw how the axiomatic system of a group allows you to study addition (and subtraction) of integers as well as multiplication (and division) of nonzero rational numbers. However, groups pose two disadvantages with regard to studying operations on *integers*:

1. You cannot study two operations *simultaneously*. For example, because a group has only one operation, there is no axiom that reflects the fact that if a, b, and c are integers, then $a(b + c) = (ab) + (ac)$.

2. You cannot study multiplication of integers in the context of a group because the inverses of some integers are not integers, but rather, rational numbers.

In this section, you will learn a new axiomatic system that overcomes both of these deficiences.

5.3.1 Rings

To address the first issue, the new axiomatic system consists of a set, say, R, together with *two* binary operations: the first is denoted by $\oplus$ (corresponding to addition) and the second is denoted by $\odot$ (corresponding to multiplication). In developing the axioms for this system, recall that the integers together with addition form an Abelian group. Generalizing to the system $(R, \oplus, \odot)$, you therefore want the set R together with $\oplus$ to form an Abelian group.

Turning to the axioms for multiplication, you know that the integers together with multiplication do *not* form an Abelian group because the inverse of an integer is not necesssarily an integer. So what properties *does* multiplication of integers satisfy? One property is associativity:

$$\text{For all } a, b, c \in \mathbb{Z},\ (ab)c = a(bc). \tag{5.40}$$

Also, as mentioned previously, multiplication of real numbers satisfies the following property *in relation to addition*:

$$\text{For all } a, b, c \in \mathbb{Z}, a(b + c) = (ab) + (ac) \text{ and } (a + b)c = (ac) + (bc). \tag{5.41}$$

Other properties of multiplication – such as commutativity and the existence of an identity element – are considered subsequently. Putting together the pieces and expressing the properties in (5.40) and (5.41) in terms of the abstract system $(R, \oplus, \odot)$ results in the axiomatic system given in the following definition.

Definition 5.9 *A* **ring** *is a set R together with two binary operations, denoted by $\oplus$ and $\odot$, that satisfy the following axioms:*

1. (Axioms of Addition) $(R, \oplus)$ *is an Abelian group, that is:*
   (a) *There is an identity element $0 \in R$ such that for all $a \in R$, $a \oplus 0 = 0 \oplus a = a$.*
   (b) *For each element $a \in R$, there is an element $-a \in R$ such that $a \oplus (-a) = (-a) \oplus a = 0$.*
   (c) *For all $a, b, c \in R$, $(a \oplus b) \oplus c = a \oplus (b \oplus c)$.*
   (d) *For all $a, b \in R$, $a \oplus b = b \oplus a$.*
2. (Axiom of Multiplication) *For all $a, b, c \in R$, $a \odot (b \odot c) = (a \odot b) \odot c$.*
3. (Distributive Axioms) *For all $a, b, c \in R$,* $(a \oplus b) \odot c = (a \odot c) \oplus (b \odot c)$ *and* $a \odot (b \oplus c) = (a \odot b) \oplus (a \odot c)$.

From here on, $(R, \oplus, \odot)$ is a ring. Several special cases of rings are presented next.

Example 5.8 – Rings of Numbers

The integers together with addition and multiplication form a ring. So do the rational numbers and the real numbers with these two operations.

Example 5.9 – The Ring of Integers Modulo n

For each positive integer n, $\mathbb{Z}_n$ (the integers mod n) forms a ring under the operations of $\oplus$ and $\odot$ in which, for integers a and b:

$$[a] \oplus [b] = [a + b] \text{ and } [a] \odot [b] = [ab].$$

In Section 5.2 it is shown that $(\mathbb{Z}_n, \oplus)$ is an Abelian group and $\odot$ is associative. To see that the distributive axioms holds, let $a, b, c \in \mathbb{Z}$. Then

$$\begin{aligned} [a] \odot ([b] \oplus [c]) &= [a] \odot ([b + c]) && \text{(definition of } \oplus) \\ &= [a(b + c)] && \text{(definition of } \odot) \\ &= [ab + ac] && \text{(distributivity of + and } \cdot) \\ &= [ab] \oplus [ac] && \text{(definition of } \oplus) \\ &= ([a] \odot [b]) \oplus ([a] \odot [c]) && \text{(definition of } \odot). \end{aligned}$$

The proof of the other distributive axiom is similar and is omitted.

Example 5.10 – The Ring of Integers Over $\sqrt{2}$

Let $\mathbb{Z}[\sqrt{2}] = \{a + b\sqrt{2} : a, b \in \mathbb{Z}\}$. Define two binary operations of addition ($\oplus$) and multiplication ($\odot$) on $a + b\sqrt{2}, c + d\sqrt{2} \in \mathbb{Z}[\sqrt{2}]$ in the following natural way on the basis of how you add and multiply those real numbers, that is:

$$(a + b\sqrt{2}) \oplus (c + d\sqrt{2}) = (a + c) + (b + d)\sqrt{2}$$

and

$$\begin{aligned}(a + b\sqrt{2}) \odot (c + d\sqrt{2}) &= ac + ad\sqrt{2} + bc\sqrt{2} + bd\left(\sqrt{2}\right)^2 \\ &= (ac + 2bd) + (ad + bc)\sqrt{2}.\end{aligned}$$

In the exercises, you are asked to show that $(\mathbb{Z}[\sqrt{2}], \oplus, \odot)$ is a ring.

Example 5.11 – The Ring of Functions

Let $M(\mathbb{R}) = \{f : \mathbb{R} \longrightarrow \mathbb{R}\}$. For $f, g \in M(\mathbb{R})$, define the two operations of $f \oplus g$ and $f \odot g$ as the corresponding arithmetic operations on these two functions, that is, for each real number x,

$$(f \oplus g)(x) = f(x) + g(x) \text{ and } (f \odot g)(x) = f(x)g(x).$$

(Do not confuse $f \odot g$, which is multiplication, with $f \circ g$, which is composition.) In the exercises, you are asked to show that $M(\mathbb{R})$ together with $\oplus$ and $\odot$ is a ring.

Properties of Rings

As mentioned in Chapter 2, one of the disadvantages of abstraction is that you move away from specific items. For instance, when you work with a ring, you lose some of the properties that apply *specifically* to the integers. For example, you know that if a and b are integers for which $ab = 0$, then either $a = 0$ or $b = 0$. This statement is *not* necessarily *true* when a and b are elements of a ring. For example, in the ring $M(\mathbb{R})$ in Example 5.11, consider the two functions $f, g : \mathbb{R} \rightarrow \mathbb{R}$ defined as follows:

$$f(x) = \begin{cases} 0, & \text{if } x \geq 0 \\ 1, & \text{if } x < 0 \end{cases} \text{ and } g(x) = \begin{cases} 1, & \text{if } x \geq 0 \\ 0, & \text{if } x < 0. \end{cases}$$

Then $f \odot g$ is the zero function because $(f \odot g)(x) = f(x)g(x) = 0$ and yet neither f nor g is the zero function. The objective of this section is to identify a number of properties that do and do not hold for a ring $(R, \oplus, \odot)$.

Because $(R, \oplus)$ is an Abelian group, all properties that hold for an Abelian group also hold for $(R, \oplus)$. For example, if a is an element of an Abelian group, then you know from Proposition 5.8(d) in Section 5.2.2 that $(a^{-1})^{-1} = a$. In the context of a ring $(R, \oplus, \odot)$, this rule translates to $-(-a) = a$. The following proposition summarizes many of these basic properties as they apply to a ring under $\oplus$. Throughout, $a - b = a \oplus (-b)$ and if n is a positive integer, then na is the sum of n copies of a, that is, $na = a \oplus \cdots \oplus a$ (n terms) and $(-n)a = -(na)$.

Proposition 5.10 *Suppose that $(R, \oplus, \odot)$ is a ring with $a, b, c \in R$.*

(a) The zero element of R is unique.

(b) Every element of R has a unique negative.

(c) If $a \oplus b = a \oplus c$, then $b = c$ (left cancellation law).

(d) If $b \oplus a = c \oplus a$, then $b = c$ (right cancellation law).

(e) $-(-a) = a$ and $-(a \oplus b) = (-a) \oplus (-b)$.

(f) The unique solution to the equations $a \oplus x = b$ and $x \oplus a = b$ for $x \in R$ is $x = b - a$.

(g) For all integers m and n, $m(a \oplus b) = (ma) \oplus (mb)$, $(m + n)a = (ma) \oplus (na)$, and $m(na) = (mn)a$.

Proof. Many of these are the translation of the properties of an Abelian group developed in Section 5.2 (see Proposition 5.8, for example) to a ring under $\oplus$. The details are left to the exercises.

QED

The next proposition presents some of the basic properties of the operation $\odot$ in a ring.

Proposition 5.11 *Suppose that $(R, \oplus, \odot)$ is a ring whose zero element is 0. The following properties hold:*

(a) For all $a \in R$, $0 \odot a = a \odot 0 = 0$.

(b) For all $a, b \in R$, $a \odot (-b) = (-a) \odot b = -(a \odot b)$.

(c) For all $a, b \in R$, $(-a) \odot (-b) = a \odot b$.

(d) For all $a, b, c \in R$, $a \odot (b - c) = (a \odot b) - (a \odot c)$ and $(a - b) \odot c = (a \odot c) - (b \odot c)$.

Proof.

(a) Let $a \in R$. Then

$$\begin{aligned}(0 \odot a) \oplus 0 &= 0 \odot a && \text{(property of 0)}\\ &= (0 \oplus 0) \odot a && \text{(property of 0)}\\ &= (0 \odot a) \oplus (0 \odot a) && \text{(distributive axiom).}\end{aligned}$$

From Proposition 5.10(c), you can cancel $0 \odot a$ from the first and last expressions to obtain $0 = 0 \odot a$. The proof that $a \odot 0 = 0$ is similar.

(b) Let $a, b \in R$. To see that $a \odot (-b) = -(a \odot b)$, note that

$$\begin{aligned}[a \odot (-b)] \oplus (a \odot b) &= a \odot (-b \oplus b) && \text{(distributive axiom)}\\ &= a \odot 0 && (-b \oplus b = 0)\\ &= 0 && \text{[from part (a)].}\end{aligned}$$

Thus, $x = a \odot (-b)$ is a solution to $x \oplus (a \odot b) = 0$, but so is $x = -(a \odot b)$. Hence, from the uniqueness property in Proposition 5.10(f), it must be that $a \odot (-b) = -(a \odot b)$.

The proof that $(-a) \odot b = -(a \odot b)$ is similar.

(c) Let $a, b \in R$. Replacing a with $-a$ in part (b) leads to

$$(-a) \odot (-b) = -[(-a) \odot b] = -[-(a \odot b)].$$

From Proposition 5.10(e), $-[-(a \odot b)] = a \odot b$, and so $(-a) \odot (-b) = a \odot b$.

(d) Let $a, b, c \in R$. Then

$$\begin{aligned}a \odot (b - c) &= a \odot [b \oplus (-c)] && \text{(definition of } -)\\ &= (a \odot b) \oplus [a \odot (-c)] && \text{(distributive axiom)}\\ &= (a \odot b) \oplus [-(a \odot c)] && \text{[from part (b)]}\\ &= (a \odot b) - (a \odot c) && \text{(definition of } -).\end{aligned}$$

The proof that $(a - b) \odot c = (a \odot c) - (b \odot c)$ is similar.

QED

Integral Domains and Subrings

Recall that one of the objectives in developing a ring R is to study the relevant properties of addition and multiplication of integers. The axiom that R together with addition is an Abelian group *does* reflect all the relevant properties of addition for integers. However, *multiplication* of integers has several properties *besides* that of associativity. For example, the integers under multiplication have an identity element – the number 1 – which a general ring might not have (for instance, the ring of even integers has no identity element under multiplication). As another example, multiplication of integers is commutative, which may *not* be the case for a general ring. Each of these properties gives rise to a special kind of ring, as summarized in the following definitions.

Definition 5.10 *An element 1 in a ring $(R, \oplus, \odot)$ is a* **unity element** *of R if and only if for all $a \in R$, $a \odot 1 = 1 \odot a = a$.*

The term *unity* element is used specifically for $\odot$ to distinguish the element 1 from the *identity* element under $\oplus$, which is 0.

Definition 5.11 *A ring $(R, \oplus, \odot)$ is a* **commutative ring** *if and only if for all $a, b \in R$, $a \odot b = b \odot a$.*

As mentioned previously, if a and b are integers with $ab = 0$, then $a = 0$ or $b = 0$. This property is *not* true for some rings. For example, in $\mathbb{Z}_8$, $[2] \odot [4] = [2(4)] = [8] = [8 \ (mod\ 8)] = [0]$, yet $[2] \neq [0]$ and $[4] \neq [0]$. By including this property together with those in Definition 5.10 and Definition 5.11, you obtain a special kind of ring that *does* capture the properties of the integers with addition and multiplication.

Definition 5.12 *An* **integral domain** *is a commutative ring $(R, \oplus, \odot)$ with a unity element $1 \neq 0$ such that if $a, b \in R$ with $a \odot b = 0$, then $a = 0$ or $b = 0$.*

Example 5.12 – The Integral Domains of the Integers, the Rationals, and the Reals

The ring of integers, the ring of rationals, and the ring of real numbers are all examples of integral domains. In contrast, the ring of even integers is *not* an integral domain because there is no unity element.

Example 5.13 – The Integral Domain of $\mathbb{Z}_n$ When n is Prime

As mentioned previously, $\mathbb{Z}_8$ is *not* an integral domain because $[2] \odot [4] = [0]$, yet $[2] \neq [0]$ and $[4] \neq [0]$. In general, though, $\mathbb{Z}_n$ is commutative and has a unity element ($[1]$), but may not be an integral domain because $[a] \odot [b] = [0]$, yet neither $[a]$ nor $[b]$ are $[0]$. However, as you are asked to show in the exercises, when n is *prime*, $\mathbb{Z}_n$ *is* an integral domain.

One additional property of an integral domain (that is not necessarily *true* in a ring) is the cancellation laws for $\odot$, as stated in the following proposition.

Proposition 5.12 *If $(R, \oplus, \odot)$ is an integral domain and $a, b, c \in R$ with $a \neq 0$ and $a \odot b = a \odot c$, then $b = c$. Likewise, if $b \odot a = c \odot a$, then $b = c$.*

Proof. Suppose that $a \odot b = a \odot c$, then $(a \odot b) - (a \odot c) = 0$ and so $a \odot (b - c) = 0$. Because $(R, \oplus, \odot)$ is an integral domain and $a \neq 0$, it follows that $b - c = 0$, and so $b = c$. A similar proof shows that if $b \odot a = c \odot a$, then $b = c$.

QED

The concept of a *subring* is analogous to that of a subgroup, as stated in the following definition.

Definition 5.13 *Let $(R, \oplus, \odot)$ be a ring and S a subset of R. Then $(S, \oplus, \odot)$ is a* **subring** *of $(R, \oplus, \odot)$ if and only if $(S, \oplus, \odot)$ is a ring.*

Example 5.14 – Subrings of Numbers

Under addition and multiplication, the set of even integers is a subring of the set of integers, which in turn is a subring of the set of rationals, which in turn is a subring of the set of real numbers.

As you learned in Chapter 1, one use of mathematics is to solve problems more efficiently. In that spirit, the following proposition provides a more efficient method than using the definition to verify if a subset S of a ring is a subring or not.

Proposition 5.13 *Let $(R, \oplus, \odot)$ be a ring and S be a nonempty subset of R. Then $(S, \oplus, \odot)$ is a subring of $(R, \oplus, \odot)$ if and only if for all elements $a, b \in S$,*

(a) $a \oplus b \in S$,

(b) $a \odot b \in S$, *and*

(c) $-a \in S$.

Proof. The proof is left to the exercises. QED

5.3.2 Fields

As you now know, for a ring $(R, \oplus, \odot)$, the pair $(R, \oplus)$ forms an Abelian group (this is one of the axioms of a ring). Is the same *true* for $(R, \odot)$? To be an Abelian group, $(R, \odot)$ must satisfy the following axioms:

1. There must be a unity element $1 \in R$ such that for all $a \in R$, $a \odot 1 = 1 \odot a = a$.

2. For each $a \in R$, there must be an inverse element, that is, an element $b \in R$ such that $a \odot b = b \odot a = 1$.

3. The operation $\odot$ must be associative.

4. The operation $\odot$ must be commutative.

In general, a ring need only satisfy (3). However, as you just learned, an *integral domain* also satisfies (1) and (4). Thus, an integral domain D can fail to be an Abelian group under $\odot$ only if (2) does not hold.

Examining (2) more carefully, consider the element 0 in an integral domain D. This element *never* has an inverse because, if b *were* such an inverse, it would mean that $0 \odot b = b \odot 0 = 1$ yet, from Proposition 5.11, you also know that $0 \odot b = b \odot 0 = 0$ and, in an integral domain, $0 \neq 1$. Therefore, if you want to form an Abelian group from an integral domain D using $\odot$, you should consider only the *nonzero* elements of D, so let

$$D_0 = \{a \in D : a \neq 0\}.$$

Even in this case, $(D_0, \odot)$ may not form an Abelian group because an element in D_0 may not have an inverse. For example, the nonzero elements of the integral domain of integers does not form an Abelian group under multiplication because the element 2, for example, has no (integer) inverse. In contrast, the nonzero *rational* numbers *do* form a group under multiplication because each nonzero rational number has an inverse that is also a nonzero rational number. The same is *true* for the nonzero real numbers. These special kinds of integral domains are formalized in the following definition.

Definition 5.14 *An integral domain $(D, \oplus, \odot)$ in which the nonzero elements form an Abelian group with respect to $\odot$ is a* **field**.

As mentioned previously, the rationals and the reals are examples of fields. So are the complex numbers. An example of a field with a *finite* number of elements is $\mathbb{Z}_5$ because the nonzero elements – namely, [1], [2], [3], and [4] – form an Abelian group with respect to the operation $[a] \odot [b] = [ab]$. In fact, *every* finite integral domain is a field, as shown in the the next proposition.

Proposition 5.14 *Every finite integral domain is a field.*

Proof. Let $(D, \oplus, \odot)$ be a finite integral domain with unity element 1 and let D_0 be the nonzero elements of D. For an element $a \in D_0$, it must be shown that a has an inverse in D_0, that is, there is an element $b \in D_0$ such that $a \odot b = b \odot a = 1$.

To construct this element b, consider the following powers of $a \neq 0$:

$$1 = a^0, a^1, a^2, a^3, \ldots.$$

None of these elements is 0 because D is an integral domain. Also, because D is *finite*, a pair of the elements in the foregoing list must repeat, that is, there is a nonnegative integer n and a positive integer k such that

$$a^n = a^{n+k},$$

or equivalently,

$$a^n \odot 1 = a^n \odot a^k. \tag{5.42}$$

By the left cancellation law in Proposition 5.12, you can cancel a^n from both sides of (5.42) to obtain

$$1 = a^k = a \odot a^{k-1}. \tag{5.43}$$

From (5.43), it is evident that $b = a^{k-1} \in D_0$. Finally, because multiplication is commutative, $b \odot a = a \odot b = a \odot a^{k-1} = 1$. Thus, every element of D_0 has an inverse in D_0 and so D is a field.

QED

As an example of Proposition 5.14, $\mathbb{Z}_n$ is a field if n is prime. This is because when n is prime, $\mathbb{Z}_n$ is a *finite integral domain* (see Example 5.13).

On a final note, a *subfield* is defined analogously to a subgroup and a subring, as stated in the following definition.

Definition 5.15 *Let $(F, \oplus, \odot)$ be a field and K be a nonempty subset of F. Then $(K, \oplus, \odot)$ is a* **subfield** *of $(F, \oplus, \odot)$ if and only if $(K, \oplus, \odot)$ is a field.*

Analogous to Proposition 5.13, you can verify more easily if $(K, \oplus, \odot)$ is a subfield of a field $(F, \oplus, \odot)$ by using the following proposition.

Proposition 5.15 *Suppose that $(F, \oplus, \odot)$ is a field with identity 0 and unity 1 and that K is a subset of F. Then $(K, \oplus, \odot)$ is a subfield of $(F, \oplus, \odot)$ if and only if the following conditions hold:*

(a) $0 \in K$ and $1 \in K$.

(b) For all $a, b \in K$, $a \oplus b \in K$ and $a \odot b \in K$.

(c) For all $a \in K$, $-a \in K$.

(d) For all $a \in K$ with $a \neq 0$, $a^{-1} \in K$.

Proof. The proof is left to the exercises. QED

In this section, you have learned about the axiomatic system of a ring that allows you to study simultaneously certain properties of addition and multiplication of integers as well as many other special cases. However, if you want to study all the important properties of *multiplication*, then you must work with integral domains. To study the rationals, the reals, and the complex numbers, the axiomatic system of a field is used. A summary of these various topics follows.

Chapter Summary

In this chapter, you have seen how the ideas of Chapters 1 and 2 are applied in abstract algebra. Various axiomatic systems are developed to study the properties of number systems, including the integers, the rationals, the reals, and even the complex numbers. A summary of these axiomatic systems is presented here.

Integer Arithmetic

Starting with the integers, you saw some of the properties of addition, subtraction, multiplication, and division. The first three operations always result in an integer, but division requires special handling because the result of dividing one integer by another is not, in general, an integer. In Section 5.1, you saw how to perform division using remainders (the division algorithm) and modulo arithmetic, which partitions the integers into equivalence classes. You also learned the special properties of *divisible* integers and that any integer greater than 1 is expressible as a product of prime numbers.

Groups

Another way to study division of integers is to work with the *rational* numbers because the result of dividing two integers (and two rationals) is a rational number. Through the process of identifying similarities and differences, it is possible to unify addition and subtraction of integers with multiplication and division of (nonzero) rationals. This is accomplished through the axiomatic system of a group. The associated axioms capture the properties of an identity element, an inverse element, and associativity. You saw some of the basic properties of groups that apply to all the special cases, including the group of equivalence classes of integers modulo n.

In each of these foregoing examples of a group, commutativity holds, but in some groups – such as functions under composition and matrices under multiplication – commutativity does *not* hold. A group in which commutativity *does* hold is an Abelian group. An Abelian group is therefore a special case of a group because any Abelian group is also a group. You have seen some of the basic properties of a group, such as the ability to solve an equation of the form $a \odot x = b$ for x, where a and b are given elements of the group. You also learned that a subgroup is a subset of a group that is itself a group.

Rings and Fields

Because a group has only one operation, you cannot use a group to study both addition and multiplication of numbers simultaneously. To do so, you need the axiomatic system of a ring, which consists of a set R together with *two* binary operations, say, $\oplus$ and $\odot$. The axioms of a ring state that $(R, \oplus)$ forms a group, that $\odot$ is associative, and that the distributive axioms hold between $\odot$ and $\oplus$. Other properties that result from these axioms are also presented, as is the concept of a subring, which is a subset of a ring that itself is a ring.

To study fully the properties of multiplication of *integers*, several additional axioms are added to those of a ring. In particular, by including the existence of a unity element for $\odot$, the axiom that $\odot$ is commutative, and the fact that if a and b are nonzero elements of a ring, then $a \odot b$ is also nonzero, you obtain an integral domain. An integral domain captures the properties of addition, subtraction, and multiplication of integers.

To study the rationals, the reals, and the complex numbers, the axiomatic system of a field is used. A field is an integral domain whose nonzero elements form an Abelian group under $\odot$ and a subfield is a subset of a field that itself is a field under $\oplus$ and $\odot$. In a field, not only does each element have an inverse with respect to $\oplus$, but also, each *nonzero* element has an inverse with respect to $\odot$. It is this property that fails to hold for the integers under multiplication.

Exercises

Exercise 5.1 Suppose that a, b, and c are integers. Complete Proposition 5.1 in Section 5.1.2 by proving each of the following facts:

(a) If $a \mid b$, then $a \mid (-b)$.

(b) If $a \mid b$ and $b \mid c$, then $a \mid c$.

(c) If $a \mid b$ and $a \mid c$, then $a \mid (b + c)$.

(d) If $a \mid b$ and $b \mid a$, then $a = \pm b$.

Exercise 5.2 Is $\mid$ an equivalence relation (see Definition 2.19 in Section 2.3.3)? If so, prove this fact. If not, give specific numerical examples to indicate which of the reflexive, symmetric, and transitive properties do not hold.

Exercise 5.3 Sequentially generalize Definition 5.2 in Section 5.1.1 for a common divisor, as follows.

(a) Define a common divisor of the integers a, b, and c. For example, 3 is a common divisior of 6, 18, and 36.

(b) By using appropriate subscript notation, generalize your definition in part (a) to a common divisor of n integers.

Exercise 5.4 Sequentially generalize Definition 5.3 in Section 5.1.1 for a greatest common divisor in the following manner.

(a) Define a greatest common divisor of the integers a, b, and c. For example, the greatest common divisor of 6, 18, and 36 is 6.

(b) By using appropriate subscript notation, generalize your definition in part (a) to a greatest common divisor of n integers.

Exercise 5.5 Apply abstraction to Definition 5.1 in Section 5.1.1 for the concept of the integer a divides the integer b. That is, by thinking of objects rather than integers, create a meaningful abstract system and a corresponding generalization of Definition 5.1. (Hint: You need to generalize the multiplication of the integers a and b to an abstract system.)

Exercise 5.6 Restate each of the following definitions from Section 5.1.1 in terms of the abstract system you developed in Exercise 5.5.

(a) Definition 5.2 of a common divisor.

(b) Definition 5.3 of a greatest common divisor.

Exercise 5.7 In this exercise, you will work with the definition of a **common multiple** of the integers a and b, which is an integer m such that $a \mid m$ and $b \mid m$. For example, 108 is a common multiple of 9 and -12 because $9 \mid 108$ and $-12 \mid 108$. Observe that 72 is also a common multiple of 9 and -12. In fact, 36 is the "least" common multiple of 9 and -12 (as is -36).

(a) Create a definition of a *least common multiple* of the two integers a and b.

(b) By using appropriate subscript notation, generalize the definition of a common multiple to n integers.

(c) By using appropriate subscript notation together with part (b), generalize your definition of the least common multiple in part (a) to n integers.

Exercise 5.8 Generalize the definitions in Exercise 5.7 according to the following guidelines.

(a) Apply abstraction to the definition of a common multiple of the integers a and b. That is, by thinking of objects rather than integers, create a meaningful abstract system and a corresponding generalization of this definition. (Hint: Use the same approach as in Exercise 5.5.)

(b) Repeat part (a) for the definition of a least common multiple, as developed in Exercise 5.7(a).

Exercise 5.9 Complete Proposition 5.3 in Section 5.1.4 by proving that if p is prime and $p \nmid b$, then $(p, b) = 1$. (Hint: Show that the only positive common divisor of p and b is 1.)

Exercise 5.10 Prove that if p and q are primes with $p \neq q$, then for any integer k, there are integers m and n such that $k = mp + nq$. (Hint: Use Exercise 5.9 and Theorem 5.3 in Section 5.1.2.)

Exercise 5.11 Prove that if p is prime and a and b are integers with $0 \leq a < p$ and $0 \leq b < p$ and $p \mid (ab)$, then $a = 0$ or $b = 0$. (Hint: Use Proposition 5.3 in Section 5.1.4.)

Exercise 5.12 Complete Example 5.13 in Section 5.3.1 by proving that if n is prime and $[a], [b] \in \mathbb{Z}_n$ with $[a] \odot [b] = [0]$, then either $[a] = [0]$ or $[b] = [0]$. (Hint: Use Exercise 5.11.)

Exercise 5.13 Suppose you want to determine if a given integer $n > 1$ is prime or not. Consider Definition 5.5 in Section 5.1.4 for a prime number and then explain why you would want to prove the following proposition:

If for each integer k with $2 \leq k \leq \sqrt{n}$, $k \nmid n$, then n is prime.

Exercise 5.14 Prove the proposition in Exercise 5.13. (Hint: Use the contrapositive method.)

Exercise 5.15 Determine whether each of the following sets of numbers forms a group under the given operation. If so, indicate the identity element and the inverse of each element in the set. If not, explain why not. (Be sure to verify that the given operation is closed.)

(a) $\{0, 1\}$ under addition.

(b) $\mathbb{Q}$ under addition.

(c) $\mathbb{Q}$ under division.

Exercise 5.16 Repeat Exercise 5.15 for each of the following.

(a) $\{\ldots, -10, -5, 0, 5, 10, \ldots\}$ under addition.

(b) $\{2^k : k \in \mathbb{Z}\}$ under multiplication.

(c) $M(\mathbb{R})$ (the set of all functions $f : \mathbb{R} \to \mathbb{R}$) under multiplication, that is, for $f, g \in M(\mathbb{R})$, define $f \odot g : \mathbb{R} \to \mathbb{R}$ by $(f \odot g)(x) = f(x)g(x)$.

Exercise 5.17 Prove that $(\mathbb{Z}_n, \odot)$ satisfies Axioms (1) and (3) of Definition 5.6 in Section 5.2.1 for a group, but not Axiom (2), where, for $[a], [b] \in \mathbb{Z}_n$, $[a] \odot [b] = [ab]$. That is, show that there is an identity element and that associativity holds, but that some element of $\mathbb{Z}_n$ has no inverse under $\odot$.

Exercise 5.18 Consider $\mathbb{Z}_n$ together with the operation $\odot$ defined by $[a] \odot [b] = [ab]$, for $[a], [b] \in \mathbb{Z}_n$.

(a) Is $(\mathbb{Z}_4 \backslash [0], \odot)$ a group? If so, indicate the identity element and the inverse of each element. If not, explain why not.

(b) Is $(\mathbb{Z}_5 \backslash [0], \odot)$ a group? If so, indicate the identity element and the inverse of each element. If not, explain why not.

Exercise 5.19 Suppose that $(G, \odot)$ is a group with identity element e and that $a, b \in G$. Complete Proposition 5.8(b) in Section 5.2.2 by proving that if $b \odot a = c \odot a$, then $b = c$.

Exercise 5.20 Suppose that $(G, \odot)$ is a group with identity element e and that $a, b \in G$. Complete Proposition 5.8(c) in Section 5.2.2 by proving that the unique solution to the equation $x \odot a = b$ for $x \in G$ is $x = b \odot a^{-1}$.

Exercise 5.21 Recall that one reason for developing an axiomatic system is that you can apply the results you obtain to each of the special cases. In that spirit, apply the results from Exercises 5.19 and 5.20 to the special case of the group consisting of all nonsingular $(n \times n)$ matrices under matrix multiplication. Use the symbols A, B, and C for matrices in the group.

Exercise 5.22 Repeat Exercise 5.21 for the special case of the symmetric group, $\text{Sym}(S) = (\overline{M(S)}, \circ)$, where S is a given set, $\overline{M(S)} = \{f : S \to S : f \text{ is bijective}\}$, and $\circ$ is composition (see Section 5.2.2). Use the symbols f, g, and h for functions in $\overline{M(S)}$.

Exercise 5.23 In developing a group for functions under composition, it is assumed that a set S serves as both the domain and codomain of each function. Explain what difficulties arise in trying to generalize this result to a set of functions $f : A \to B$, where A and B are sets.

Exercise 5.24 In developing a group for matrices under multiplication, it is assumed that each matrix has n rows and n columns. Explain what difficulties arise in trying to generalize this result to a set of matrices with m rows and n columns.

Exercise 5.25 Suppose that $(H, \odot)$ is a subgroup of a group $(G, \odot)$ whose identity element is e. Complete Proposition 5.9 in Section 5.2.3 by proving each of the following statements.

(a) If $a, b \in H$, then $a \odot b \in H$.

(b) $e \in H$.

(c) If $a \in H$, then $a^{-1} \in H$.

Exercise 5.26 Suppose that $(G, \odot)$ is a group whose identity element is e and that $H \subseteq G$.

(a) Looking at Proposition 5.9, explain why you would want to prove the following proposition:

$(H, \odot)$ is a subgroup of $(G, \odot)$ if and only if (1) $e \in H$ and (2) if $a, b \in H$, then $a \odot b^{-1} \in H$.

(b) Prove the proposition in part (a). (Hint: Use Proposition 5.9 in Section 5.2.3.)

Exercise 5.27 Complete Example 5.10 in Section 5.3.1 by proving that $\mathbb{Z}[\sqrt{2}] = \{a + b\sqrt{2} : a, b \in \mathbb{Z}\}$ is a ring under the following operations of $\oplus$ and $\odot$ for $a + b\sqrt{2}, c + d\sqrt{2} \in \mathbb{Z}[\sqrt{2}]$:

$$(a + b\sqrt{2}) \oplus (c + d\sqrt{2}) = (a + c) + (b + d)\sqrt{2}$$
$$(a + b\sqrt{2}) \odot (c + d\sqrt{2}) = (ac + 2bd) + (ad + bc)\sqrt{2}.$$

Exercise 5.28 Complete Example 5.11 in Section 5.3.1 by proving that $M(\mathbb{R}) = \{f : \mathbb{R} \to \mathbb{R}\}$ together with arithmetic addition and multiplication of functions is a ring.

Exercise 5.29 Prove Proposition 5.10 in Section 5.3.1.

Exercise 5.30 Prove Proposition 5.13 in Section 5.3.1.

Exercise 5.31 Prove Proposition 5.15 in Section 5.3.2.

Chapter 6

SELECTED TOPICS IN REAL ANALYSIS

You will now see how the techniques of Chapters 1 and 2 are applied in the area of **real analysis**, which is the study of the properties of real numbers and related concepts. The following sets are used throughout this chapter:

$$\begin{aligned}
\mathbb{N} &= \{1, 2, 3, \ldots\}, \\
\mathbb{Z} &= \{\ldots, -2, -1, 0, 1, 2, \ldots\}, \\
\mathbb{Q} &= \left\{ \frac{p}{q} : p, q \in \mathbb{Z} \right\}, \\
\mathbb{R} &= \{\text{real numbers}\}.
\end{aligned}$$

Other material from this book that you should review is indicated when necessary.

6.1 Algebraic and Order Properties of Real Numbers

One of the goals in this book is to identify and understand the similarities and differences between the sets $\mathbb{Z}$, $\mathbb{Q}$, and $\mathbb{R}$. How many similarities and differences can you find? Compare your list with the following one.

Similarities Between the Sets $\mathbb{Z}$, $\mathbb{Q}$, and $\mathbb{R}$

1. You can apply the binary operations of addition, subtraction, multiplication, and division to the numbers in each of these sets. In each case, these operations satisfy certain properties, such as associativity, commutativity, and the like (see Chapter 5).

2. You can compare two numbers in each of these sets with the binary relations $=, >, \geq, <$, and $\leq$ to determine their relative *order*, that is, their relationship to each other.

3. All three sets contain an infinite number of elements.

Differences Between the Sets $\mathbb{Z}$, $\mathbb{Q}$, and $\mathbb{R}$

1. The result of dividing two rationals (or two reals) is a rational (or a real) if the divisor is not 0, but the result of dividing two integers is *not* necessarily an integer.

2. You know that $\mathbb{Z} \subseteq \mathbb{Q} \subseteq \mathbb{R}$, but there are elements of $\mathbb{Q}$ (such as $\frac{1}{2}$) that are not in $\mathbb{Z}$. Similarly, but perhaps not so obvious, is the fact that there are elements of $\mathbb{R}$ (such as $\sqrt{2}$) that are not in $\mathbb{Q}$.

3. Between any two distinct real numbers, you can find another real number – for example, between 1.5 and 1.6 is the real number 1.55. The same is *true* for the rational numbers. However, this property is *not true* for the integers – for example, between the integers 1 and 2, you *cannot* find another integer.

4. Each set contains an infinite number of elements but, as you will learn in Section 6.2, the number of elements in $\mathbb{Z}$ is *not* the same as the number of elements in $\mathbb{R}$, which has many more elements than does $\mathbb{Z}$. Surprisingly, $\mathbb{Z}$ and $\mathbb{Q}$ *do* contain the same number of elements.

5. As described in Section 5.1, the Least-Integer Principle is an axiom stating that a nonempty set of positive integers has a least element. This property is *not* satisfied by the reals. For example,

$$T = \{x \in \mathbb{R} : 0 < x\}$$

is a nonempty set of positive real numbers, but T has no least element (because $0 \notin T$). Likewise, a nonempty set of rational numbers may not have a least element. In this sense, the rationals and the reals are the same. However, certain nonempty subsets of $\mathbb{R}$ satisfy a property (described in Section 6.2) that is *not* shared by those same nonempty subsets of rationals, and this fact is what differentiates the real numbers from the rational numbers.

The objective of Section 6.1 and Section 6.2 is to develop these similarities and differences in a formal way through the use of appropriate axioms and proofs. In this section, algebraic and order properties of the real numbers are developed.

6.1.1 Algebraic Properties of Real Numbers

You already know the four binary operations of adding, subtracting, multiplying, and dividing real numbers. As you learned in Section 5.1, to understand these operations, you need only study addition and multiplication because, with the concept of an *inverse*, you can perform subtraction and division by using addition and multiplication, respectively. For example, if a and b are real numbers, then you can subtract b from a by adding the *negative* of b to a, that is:

$$a - b = a + (-b).$$

Likewise, if $b \neq 0$, then you can divide a by b by multiplying a by the *inverse* of b, that is:

$$\frac{a}{b} = ab^{-1}.$$

These operations and the fundamental properties they satisfy are stated succinctly in the axiomatic system of a *field*, as presented in Section 5.3.2. Those axioms, as they apply to the real numbers, are repeated here without further explanation. They are the laws you are familiar with from your previous experience with addition and multiplication.

Axioms Pertaining to Addition

1. There exists an element $0 \in \mathbb{R}$ such that for all $a \in \mathbb{R}$, $a + 0 = 0 + a = a$.

2. For all $a \in \mathbb{R}$, there exists an element $-a \in \mathbb{R}$ such that $a + (-a) = (-a) + a = 0$.

3. For all $a, b, c \in \mathbb{R}$, $(a + b) + c = a + (b + c)$.

4. For all $a, b \in \mathbb{R}$, $a + b = b + a$.

Axioms Pertaining to Multiplication

5. There exists an element $1 \in \mathbb{R}$ such that for all $a \in \mathbb{R}$, $a(1) = 1(a) = a$.

6. For all $a \in \mathbb{R}$ with $a \neq 0$, there exists an element $a^{-1} \in \mathbb{R}$ such that $aa^{-1} = a^{-1}a = 1$. (The number a^{-1} is also written $1/a$.)

7. For all $a, b, c \in \mathbb{R}$, $(ab)c = a(bc)$.

8. For all $a, b \in \mathbb{R}$, $ab = ba$.

Distributive Axioms

9. For all $a, b, c \in \mathbb{R}$, $a(b + c) = ab + ac$ and $(a + b)c = ac + bc$.

The integers are a subset of the reals in that you can think of a positive integer n as the n-fold sum of the real number 1, the integer 0 is the same as the real number 0, and the negative integer $-n$ is the n-fold sum of the real number -1. Note, however, that Axiom (6) does *not* hold for the integers – for example, the integer 2 has no *integer* inverse. Thus, this property distinguishes the reals from the integers.

The rational numbers $\mathbb{Q}$ are also a subset of $\mathbb{R}$ and satisfy the property that the sum and product of two rational numbers are also rational numbers (prove this for yourself). In fact, $\mathbb{Q}$ together with addition and multiplication satisfy the nine foregoing axioms. Thus, some *additional* property is needed to differentiate the rationals from the reals. You will learn about that property in Section 6.2.

Several other important properties that follow from these axioms are presented in Proposition 5.10 and Proposition 5.11 and are repeated here for the real numbers. See Section 5.3.1 for the proofs.

Proposition 6.1 *Suppose that $a, b, c \in \mathbb{R}$.*

(a) The zero element of $\mathbb{R}$ is unique.

(b) Every element of $\mathbb{R}$ has a unique negative.

(c) If $a + b = a + c$, then $b = c$ (left cancellation law).

(d) If $b + a = c + a$, then $b = c$ (right cancellation law).

(e) $-(-a) = a$ and $-(a + b) = (-a) + (-b)$.

(f) The unique solution to the equations $a + x = b$ and $x + a = b$ for $x \in \mathbb{R}$ is $x = b - a$.

(g) For all integers m and n, $m(a + b) = ma + mb$, $(m + n)a = ma + na$, and $m(na) = (mn)a$.

Proposition 6.2 *The following conditions hold for all $a, b, c \in \mathbb{R}$:*

(a) $0(a) = a(0) = 0$.

(b) $a(-b) = (-a)b = -(ab)$.

(c) $(-a)(-b) = ab$.

(d) $a(b - c) = ab - ac$ and $(a - b)c = ac - bc$.

6.1.2 Order Properties of Real Numbers

You already know how to compare two real numbers to determine their relative *order* with the binary relations $=, >, \geq, <$, and $\leq$. The objective of this section is to generalize these relations through the development of an axiomatic system. You could do so for an *arbitrary* set of objects, but here,

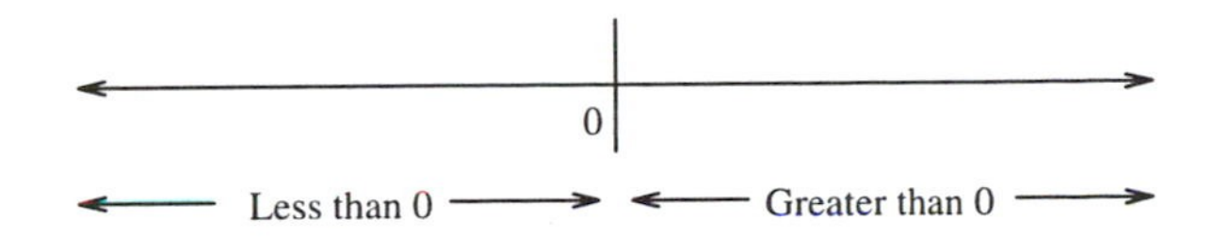

Figure 6.1: The Real Line

only the set $\mathbb{R}$ of real numbers is used together with the binary operations of addition and multiplication.

Using Binary Relations to Compare Real Numbers

The approach mathematicians have developed is to begin by comparing real numbers to 0. In this regard, you know that every real number is either less than 0, equal to 0, or greater than 0, as illustrated by the visual image of the **real line** in Figure 6.1. To express this fact *without* using the words "less than" or "greater than" (which are going to be defined formally), consider first the set of *positive* real numbers, denoted by P. The idea is to identify *axioms* that the numbers in P satisfy. For example, the positive real numbers satisfy the following property with regard to addition:

1. If $a, b \in P$, then $a + b \in P$.

Observe, however, that the negative real numbers *also* satisfy (1), that is, the sum of two negative real numbers is a negative real number. This fact indicates that another axiom is needed to distinguish the positives from the negatives. One such property is that when you *multiply* two positive real numbers, you get a positive real number, but when you multiply two *negative* real numbers, you do *not* get a negative real number. You can express this observation in terms of the elements in the set P as follows:

2. If $a, b \in P$, then $ab \in P$.

The third property to be included in the axiomatic system is the one mentioned previously: for each real number a, either a is positive, a is 0, or a is negative. You can express this observation in terms of the set P and the number 0 as follows:

3. For each $a \in P$, exactly one of the following holds:

$$a \in P,\ a \in \{0\},\ -a \in P.$$

These three foregoing properties are summarized in the following axiomatic system.

The Axiom of the Existence of the Positive Real Numbers

There exists a nonempty subset P of $\mathbb{R}$, called the **positive real numbers**, that satisfies the following three properties:

1. For all $a, b \in P$, $a + b \in P$.
2. For all $a, b \in P$, $ab \in P$.
3. For all $a \in \mathbb{R}$, exactly one of the following is *true*:
   (a) $a \in P$,
   (b) $a \in \{0\}$, or
   (c) $-a \in P$.

The **negative real numbers** is the set $-P = \{-a : a \in P\}$.

Property (3) is called the **trichotomy property** which, in the terminology of Section 2.2.1, means that the real numbers are *partitioned* into three mutually exclusive sets: the positive real numbers, the negative real numbers, and $\{0\}$.

Now it is possible to define the meaning of the symbols $=, >, \geq, <$, and $\leq$ when comparing real numbers to 0.

Definition 6.1 *Let P be the set of positive real numbers and $a \in \mathbb{R}$.*

1. *a **equals** 0, written $a = 0$, if and only if $a \in \{0\}$.*
2. *a is **positive** (or **strictly positive**), written $a > 0$, if and only if $a \in P$.*
3. *a is **nonnegative**, written $a \geq 0$, if and only if $a \in P \cup \{0\}$.*
4. *a is **negative** (or **strictly negative**), written $a < 0$, if and only if $-a \in P$.*
5. *a is **nonpositive**, written $a \leq 0$, if and only if $-a \in P \cup \{0\}$.*

With the use of Definition 6.1, you can define what it means to compare *any* two real numbers with the symbols $=, >, \geq, <$, and $\leq$. The idea is to compare the *difference* of the two real numbers with 0, as described in the following definition.

Definition 6.2 *Let P be the set of positive real numbers and $a, b \in \mathbb{R}$.*

1. *a **equals** b, written $a = b$, if and only if $a - b = 0$.*
2. *a is **greater than** (or **strictly greater than**) b, written $a > b$, if and only if $a - b \in P$.*
3. *a is **greater than or equal to** b, written $a \geq b$, if and only if $a - b \in P \cup \{0\}$.*
4. *a is **less than** (or **strictly less than**) b, written $a < b$, if and only if $b - a \in P$.*
5. *a is **less than or equal to** b, written $a \leq b$, if and only if $b - a \in P \cup \{0\}$.*

Note that $a < b$ is the same as $b > a$ and that $a \leq b$ is the same as $b \geq a$. Also, for $a, b, c \in \mathbb{R}$, the notation

$$a < b < c$$

means that $a < b$ and $b < c$. Similarly,

$$a \leq b \leq c$$

means that $a \leq b$ and $b \leq c$. These same notations apply to $>$ and $\geq$.

As with any axiomatic system, you can prove that certain properties follow from the basic axioms. Such a list of many of the standard properties of inequalities is presented in the next proposition.

Proposition 6.3 *Suppose that $a, b, c \in \mathbb{R}$.*

(a) If $a < b$ and $b < c$, then $a < c$, that is, $<$ is a transitive *relation (see Definition 2.18 in Section 2.3.3). The same is true when $<$ is replaced, respectively, by $\leq$, $>$, and $\geq$.*

(b) One and only one of the following holds: $a < b$, $a = b$, $a > b$.

(c) If $a \leq b$ and $b \leq a$, then $a = b$.

(d) If $a \neq 0$, then $a^2 > 0$.

(e) If $a, b \geq 0$ and $a^2 > b^2$, then $a > b$.

Proof. Let $a, b, c \in \mathbb{R}$ and P be the set of positive real numbers.

(a) Suppose that $a < b$ and $b < c$. By (4) of Definition 6.2, this means that $b - a \in P$ and $c - b \in P$. Adding $b - a$ and $c - b$ and applying Property (1) of the positive real numbers yields that $(b - a) + (c - b) \in P$, or equivalently, by the laws of algebra, that $c - a \in P$. The desired conclusion that $a < c$ now follows from (4) of Definition 6.2. This proof remains valid when $<$ is replaced, respectively, by $\leq, >$, and $\geq$.

(b) Applying the trichotomy property to the real number $b-a$ means that exactly one of the following conditions holds: $b - a \in P, b - a \in \{0\}$, or $-(b - a) \in P$. Proceeding by cases, if $b - a \in P$, then, from (4) of Definition 6.2, $a < b$. Alternatively, if $b - a \in \{0\}$, then $b - a = 0$ and from (1) of Definition 6.2, $a = b$. Finally, if $-(b-a) \in P$, then $a-b \in P$ and so, from (2) of Definition 6.2, $a > b$. Thus, either $a < b, a = b$, or $a > b$.

(c) Because $a \leq b$, by (5) of Definition 6.2, $b - a \in P \cup \{0\}$. Proceed by cases. If $b - a \in \{0\}$, then, by (1) of Definition 6.2, $a = b$, as desired. The other possibility that $b-a \in P$ cannot happen because otherwise, $b > a$, which contradicts the hypothesis that $b \leq a$.

(d) Suppose that $a \neq 0$. By the trichotomy property, this means that $a \in P$ or $-a \in P$. Proceed by cases. If $a \in P$ then, by Property (2) of the positive real numbers, $a^2 = aa \in P$, so $a^2 > 0$. On the other hand, if $-a \in P$, then $a^2 = aa = (-a)(-a) \in P$. In either case, $a^2 \in P$ so, by (2) of Definition 6.1, $a^2 > 0$.

(e) By contradiction, suppose that $a, b \geq 0$, $a^2 > b^2$, and that $a \leq b$. It then follows that $a - b \leq 0$ and $a + b \geq 0$, so

$$\begin{aligned} 0 &< a^2 - b^2 \\ &= (a - b)(a + b) \\ &\leq 0. \end{aligned}$$

The contradiction that $0 < 0$ means that $a > b$.

QED

Several properties of $<$ in relation to the operations of addition and multiplication are summarized in the following proposition.

Proposition 6.4 *Suppose that $a, b, c, d \in \mathbb{R}$.*

(a) If $a < b$, then $a + c < b + c$.

(b) If $a < b$ and $c < d$, then $a + c < b + d$.

(c) If $a < b$ and $c > 0$, then $ac < bc$.
If $a < b$ and $c < 0$, then $ac > bc$.

(d) If $a < 0$, then $a^{-1} < 0$.
If $a > 0$, then $a^{-1} > 0$.

Proof.

(a) If $a < b$, then $b - a > 0$, so $b - a = (c + b) - (a + c) > 0$ and hence $a + c < b + c$.

(b) If $a < b$, then, from part (a), $a + c < b + c$. Similarly, because $c < d$, $b + c < b + d$. From the transitive property in Proposition 6.3(a), it now follows that $a + c < b + c < b + d$, so $a + c < b + d$.

(c) Suppose that $a < b$ and $c > 0$. Then $b - a > 0$ and $c > 0$ so, by Property (2) of the positive real numbers, $(b - a)c > 0$. This means that $bc - ac > 0$, or equivalently, that $ac < bc$.

Now suppose that $a < b$ and $c < 0$. Then $b - a > 0$ and $-c > 0$ so, by Property (2) of the positive real numbers, $(b - a)(-c) > 0$. This means that $ac - bc > 0$, or equivalently, that $ac > bc$.

(d) Suppose that $a < 0$. By the trichotomy property, $a^{-1} > 0, a^{-1} < 0$, or $a^{-1} = 0$. To show that $a^{-1} < 0$, you can rule out the two undesirable cases because if $a^{-1} > 0$, then, by part (c), $1 = aa^{-1} < 0$, which is a contradiction. Likewise, if $a^{-1} = 0$, then $1 = aa^{-1} = a(0) = 0$, which is also a contradiction. The only alternative is that $a^{-1} < 0$.

Suppose now that $a > 0$. As before, you can rule out the two cases that $a^{-1} = 0$ and $a^{-1} < 0$. The only alternative is that $a^{-1} > 0$.

QED

The next proposition confirms that the natural numbers, $\mathbb{N}$, are positive real numbers.

Proposition 6.5 *For every* $n \in \mathbb{N}$, $n > 0$.

Proof. By induction on n, consider first $n = 1$. Because $1 = 1^2$, by Proposition 6.3(d), $1 > 0$. For the induction step, suppose that $n \in \mathbb{N}$ and that $n > 0$. It must be shown that $n + 1 > 0$. But because $n > 0$ and $1 > 0$, by adding these two inequalities you have that $n + 1 > 0 + 0 = 0$. The conclusion now follows by induction. QED

The Absolute Value

From your previous experience with real numbers, you know that the *absolute value* of a real number a is the nonnegative number specified in the following definition.

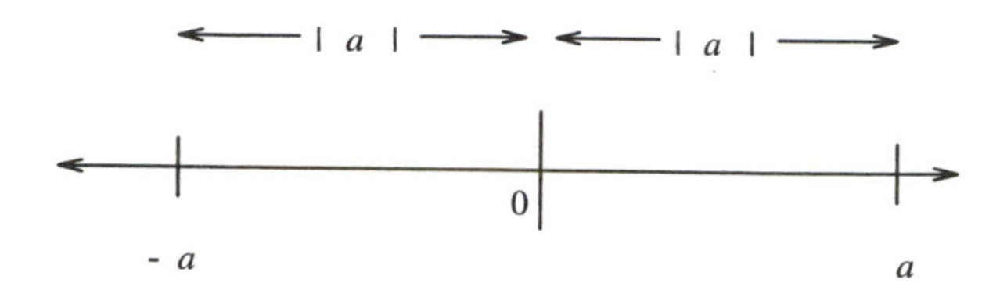

Figure 6.2: The Absolute Value of a Real Number

Definition 6.3 *For $a \in \mathbb{R}$, the* **absolute value** *of a, denoted by $|a|$, is computed as follows:*

$$|a| = \begin{cases} a, & \text{if } a \geq 0 \\ -a, & \text{if } a < 0. \end{cases}$$

The absolute value is a unary operation that associates to each real number a, the nonnegative real number $|a|$. For example, $|5| = 5$, $|0| = 0$, and $|-3| = 3$. As you learned in Section 2.1, it is useful to develop a visual image of a definition. An appropriate picture for $|a|$ is the *distance* on the real line from a to 0, as shown in Figure 6.2. Several common properties of the absolute value are established in the following proposition.

Proposition 6.6 *Suppose that $a, b, c \in \mathbb{R}$ with $c \geq 0$.*

(a) $|a| = 0$ *if and only if* $a = 0$.

(b) $|-a| = |a|$.

(c) $|ab| = |a||b|$.

(d) $|a| \leq c$ *if and only if* $-c \leq a \leq c$.

(e) $-|a| \leq a \leq |a|$.

Proof.
Definition 6.3 is used in the proof of each part and, because the absolute value depends on the sign of the number, a proof by cases (see Section 1.6.11) is usually appropriate.

(a) Suppose first that $|a| = 0$. To show that $a = 0$, the possibilities that $a > 0$ and $a < 0$ are ruled out by contradiction. If $a > 0$, then $|a| = a > 0$, which contradicts the assumption that $|a| = 0$. Likewise, if $a < 0$, then $|a| = -a > 0$, again a contradiction to $|a| = 0$. The only alternative is $a = 0$.

Now suppose that $a = 0$. From Definition 6.3, $|a| = |0| = 0$.

(b) The proof is by cases, depending on the sign of a. If $a = 0$, then $|-a| = |-0| = 0 = |a|$. On the other hand, if $a > 0$, then $-a < 0$, so $|-a| = -(-a) = a = |a|$. Finally, if $a < 0$, then $-a > 0$, so $|-a| = -a = |a|$. Thus, in all cases, $|-a| = |a|$.

(c) The proof is by cases, depending on the signs of a and b. If $a = 0$ or $b = 0$, then $ab = 0$ and also $|a| = 0$ or $|b| = 0$. Thus, $|ab| = |0| = 0 = |a||b|$. If $a > 0$ and $b > 0$, then $ab > 0$, so $|ab| = ab = |a||b|$. If $a > 0$ and $b < 0$, then $ab < 0$, so $|ab| = -(ab) = a(-b) = |a||b|$. Similar proofs cover the cases $a < 0,\ b > 0$ and $a < 0,\ b < 0$.

(d) Suppose first that $|a| \leq c$ and recall from the hypothesis that $c \geq 0$. The proof is by cases, depending on the sign of a. If $a \geq 0$, then $-c \leq 0 \leq a = |a| \leq c$. Alternatively, if $a < 0$, then $-a > 0$, so $-c \leq 0 \leq -a = |a| \leq c$. Multiplying through by -1 results in $c \geq 0 \geq a \geq -c$, or equivalently, $-c \leq a \leq c$. Thus, in either case, $-c \leq a \leq c$.

Suppose now that $-c \leq a \leq c$. The proof is by cases, depending on the sign of a. If $a \geq 0$, then $|a| = a \leq c$. Alternatively, if $a < 0$, then $|a| = -a \leq c$. Thus, in either case, $|a| \leq c$.

(e) The result follows from part (d) with $c = |a|$.

QED

One of the most useful properties of the absolute value is the **triangle inequality**, as stated in the next theorem.

Theorem 6.1 *For any* $a, b \in \mathbb{R}$, $|a + b| \leq |a| + |b|$.

Proof. From Proposition 6.6(e),

$$-|a| \leq a \leq |a| \tag{6.1}$$

and similarly,

$$-|b| \leq b \leq |b|. \tag{6.2}$$

Adding (6.1) and (6.2) and using Proposition 6.4(b) results in the following:

$$-(|a| + |b|) \leq a + b \leq |a| + |b|. \tag{6.3}$$

The result that $|a + b| \leq |a| + |b|$ follows from (6.3) by Proposition 6.6(d) with $c = |a| + |b|$.

QED

In this section, you have seen some of the algebraic and order properties satisfied by the real numbers. These properties are stated in the form of axioms from which further relationships are derived. The rational numbers *also* satisfy all of the algebraic and order properties identified in this section. Therefore, some other property is needed to differentiate $\mathbb{R}$ from $\mathbb{Q}$. This additional property is developed in Section 6.2

6.2 Some Differences Between $\mathbb{Z}$, $\mathbb{Q}$, and $\mathbb{R}$

Now that you know the basic algebraic and order properties of the reals, you will learn some of the ways in which these numbers are different from both the integers and the rationals. For instance, the result of dividing two integers is not necessarily an integer – for example, 3 divided by 2 is not an integer. In contrast, the result of dividing two rationals (or two reals) *is* a rational (or a real). In this section, the remaining differences between $\mathbb{Z}$, $\mathbb{Q}$, and $\mathbb{R}$, as identified at the beginning of Section 6.1, are established.

6.2.1 The Infimum Property of Real Numbers

Up to this point, the rationals and reals have satisfied all of the same properties, including the nine axioms for addition and multiplication given at the beginning of Section 6.1 and the order properties of the binary relations $=, >, \geq, <$ and $\leq$. So what is it that distinguishes the rationals from the reals?

One answer is that there are real numbers that are *not* rationals – for example, $\sqrt{2}$. First of all, how do you know that $\sqrt{2}$ *is*, in fact, a real number? Equivalently stated, how do you know that there is a real number x such that $x^2 = 2$? The existence of such a number is established in this section on the basis of a unique axiom of the reals that is *not* satisfied by the rationals. It is this property that distinguishes the reals from the rationals.

Lower Bounds for Sets of Real Numbers

To understand this property, recall, from Section 5.1.1, the Least-Integer Principle:

Any nonempty set of positive integers has a least element.

For example, the least element of

$$\{n \in \mathbb{Z} : n > 0\}$$

is 1. In contrast, consider the same set of *real* numbers, that is:

$$T = \{x \in \mathbb{R} : x > 0\}. \tag{6.4}$$

In the following proposition, it is proved that T has no least element.

Proposition 6.7 *The set $T = \{x \in \mathbb{R} : x > 0\}$ has no least element.*

Proof. By contradiction, suppose that $a > 0$ *is* the least element of T. A contradiction is reached by producing a *smaller* element of T. In fact, $a/2 = a(2^{-1})$ is such a number. You can see that $a/2 > 0$ because $2 > 0$ so, by Proposition 6.4(d), $\frac{1}{2} = 2^{-1} > 0$. But then, by Property (2) of the positive real numbers, because $a > 0$, $a/2 = a(2^{-1}) > 0$.

It remains to show that $a/2 < a$. But $0 < a$ so, by Proposition 6.4(a),

$$a = 0 + a < a + a = 2a. \tag{6.5}$$

On multiplying both sides of (6.5) by the positive number 2^{-1} and using the algebraic properties of real numbers, it follows that $a/2 < a$. The contradiction that $a/2$ is a smaller element of T than a establishes that T has no least element.

QED

Even though $T = \{x \in \mathbb{R} : x > 0\}$ has no least element (because $0 \notin T$), 0 is an important number relative to T because *every* element of T is greater than or equal to 0. This motivates the following definition.

Definition 6.4 *A real number t is a* **lower bound for a set T of real numbers** *if and only if for each $x \in T$, $t \le x$.*

As you learned in Section 2.1, you should assoicate a visual image with a definition. For Definition 6.4, such a picture arises by realizing that if t is a lower bound for T, then all of T is to the *right* of t, as illustrated in Figure 6.3. An advantage of a symbolic definition is that you can write the negation more easily. For example, applying the rule for negating the quantifier *for all* in Definition 6.4 results in the following statement:

> t is *not* a lower bound for the set T of real numbers if and only if there is an element $x \in T$ such that $x < t$.

The Infimum of a Set of Real Numbers

Some sets of real numbers *have* lower bounds – such as the set T in (6.4) – but other sets do not. For example,

$$\{x \in \mathbb{R} : x < 2\} \tag{6.6}$$

has no lower bound. These two types of sets of real numbers are distinguished by the following definition.

Definition 6.5 *A set T of real numbers is* **bounded below** *if and only if there is a real number t such that t is a lower bound for T.*

Observe that the empty set ($\emptyset$) *is* bounded below because, according to Definition 6.4 and Definition 6.5, *every* real number is a lower bound for $\emptyset$.

If a set T *has* a lower bound, say, t, then T has *infinitely* many lower bounds because any real number $s < t$ is also a lower bound for T (see Figure 6.3). As an example, consider again the set $T = \{x \in \mathbb{R} : x > 0\}$ in (6.4), for which you know that 0 is a lower bound. Then -1 is also a lower bound, as is any real number $s < 0$. In this regard, 0 is a *special*

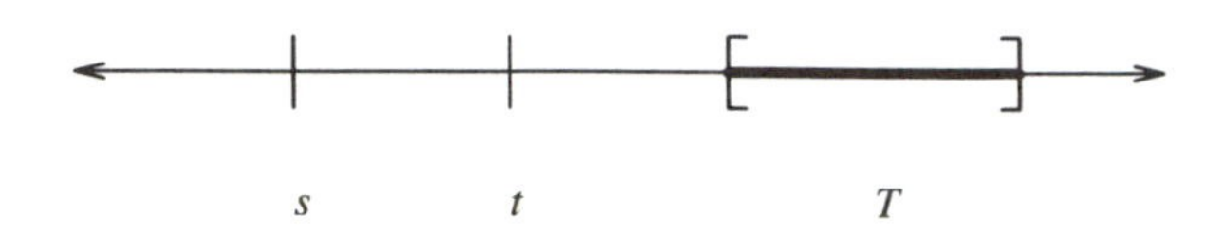

Figure 6.3: Lower Bounds for a Set T of Real Numbers

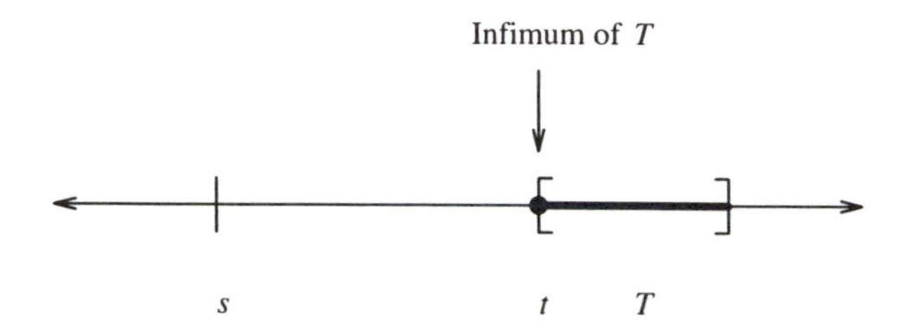

Figure 6.4: The Infimum of a Set T of Real Numbers

lower bound for this set T because 0 is the *largest* of the lower bounds. This concept is formalized in the following definition.

Definition 6.6 *A real number t is an* **infimum** *(or* **greatest lower bound***) of a set T of real numbers, written $t = \inf T$, if and only if both of the following conditions hold:*

1. *t is a lower bound for T.*
2. *For any lower bound s for T, $s \leq t$.*

The visual image associated with an infimum is illustrated in Figure 6.4.

As with lower bounds, some sets *have* an infimum and others do not. For example, 0 is an infimum of the set T in (6.4), but the set in (6.6) *has no* infimum. Also, when a set *does* have an infimum, that infimum may or may not be an element of the set. For example, 0 is the infimum of the set T in (6.4), but $0 \notin T$. In contrast, 0 is also an infimum of the set

$$T = \{x \in \mathbb{R} : x \geq 0\},$$

but here, $0 \in T$. In any case, it is possible to prove that *if* a set of real numbers *has* an infimum, then that infimum is unique, as shown in the following proposition.

Proposition 6.8 *If t is the infimum of a set T of real numbers, then t is unique.*

Proof. According to the uniqueness method (see Section 1.6.12), you can assume that s is also an infimum of T and must then show that $t = s$. According to Proposition 6.3(c), you can accomplish this by using Definition 6.6 to establish that both $t \le s$ and $s \le t$. In particular, because both s and t are infimums of T, by condition (1) of Definition 6.6, both s and t are lower bounds for T. Consequently, by condition (2) in the definition of t being an infimum of T, it follows that $s \le t$. Likewise, by condition (2) in the definition of s being an infimum of T, it follows that $t \le s$. This shows that $s = t$ and hence that t is unique.

QED

There is another useful way to think of the infimum of a set, as described in the following proposition.

Proposition 6.9 *A lower bound t is the infimum of a set T of real numbers if and only if for each real number $\varepsilon > 0$, there is an element $u \in T$ such that $u < t + \varepsilon$.*

Proof. Suppose first that $t = \inf T$ and let $\varepsilon > 0$. It must be shown that there is an element $u \in T$ such that $u < t + \varepsilon$. To construct this element, observe that because t is the *greatest* lower bound for T and $t + \varepsilon > t$, $t + \varepsilon$ cannot be a lower bound for T. By negating Definition 6.4 for $t + \varepsilon$, it follows that there is an element $u \in T$ such that $u < t + \varepsilon$.

Conversely, suppose that for every real number $\varepsilon > 0$, there is an element $u \in T$ such that $u < t + \varepsilon$. Because t is a lower bound for T, according to Definition 6.6, it remains only to show that for any lower bound s for T, $s \le t$. So, let s be a lower bound for T. By contradiction, if $s > t$ then, letting $\varepsilon = s - t > 0$, it follows that there is an element $u \in T$ such that $u < t + \varepsilon = t + (s - t) = s$. In other words, $u < s$, which contradicts the fact that s is a lower bound for T and this completes the proof.

QED

The Infimum Property of Real Numbers

The property that differentiates the reals from the rationals arises by trying to identify *when* – that is, under what conditions – a set T of real numbers *has* an infimum. By working with numerous specific examples, you should come to the conclusion that there are two conditions under which T does *not* have an infimum:

1. When T has no lower bound at all. For example, $T = \{x \in \mathbb{R} : x < 2\}$ has no lower bound.

2. When the set T *has* a lower bound but no *greatest* lower bound. The only such set is the empty set.

Any set of real numbers, other than the foregoing two types, *has* an infimum, as stated formally in the following.

The Infimum Property of $\mathbb{R}$

Every nonempty set of real numbers that is bounded below (that is, that has a lower bound) has an infimum.

6.2.2 Using the Infimum Property to Distinguish the Reals from the Rationals

The specific way in which the infimum property differentiates the reals from the rationals is explained now. To that end, consider the following set of *rational* numbers (which is also a set of *real* numbers):

$$T = \{q \in \mathbb{Q} : q > 0 \text{ and } q^2 > 2\}.$$

T is nonempty ($2 \in T$) and has a lower bound (0, for example). By the infimum property, this set of real numbers has an infimum, which is a real number. However, this infimum is *not* a rational number, as you will see shortly. In other words, the following is the primary difference between the rationals and the reals.

The Difference Between the Rationals and the Reals

What differentiates the rationals from the reals is that the infimum of a set of rational numbers need not be a rational number. In contrast, the infimum of a set of real numbers *is* a real number.

This and other consequences of the infimum property of real numbers are developed now.

The Archimedean Property

In the next proposition, called the **Archimedean property**, the obvious fact that for any real number z, there is a natural number $n > z$ is proved. What is interesting is that the proof is based on the infimum property and cannot be established by the algebraic or order properties of the real numbers presented in Section 6.1.

Proposition 6.10 *For every real number z, there is an $n \in \mathbb{N}$ such that $n > z$.*

Proof. By contradiction, suppose that z is a real number such that for all $n \in \mathbb{N}$, $n \leq z$. Consider now the following nonempty set of real numbers:

$$T = \{z - n : n \in \mathbb{N}\}.$$

Because $n \leq z$ for all $n \in \mathbb{N}$, $z - n \geq 0$ and so 0 is a lower bound for T. By the infimum property, T has an infimum, say, the real number t. A contradiction is reached by showing that t is *not* a lower bound for T. This, in turn, is accomplished by producing an element $s \in T$ such that $s < t$.

By Proposition 6.9 with $\varepsilon = 1$, there is an $r \in T$ such that $r < t + 1$, so, $r - 1 < t$. But because $r \in T$, so is $r - 1$. Hence, $s = r - 1$ is an element of T that is less than t. This contradicts the fact that t is a lower bound for T, thus establishing the claim.

QED

The Archimedean property provides several other useful relationships between the natural numbers and the reals, including the fact that between any two distinct positive real numbers, there is another real number, as stated in the following proposition.

Proposition 6.11 *Let x and z be positive real numbers.*

(a) There is an $n \in \mathbb{N}$ such that $z < nx$.

(b) There is an $n \in \mathbb{N}$ such that $0 < \dfrac{1}{n} < x$.

(c) There is an $n \in \mathbb{N}$ such that $n - 1 \leq x < n$.

(d) If $x < z$, then there is a real number y with $x < y < z$.

Proof.

(a) Let $w = z/x$. By the Archimedean property, there is an $n \in \mathbb{N}$ such that $w < n$, that is

$$\frac{z}{x} < n. \tag{6.7}$$

The result follows on multiplying both sides of (6.7) by the positive real number x.

(b) By setting the value of z to 1 in part (a), it follows that there in an $n \in \mathbb{N}$ such that $0 < 1 < nx$. On dividing through by the positive integer n, you have that $0 < 1/n < x$.

(c) Consider the following set of positive integers:

$$T = \{m \in \mathbb{N} : x < m\}.$$

This set is not empty by the Archimedean property so, from the Least-Integer Principle, this set has a least element, say, $n \in T$, so $x < n$. Now $n - 1 < n$ and because n is the *least* element of T, $n - 1 \notin T$, which means that $n - 1 \leq x$. You therefore have that $n - 1 \leq x < n$.

(d) If $x < z$, then $z - x > 0$ so, by part (b), there is an $n \in \mathbb{N}$ such that

$$0 < \frac{1}{n} < z - x,$$

or, equivalently, by adding x to each term,

$$x < x + \frac{1}{n} < z. \tag{6.8}$$

From (6.8), you can see that the desired value of y is

$$y = x + \frac{1}{n}.$$

QED

The Existence of Real Numbers that Are Not Rational

The infimum property also makes it possible to prove that certain real numbers exist. When applied in a particular way, this establishes that there are real numbers that are not rational. For example, the next theorem uses the infimum property to show that $\sqrt{2}$ is a real number that is not rational.

Theorem 6.2 *There is a real number x such that $x^2 = 2$ and furthermore, x is not rational.*

Proof. The infimum property is used to construct this number by establishing that x is in fact the infimum of the following set of real numbers:

$$T = \{s \in \mathbb{R} : s > 0 \text{ and } s^2 > 2\}.$$

To apply the infimum property, you must show that T is not empty and that T has a lower bound. You can see that $2 \in T$ because $2 > 0$ and $2^2 = 2(2) > 2(1) = 2$, so T is not empty. You can also see that 1 is a lower bound for T because, for any element $s \in T$, $s^2 > 2 > 1 = 1^2$ so, by Proposition 6.3(e), $s > 1$. Consequently, by the infimum property, T has an infimum, say, the real number $x = \inf T$. Note that $x > 0$ because 1 is a lower bound and x is the *greatest* lower bound, so $x \geq 1 > 0$.

It is now necessary to show that $x^2 = 2$. This, in turn, is accomplished by using contradiction to rule out the other two possibilities: $x^2 > 2$ and

$x^2 < 2$. The proof proceeds by cases.

Case 1: $x^2 > 2$.

In this case, a contradiction is reached by showing that x is not a lower bound for T. To do so, you need to construct an $n \in \mathbb{N}$ such that the real number $x - (1/n)$ (which is less than x) is in T.

To construct this n, use the assumption that $x^2 > 2$ and the fact that $x > 0$, from which it follows that

$$\frac{x^2 - 2}{2x} > 0.$$

Now you can construct the desired value for n. In particular, by Proposition 6.11(b), there is an $n \in \mathbb{N}$ such that

$$\frac{1}{n} < \frac{x^2 - 2}{2x}. \tag{6.9}$$

To see that $x - (1/n) \in T$, apply the rules of algebra and inequalities to obtain that

$$\begin{aligned}
\left(x - \frac{1}{n}\right)^2 &= x^2 - \frac{2x}{n} + \frac{1}{n^2} && \text{(algebra)} \\
&> x^2 - \frac{2x}{n} && (1/n^2 > 0) \\
&> x^2 - (x^2 - 2) && \text{[from (6.9)]} \\
&= 2.
\end{aligned}$$

Also, from (6.9),

$$x - \frac{1}{n} > x - \frac{x^2 - 2}{2x} = \frac{x^2 + 2}{2x} > 0.$$

In other words,

$$x - \frac{1}{n} \in T \text{ and } x - \frac{1}{n} < x,$$

so x is not a lower bound for T. This contradiction rules out the possibility that $x^2 > 2$.

Case 2: $x^2 < 2$.

In this case, a contradiction is reached by showing that x is *not* the greatest lower bound for T. This, in turn, is accomplished by constructing an $n \in \mathbb{N}$ such that the real number $x + (1/n)$ (which is greater than x) is also a lower bound for T.

To construct this n, use the assumption that $x^2 < 2$ and the fact that $2x + 1 > 0$, from which it follows that

$$\frac{2 - x^2}{2x + 1} > 0.$$

Now you can construct the desired value for n. In particular, by Proposition 6.11(b), there is an $n \in \mathbb{N}$ such that

$$\frac{1}{n} < \frac{2 - x^2}{2x + 1}. \tag{6.10}$$

The following result is used to show subsequently that $x + (1/n)$ is a lower bound for T:

$$\begin{aligned}
\left(x + \frac{1}{n}\right)^2 &= x^2 + \frac{2x}{n} + \frac{1}{n^2} && \text{(algebra)} \\
&= x^2 + \frac{1}{n}\left(2x + \frac{1}{n}\right) && \text{(algebra)} \\
&\leq x^2 + \frac{1}{n}(2x + 1) && (1/n \leq 1) \\
&< x^2 + (2 - x^2) && \text{[from (6.10)]} \\
&= 2.
\end{aligned}$$

To see that

$$x + \frac{1}{n}$$

is a lower bound for T, let $s \in T$. Then

$$\left(x + \frac{1}{n}\right)^2 \leq 2 < s^2,$$

and so, from Proposition 6.3(e),

$$x + \frac{1}{n} < s.$$

The fact that

$$x + \frac{1}{n} \text{ is a lower bound for } T \text{ and } x + \frac{1}{n} > x$$

contradicts the fact that x is the *greatest* lower bound for T, so it cannot be the case that $x^2 < 2$.

Having ruled out the possibilities that $x^2 > 2$ and $x^2 < 2$, it must that $x^2 = 2$.

To complete the proof, it remains to show that x is not rational, and this is established in Proposition 1.8 in Section 1.6.9.

QED

The set T in Theorem 6.2 shows that the rationals do not satisfy the same infimum property as the reals. That is, when a set S of rationals *does*

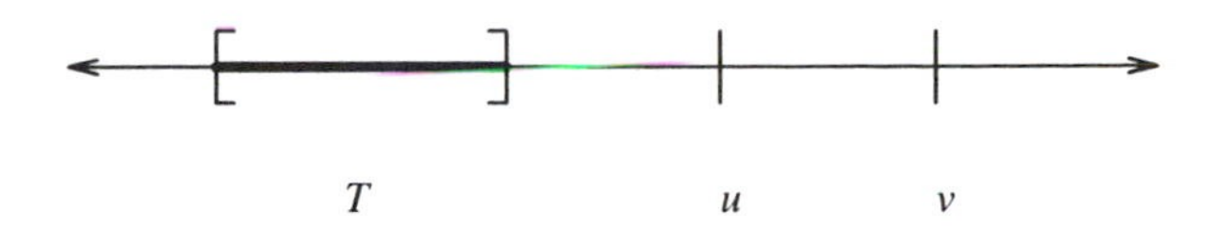

Figure 6.5: Upper Bounds for a Set T of Real Numbers

have an infimum, it is *not* always *true* that inf S is a rational number. For example,

$$S = \{q \in \mathbb{Q} : q > 0 \text{ and } q^2 > 2\}$$

is a nonempty set of rational numbers that is bounded below. Even though S *has* an infimum, that infimum is *not* a rational number. In particular, $\inf S = \sqrt{2} \notin \mathbb{Q}$.

6.2.3 Upper Bounds and Suprema

Until now, you have worked with lower bounds and the infimum of a set of real numbers. According to Definition 6.4, a lower bound for a set T is a real number t such that all of T is to the *right* of t (see Figure 6.3). It is also possible to have a situation in which all of T is to the *left* of a particular real number, as seen in Figure 6.5 and as stated in the following definition.

Definition 6.7 *A real number u is an* **upper bound for a set T of real numbers** *if and only if for each $x \in T$, $x \leq u$.*

As with lower bounds, some sets of real numbers *have* upper bounds and others do not, which motivates the following definition corresponding to Definition 6.5.

Definition 6.8 *A set T of real numbers is* **bounded above** *if and only if there is a real number u such that u is an upper bound for T.*

A set T of real numbers is *bounded* if T has both a lower and an upper bound. An equivalent way to say this is presented in the next definition.

Definition 6.9 *A set T of real numbers is* **bounded** *if and only if there is a real number $M > 0$ such that for all $x \in T$, $|x| \leq M$.*

Corresponding to the largest of the lower bounds is the smallest of the upper bounds, as defined next and illustrated in Figure 6.6.

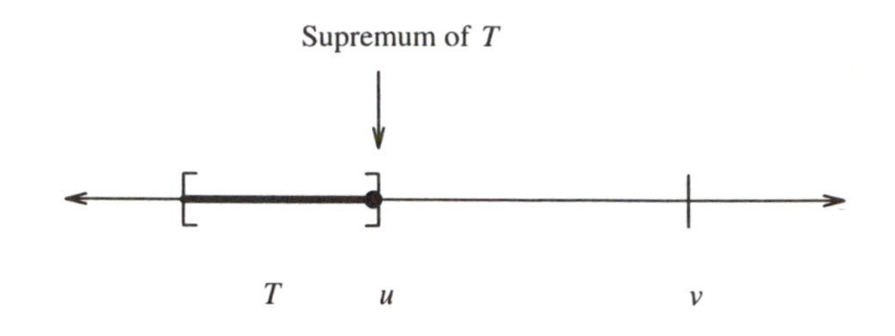

Figure 6.6: The Supremum of a Set T of Real Numbers

Definition 6.10 *A real number u is a* **supremum** *(or* **least upper bound***) of a set T of real numbers, written $u = \sup T$, if and only if both of the following conditions hold:*

1. *u is an upper bound for T.*
2. *For any upper bound v for T, $v \geq u$.*

Finally, corresponding to the infimum property is the following *supremum* property for real numbers.

The Supremum Property of $\mathbb{R}$

Every nonempty set of real numbers that is bounded above (that is, that has an upper bound) *has* a supremum.

In the exercises, you are asked to establish properties of upper bounds and suprema corresponding to those of lower bounds and infima.

6.2.4 The Number of Elements in a Set

The existence of $\sqrt{2}$, as established in Theorem 6.2, shows that there are real numbers that are not rational. Collectively, the set of all real numbers that are not rational is called the set of **irrational numbers**. As you are about to learn, there are many more irrational numbers (and hence real numbers) than rational numbers. Surprisingly, however, the number of rationals and the number of integers are the *same*. This raises the next issue: How, exactly, do you determine the number of elements in a set, especially when the set has *infinitely* many elements?

Finite Sets

To begin with, how do you know that the sets $S = \{2, 4, 7\}$ and $T = \{3, 8, 9\}$ have the same number of elements? Quite simply, you *count* them, thus leading you to the conclusion that both S and T have three elements. One of the problems with "counting" is that you cannot generalize this concept to *infinite* sets, that is, you cannot *count* the number of elements in sets such as the integers, the rationals, and the reals.

What is needed, therefore, is another method for determing how many elements are in a finite set – a method that you *can* generalize to infinite sets. Mathematicians have discovered that one approach to doing so is to use *bijective functions* (see Section 3.2.3). After reviewing that material, you can understand the following definition of what it means for a set to have n elements (where $n \in \mathbb{N}$).

Definition 6.11 *Let $n \in \mathbb{N}$ and define*

$$\mathbb{N}_n = \{1, 2, \ldots, n\}.$$

A **set** S **has** n **elements**, *written* $|S| = n$, *if and only if there is a bijective function* $f : \mathbb{N}_n \to S$. *A set* S *is* **finite** *if and only if* S *is empty or if there is an* $n \in \mathbb{N}$ *such that* $|S| = n$. *The set* S *is* **infinite** *if and only if* S *is not finite.*

Throughout the rest of this section, whenever it is necessary to create a bijective function, the specific method for computing that function is described but the details of showing that the function is bijective are left to the reader, as in the following examples.

Example 6.1 – A Finite Set with Three Elements

The set $S = \{2, 4, 7\}$ has $n = 3$ elements because the following function $f : \mathbb{N}_3 \to S$ is bijective:

$$f(1) = 2,\ f(2) = 4,\ f(3) = 7.$$

Example 6.2 – The Finite Set $\{1, \ldots, n\}$

For each $n \in \mathbb{N}$, the set $\mathbb{N}_n$ is finite because the *identity* function $i : \mathbb{N}_n \to \mathbb{N}_n$ defined by $i(k) = k$, for each $k \in \mathbb{N}_n$, is bijective.

One visual image you can associate with a finite set S in which $|S| = n$ is that you can picture the bijective function $f : \mathbb{N}_n \to S$ in Definition 6.11

as numbering the elements of S from 1 to n, that is,

$$S = \{s_1, s_2, \ldots, s_n\}.$$

The next issue is how to *prove* that a particular set S is finite. One approach is to use Definition 6.11 but another alternative is to show that there is a bijective function to S from another set T that you already *know* is finite, as shown in the next proposition.

Proposition 6.12 *If T is a nonempty finite set and S is a set for which there is a bijective function $g : T \to S$, then S is finite. Moreover, $|S| = |T|$.*

Proof. By hypothesis, T is nonempty and finite, so there is a natural number $n = |T|$ and a bijective function $f : \mathbb{N}_n \to T$. To see that S is finite, observe first that the bijective function $g : T \to S$ in the hypothesis insures that $S \neq \emptyset$. You can now create the bijective function $h : \mathbb{N}_n \to S$ defined by $h(k) = (g \circ f)(k) = g(f(k))$, for each $k \in \mathbb{N}_n$. Thus, S is finite and $|S| = |T| = n$. QED

Another way to prove that a set S is finite is to show that S is a subset of a *known* finite set T, as stated in the next proposition.

Proposition 6.13 *If S is a subset of a set T and T is finite, then S is finite.*

Proof. If $T = \emptyset$, then $S = \emptyset$ and so S is finite. Thus, you can assume that $T \neq \emptyset$. The proof now proceeds by induction on $n = |T|$.

If $n = 1$, then $S = \emptyset$ or S contains the one element of T – in either case, S is finite. The induction hypothesis is that any subset of a set having n elements is finite. So suppose now that T is a finite set with $|T| = n+1 > 1$ and that $S \subseteq T$. Because T is finite, there is a bijective function $f : \mathbb{N}_{n+1} \to T$. The proof is based on the following two cases.

Case 1: $S = T$.

In this case, the bijective function $g : \mathbb{N}_{n+1} \to S$ defined by $g(k) = f(k)$, for each $k \in \mathbb{N}_{n+1}$, shows that S is finite.

Case 2: $S \neq T$.

In this case, there is an element $x \in T$ such that $x \notin S$. To use the induction hypothesis, consider the set $T' = T - \{x\} \neq \emptyset$ and let $1 \leq j \leq n+1$ be the integer for which $f(j) = x$. Then the set T' is finite because of the bijective function $g : \mathbb{N}_n \to T'$ defined, for each $k \in \mathbb{N}_n$, as follows:

$$g(k) = \begin{cases} f(k), & \text{if } k < j \\ f(k+1), & \text{if } k \geq j. \end{cases}$$

Accordingly, $|T'| = n$. Furthermore, because $x \notin S$, $S \subseteq T'$. Therefore, by the induction hypothesis, S is finite.

The result now follows by induction. QED

Infinite Sets

As mentioned previously, the advantage of using a bijective function for determining the number of elements in a finite set is that you *can* generalize this concept to *infinite* sets. But the first question is, How do you know that *there is* an infinite set? The answer is given in the next proposition.

Proposition 6.14 *The set* $\mathbb{N}$ *is infinite.*

Proof. By contradiction, suppose that the nonempty set $\mathbb{N}$ is *finite*. Thus, by definition, there is an $n \in \mathbb{N}$ and a bijective function $f : \mathbb{N}_n \to \mathbb{N}$. Now let $m = \max\{f(1), \ldots, f(n)\}$. Then you can see that the function f is not surjective (see Section 3.2.3) because there is no value of $k \in \mathbb{N}_n$ for which $f(k) = m + 1$. QED

As a result of Proposition 6.14, you know that *there are* infinite sets. In fact, any set that contains an infinite set is also infinite, as stated in the next proposition.

Proposition 6.15 *If S and T are sets for which $S \subseteq T$ and S is infinite, then T is infinite.*

Proof. The proof is left to the exercises. QED

From Proposition 6.15, $\mathbb{Z}$ is infinite because $\mathbb{N} \subseteq \mathbb{Z}$ and $\mathbb{N}$ is infinite. For the same reason, $\mathbb{Q}$ and $\mathbb{R}$ are also infinite.

The next issue is how to *count* the number of elements in an infinite set. One approach is to generalize Definition 6.11 by replacing the finite set $\mathbb{N}_n$ with the *infinite* set $\mathbb{N}$, as done in the following definition.

Definition 6.12 *A set S is* **countably infinite** *(or* **denumerable***) if and only if there is a bijective function $f : \mathbb{N} \to S$. A set that is finite or countably infinite is* **countable***. A set that is not countable is* **uncountable**.

Example 6.3 – The Countably Infinite Set $\mathbb{N}$

The set $\mathbb{N}$ is countably infinite because the identity function $i : \mathbb{N} \to \mathbb{N}$ defined by $i(n) = n$, for each $n \in \mathbb{N}$, is bijective.

Example 6.4 – The Countably Infinite Set of Positive Even Integers

The set $S = \{2, 4, 6, \ldots\}$ is countably infinite because the function $f : \mathbb{N} \to S$ defined by $f(n) = 2n$, for each $n \in \mathbb{N}$, is bijective.

Example 6.5 – The Countably Infinite Set of Integers

The set $\mathbb{Z}$ is countably infinite because the following function $f : \mathbb{N} \to \mathbb{Z}$, defined for each $n \in \mathbb{N}$, is bijective:

$$f(n) = \begin{cases} \dfrac{1-n}{2}, & \text{if } n \text{ is odd} \\ \dfrac{n}{2}, & \text{if } n \text{ is even.} \end{cases}$$

The three sets $\mathbb{N}, S$, and $\mathbb{Z}$ in Examples 6.3, 6.4, and 6.5 all have the *same* number of elements, namely, the number of elements in $\mathbb{N}$. This is because there is a bijective function from $\mathbb{N}$ to each of these sets. You will see shortly that, by this same reasoning, the number of elements in $\mathbb{Q}$ is the same as in $\mathbb{Z}$, which is the same as in $\mathbb{N}$.

The next issue is how to *prove* that a particular set S *is* countably infinite. One approach is to use Definition 6.12 but another alternative is to show, analogous to Proposition 6.12, that there is a bijective function to S from another set T that you already know is countably infinite, as shown in the next proposition.

Proposition 6.16 *If T is a countably infinite set and S is a set for which there is a bijective function $g : T \to S$, then S is countably infinite.*

Proof. To show that S is countably infinite, use the bijective function $g : T \to S$ in the hypothesis together with the bijective function $f : \mathbb{N} \to T$ that exists because T is countably infinite. You can then create the bijective function $h : \mathbb{N} \to S$ defined by $h(n) = (g \circ f)(n) = g(f(n))$, for each $n \in \mathbb{N}$. Thus, S is countably infinite. QED

Another way to show that a set is countably infinite is to show that the set is the union of two disjoint sets, S and T, each of which is known to be countably infinite, as shown in the following proposition.

Proposition 6.17 *If S and T are countably infinite disjoint sets, then so is $S \cup T$.*

Proof. Because S and T are countably infinite, there are bijective functions $f : \mathbb{N} \to S$ and $g : \mathbb{N} \to T$. To show that $S \cup T$ is countably infinite, create the function $h : \mathbb{N} \to S \cup T$ defined as follows for each $n \in \mathbb{N}$:

$$h(n) = \begin{cases} f(\frac{n+1}{2}), & \text{if } n \text{ is odd} \\ g(\frac{n}{2}), & \text{if } n \text{ is even.} \end{cases}$$

The fact that S and T are disjoint insures that h is bijective.

QED

Analogous to Proposition 6.13, you might think that another way to prove that a set S is countably infinite is to show that S is a subset of a known countably infinite set. However, this is *not* correct because S might be *finite*. The correct analogue to Proposition 6.13 follows.

Proposition 6.18 *If T is a countable set and $S \subseteq T$, then S is countable.*

Proof. If T is finite, then so is S by Proposition 6.12 and hence S is countable. So, assume that T is infinite. Likewise, if S is finite, then S is countable, so assume that S is also infinite.

To show that S is countable, it is necessary to construct a bijective function $h : \mathbb{N} \to S$. However, because T is countable and infinite, there is a bijective function $f : \mathbb{N} \to T$. The desired bijective function h is the composition of f with a new function $g : \mathbb{N} \to \mathbb{N}$, defined inductively as follows.

Let $g(1) = a_1$, where a_1 is the least element of the nonempty set $\widehat{f}^{-1}(S) = \{n \in \mathbb{N} : f(n) \in S\}$, which exists by the Least-Integer Principle. Observe that $\widehat{f}^{-1}(S) - \{a_1\} \neq \emptyset$ because S is infinite, so now define $g(2) = a_2$, the least element of the nonempty set $\widehat{f}^{-1}(S) - \{a_1\}$, which again exists by the Least-Integer Principle. Continuing in this manner, let $g(k) = a_k$, the least element of the nonempty set $\widehat{f}^{-1}(S) - \{a_1, \ldots, a_{k-1}\}$. Note that $1 \leq g(1), 2 \leq g(2), \ldots, k \leq g(k)$.

Now $f \circ g : \mathbb{N} \to T$, but you can also think of $f \circ g$ as a function from $\mathbb{N}$ to S because, for each $k \in \mathbb{N}$, $(f \circ g)(k) = f(g(k)) = f(a_k)$ and each $f(a_k) \in S$. Therefore, $f \circ g : \mathbb{N} \to S$.

It remains to show that the function $f \circ g : \mathbb{N} \to S$ is bijective, that is, that $f \circ g$ is both injective and surjective. To see that $f \circ g$ is injective, suppose that j and k are integers for which $(f \circ g)(j) = (f \circ g)(k)$, that is,

$$(f \circ g)(j) = f(g(j)) = f(a_j) = f(a_k) = f(g(k)) = (f \circ g)(k).$$

Because f is injective, it follows that $a_j = a_k$. Then, because $a_{k+1} \notin \{a_1, \ldots, a_k\}$, it must be the case that $j = k$, thus establishing that $f \circ g$ is injective.

To see that $f \circ g$ is surjective, let $s \in S \subseteq T$, for which you must show that there is an $m \in \mathbb{N}$ such that $s = (f \circ g)(m)$. However, the fact that $f : \mathbb{N} \to T$ is surjective means that there is an $n \in \mathbb{N}$ such that $f(n) = s$. In fact, $n \in \widehat{f}^{-1}(S)$ and because $k \le a_k = g(k)$ for each $k \in \mathbb{N}$, it follows that $n = g(m)$ for some $m \in \mathbb{N}$ with $m \le n$. Therefore, $s = f(n) = f(g(m)) = (f \circ g)(m)$, which shows that $f \circ g$ is surjective. Thus, $f \circ g$ is bijective and so S is countable.

QED

It is now possible to show that $\mathbb{Q}$ is countably infinite and therefore that $\mathbb{Q}$ and $\mathbb{Z}$ have the same number of elements. This is because there are bijective functions from $\mathbb{N}$ to $\mathbb{Q}$ and from $\mathbb{N}$ to $\mathbb{Z}$, so both $\mathbb{Q}$ and $\mathbb{Z}$ have the same number of elements as in the set $\mathbb{N}$.

Theorem 6.3 *The set $\mathbb{Q}$ of rational numbers is countably infinite.*

Proof. Because $\mathbb{Q} = \mathbb{Q}^+ \cup \{0\} \cup \mathbb{Q}^-$ (where $\mathbb{Q}^+$ is the set of positive rationals and $\mathbb{Q}^-$ is the set of negative rationals), by Proposition 6.17, it suffices to show that $\mathbb{Q}^+$ is countably infinite. This, in turn, is done by showing that $\mathbb{Q}^+$ is a subset of the countable set F of all positive fractions for then, by Proposition 6.18, $\mathbb{Q}^+$ is countable, that is, $\mathbb{Q}^+$ is either finite or countably infinite. However, because $\mathbb{N} \subseteq \mathbb{Q}^+$ and $\mathbb{N}$ is infinite, by Proposition 6.15, $\mathbb{Q}$ is also infinite.

To show that F is countable, write the fractions as a matrix in which the denominator of all fractions in row n is n, as follows:

$$\begin{array}{lllll} \frac{1}{1} & \frac{2}{1} & \frac{3}{1} & \frac{4}{1} & \cdots \longleftarrow \text{row } n = 1 \\[1em] \frac{1}{2} & \frac{2}{2} & \frac{3}{2} & \frac{4}{2} & \cdots \longleftarrow \text{row } n = 2 \\[1em] \frac{1}{3} & \frac{2}{3} & \frac{3}{3} & \frac{4}{3} & \cdots \longleftarrow \text{row } n = 3 \\[1em] \frac{1}{4} & \frac{2}{4} & \frac{3}{4} & \frac{4}{4} & \cdots \longleftarrow \text{row } n = 4 \\[1em] \vdots & \vdots & \vdots & \vdots & \ddots \end{array}$$

What is needed now is a bijective function $f : \mathbb{N} \to F$. To insure that f is surjective, f must cover every fraction in the foregoing matrix, so first set

$$f(1) = \frac{1}{1}.$$

Next, consider all fractions in the foregoing matrix whose numerator and denominator sum to 3 (that is, $\frac{1}{2}$ and $\frac{2}{1}$) and set

$$f(2) = \frac{1}{2} \text{ and } f(3) = \frac{2}{1}.$$

Next, consider all fractions whose numerator and denominator sum to 4 (that is, $\frac{1}{3}$, $\frac{2}{2}$, and $\frac{3}{1}$) and set

$$f(4) = \frac{1}{3},\ f(5) = \frac{2}{2},\ \text{and } f(6) = \frac{3}{1}.$$

In general, consider all fractions whose numerator, m, and denominator, n, sum to the positive integer p, that is, $m+n = p$. There are $p-1$ such fractions and so you set the next $p - 1$ values of f to those $p - 1$ fractions. More specifically, at this point, you have already identified $f(k)$ for all values of k from 1 up to the number of fractions whose numerator and denominator sum to $p - 1$. There are a total of

$$1 + 2 + \cdots + (p - 2) = \frac{(p-2)(p-3)}{2}$$

such fractions. You then set the next $p-1$ values of f to those $p-1$ fractions whose numerator and denominator sum to p.

On the basis of this approach, it is possible to derive a closed-form expression (see Section 1.1.1) for $f(k)$. In particular, the desired bijective function $f : \mathbb{N} \to F$ is defined as follows, for each $k \in \mathbb{N}$:

$$f(k) = \frac{m}{n},$$

where m and n are the unique integers for which

$$k = \frac{1}{2}(m + n - 2)(m + n - 1) + m.$$

For example, for $k = 7$, $m = 1$ and $n = 4$, so $f(7) = \frac{1}{4}$.

QED

You now know that there are many countably infinite sets – $\mathbb{N}, \mathbb{Z}$, and $\mathbb{Q}$ – but, are there sets that are *uncountable*? The final result in this section provides the answer by establishing that the reals are *uncountable*. In fact, the next theorem shows that even a small portion of the reals is uncountable.

Theorem 6.4 *The set* $S = \{x \in \mathbb{R} : 0 \leq x \leq 1\}$ *is uncountable.*

Proof. The proof is by contradiction, so assume that S *is* countable. Then S is countably infinite because S contains $\{1/n : n \in \mathbb{N}\}$, which is countably infinite. Thus, there is a bijective function $f : \mathbb{N} \to S$, so you can list all the real numbers in S using their decimal expansions, as follows:

$$\begin{array}{lcl} x_1 &=& 0.a_{11}a_{12}a_{13}\cdots a_{1n}\cdots \\ x_2 &=& 0.a_{21}a_{22}a_{23}\cdots a_{2n}\cdots \\ x_3 &=& 0.a_{31}a_{32}a_{33}\cdots a_{3n}\cdots \\ \vdots & \vdots & \qquad\vdots \\ x_n &=& 0.a_{n1}a_{n2}a_{n3}\cdots a_{nn}\cdots \\ \vdots & \vdots & \qquad\vdots \end{array} \tag{6.11}$$

Each a_{ij} in (6.11) is an integer between 0 and 9. A contradiction is reached by creating a real number $x \in S$ that is *not* in this list.

To prove that this x is not in the list in (6.11), it is shown that the decimal expansion of x is different from all the ones listed in (6.11). However, it is important to note that some real numbers have *two* decimal expansions, only one of which is in (6.11). For example, the real number 0.5 has the two decimal expansions $0.5000\cdots$ and $0.4999\cdots$, only one of which is on the right side of the equality corresponding to 0.5 in (6.11). To avoid having to check two decimal expansions, the value for x that provides the contradiction has only *one* decimal expansion because no digit of the decimal expansion of x is 0 or 9. In particular, the desired real number x that has only one decimal expansion is

$$x = 0.b_1 b_2 b_3 \cdots b_n \cdots,$$

where, for each $k \in \mathbb{N}$,

$$b_k = \begin{cases} 3, & \text{if } a_{kk} \geq 5 \\ 6, & \text{if } a_{kk} \leq 4. \end{cases} \tag{6.12}$$

Now $0 \leq x \leq 1$, so x must appear on the left side of the equality in (6.11). However, $x \neq x_1$ because the first decimal position of x is $b_1 = 3$, if $a_{11} \geq 5$ and 6, if $a_{11} \leq 4$. In either case, $b_1 \neq a_{11}$. Thus, the first decimal position of x is not equal to that of x_1 and hence $x \neq x_1$. Likewise, $x \neq x_2$ because the *second* decimal position of x is $b_2 = 3$, if $a_{22} \geq 5$ and 6, if $a_{22} \leq 4$. In either case, $b_2 \neq a_{22}$, so $x \neq x_2$. In a similar fashion, you can see that x is not equal to any of the real numbers in the list in (6.11) because the unique decimal expansion of x is different in position k from that position in the decimal expansion of x_k, that is, $b_k \neq a_{kk}$, so $x \neq x_k$. This contradiction shows that S is not countable. QED

Theorem 6.3 and Theorem 6.4 establish another difference between the rationals and the reals – the rationals are countable but the reals are not. Also, the *irrationals* are uncountable because otherwise, the reals, being the union of the rationals and the irrationals, would be countable which, according to Theorem 6.4, is not the case.

Throughout this section, you have seen numerous differences between the rationals and the reals. These differences are based on the infimum property for real numbers that is not satisfied by the rational numbers and on the number of elements in a set. In Section 6.3, you will learn another concept of real numbers that is useful in solving many problems in applied mathematics.

6.3 Sequences of Real Numbers

Now that you know some of the basic properties of real numbers, you will learn about a concept in real analysis that is used to solve many problems in engineering, computer science, and applied mathematics. To illustrate, recall, from Section 6.2, that the $\sqrt{2}$ is a real number between 1 and 2 that is not rational – but how can you find the value of the $\sqrt{2}$ in terms of a decimal expansion? Mathematicians have discovered the following *algorithm* as one way for doing so:

An Algorithm for Computing the Value of $\sqrt{2}$

Step 0. Set $x_1 = 2,\ k = 1$, and go to Step 1.

Step 1. Use the value of x_k to compute x_{k+1}, as follows:

$$x_{k+1} = \frac{1}{2}\left(x_k + \frac{2}{x_k}\right).$$

Increase k by 1 and go to Step 1.

The first five numbers obtained from this algorithm are

$$\begin{aligned} x_1 &= 2.00000000 \\ x_2 &= 1.50000000 \\ x_3 &= 1.41666667 \\ x_4 &= 1.41421569 \\ x_5 &= 1.41421356. \end{aligned}$$

This algorithm will continue to generate values, in fact, *infinitely many* of them, namely: $x_1, x_2, x_3, \ldots$. The solution to this – and to many other problems of science and engineering – is obtained by using an algorithm that generates an *infinite list of real numbers*, the study of which is the focus of this section.

6.3.1 What is a Sequence of Real Numbers?

As you learned in Section 2.2, after identifying a mathematical concept of interest that you want to study, you should create a definition in symbolic form. You will now see how to do so for an *infinite list of real numbers* and also how to work with this concept.

Perhaps an infinite list reminds you of an n-vector $\mathbf{x} = (x_1, \ldots, x_n)$ which, as you will recall from Section 4.1, is a *finite ordered* list of n real numbers. You might therefore try to define an *infinite list* by generalizing the definition of an n-vector.

Proceeding along this line of reasoning, the next step, as described in Section 2.2.2, is to choose an appropriate name for the concept being defined. In this case, rather than use the term *infinite list of real numbers*, mathematicians have chosen that term to be a *sequence of real numbers*, for which you might now create the following definition:

A *sequence of real numbers*, X, is an infinite ordered list of real numbers.

According to this definition, the following is a sequence:

$$X = (1, 3, 5, \ldots). \tag{6.13}$$

As you learned in Section 2.2.2, you must *verify* the definition to make sure that all desirable objects – such as the sequence in (6.13) – *do* satisfy the property (of being an "infinite ordered list of real numbers," in this case). However, you must also verify that *any object that satisfies the property in the definition is what you consider to be a desirable object*. For example, what about the set of *all* real numbers? That, too, is an infinite ordered list of real numbers and therefore satisfies the property in the foregoing definition. Do you want to consider such an ordered list of real numbers to be a *sequence*? The answer depends on what types of infinite ordered lists you want to study. For the purposes of *this* chapter, the answer is *no*. You must therefore modify the foregoing definition in such a way as to exclude the reals from being a sequence. One way to do so is to define a sequence as a *countably infinite* ordered list of real numbers. An equivalent approach is to use a function from the natural numbers, as follows.

Definition 6.13 *A* **sequence of real numbers** *is a function* $X : \mathbb{N} \rightarrow \mathbb{R}$. *For* $k \in \mathbb{N}$, *the notation* x_k *is used instead of* $X(k)$ *and the sequence is written as follows:*

$$X = (x_1, x_2, x_3, \ldots),$$

or equivalently,

$$X = (x_k : k \in \mathbb{N}) \text{ or } X = (x_k).$$

Example 6.6 – Examples of Sequences of Real Numbers

The **zero sequence** is the sequence in which each element is 0, that is:

$$0 = (0, 0, 0, \ldots).$$

A generalization is the **constant sequence** which, for a given value of $x \in \mathbb{R}$, is

$$X = (x, x, x, \ldots).$$

When specifying a sequence, you can list the elements in order, but another common approach is to give a closed-form expression (see Section 1.1.1) for computing the value of x_k, as shown in some of the following examples.

Example 6.7 – A Closed-Form Expression for the Sequence of Even Natural Numbers

The sequence of even natural numbers is

$$X = (2, 4, 6, \ldots),$$

or equivalently, in closed form:

$$X = (2k : k \in \mathbb{N}).$$

Example 6.8 – A Closed-Form Expression for the Reciprocals of the Natural Numbers

The sequence of reciprocals of the natural numbers is

$$X = \left(\frac{1}{1}, \frac{1}{2}, \frac{1}{3}, \ldots\right),$$

or equivalently, in closed form:

$$X = \left(\frac{1}{k} : k \in \mathbb{N}\right).$$

Example 6.9 – The Inductive Form of the Fibonacci Sequence

The Fibonacci sequence is

$$X = (1, 1, 2, 3, 5, 8, 13, \ldots),$$

in which each subsequent term, after the first two, is the sum of the preceding two terms. You can also use the *inductive form* to describe this sequence, as follows:

$$x_1 = 1,\ x_2 = 1,\ x_k = x_{k-1} + x_{k-2}\ (k = 3, 4, \ldots).$$

6.3.2 Operations on Sequences

When you encounter a new mathematical object, such as a sequence, you should learn how to perform unary and binary operations on those objects to create new objects. Several such operations on sequences are described next.

Unary Operations on Sequences

The unary operation of taking the negative of the sequence $X = (x_1, x_2, \ldots)$ results in the following sequence:

$$-X = (-x_k : k \in \mathbb{N}).$$

By considering any real number c instead of -1, you can generalize this operation to that of creating a **multiple of a sequence** X, as follows:

$$cX = (cx_k : k \in \mathbb{N}).$$

Observe that when $c = -1$, the sequence $cX = (-1)X = -X$ is the negative of X, so the negative of a sequence is a special case of a multiple of a sequence. Another example of the multiple of a sequence follows.

Example 6.10 – The Multiple of a Sequence

For the sequence

$$X = (1, 3, 5, \ldots, 2k - 1, \ldots),$$

you have that

$$\begin{aligned} 0X &= (0, 0, 0, \ldots, 0, \ldots) \text{ and} \\ 2X &= (2, 6, 10, \ldots, 4k - 2, \ldots). \end{aligned}$$

Binary Operations on Sequences

Turning now to *binary* operations, you can combine two sequences, say, $X = (x_1, x_2, \ldots)$ and $Y = (y_1, y_2, \ldots)$, to create a new sequence in any of the following ways:

Binary Operation	Symbol	Resulting Sequence
Addition	$+$	$X + Y = (x_k + y_k : k \in \mathbb{N})$
Subtraction	$-$	$X - Y = (x_k - y_k : k \in \mathbb{N})$
Multiplication	$\cdot$	$X \cdot Y = (x_k y_k : k \in \mathbb{N})$
Division	$/$	$X/Y = (x_k / y_k : k \in \mathbb{N})$

To perform the operation of division, it is necessary that for each $k \in \mathbb{N}$, $y_k \neq 0$. You might also notice that addition and subtraction of sequences are a generalization of the corresponding operations on n-vectors, as defined in Section 4.1.3, but multiplication and division are *not*. These various binary operations on sequences are illustrated in the following example.

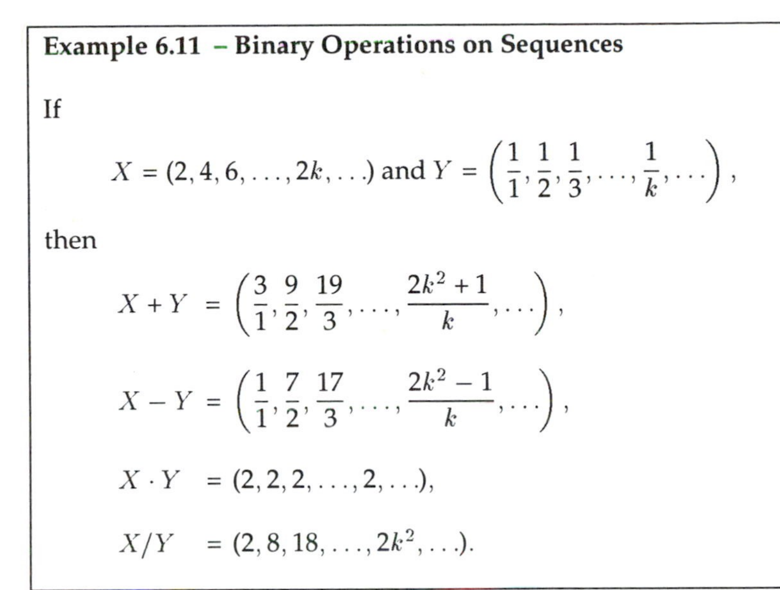

Example 6.11 – Binary Operations on Sequences

If

$$X = (2, 4, 6, \ldots, 2k, \ldots) \text{ and } Y = \left(\frac{1}{1}, \frac{1}{2}, \frac{1}{3}, \ldots, \frac{1}{k}, \ldots\right),$$

then

$$\begin{aligned} X + Y &= \left(\frac{3}{1}, \frac{9}{2}, \frac{19}{3}, \ldots, \frac{2k^2+1}{k}, \ldots\right), \\ X - Y &= \left(\frac{1}{1}, \frac{7}{2}, \frac{17}{3}, \ldots, \frac{2k^2-1}{k}, \ldots\right), \\ X \cdot Y &= (2, 2, 2, \ldots, 2, \ldots), \\ X/Y &= (2, 8, 18, \ldots, 2k^2, \ldots). \end{aligned}$$

6.3.3 Sequences with Special Properties: Bounded and Monotone Sequences

The objective of this section is to identify several properties of sequences that are exploited subsequently in Section 6.4. This is accomplished by looking at numerous sequences to see what similarities and differences they share. With this objective in mind, put the following sequences into groups in such a way that the sequences in each group all share a common property that is *not* shared by any of the sequences in the other groups:

$$\begin{aligned} U &= \left(\frac{1}{1}, \frac{1}{2}, \frac{1}{3}, \frac{1}{4}, \ldots, \frac{1}{k}, \ldots\right), \\ V &= \left(1 - \frac{1}{1}, 1 - \frac{1}{2}, 1 - \frac{1}{3}, 1 - \frac{1}{4}, \ldots, 1 - \frac{1}{k}, \ldots\right), \\ W &= (1, 2, 3, 4, \ldots, k, \ldots), \\ X &= (-2, -4, -6, -8, \ldots, -2k, \ldots), \\ Y &= (-1, 1, -1, 1, \ldots, (-1)^k, \ldots), \\ Z &= \left(1 - \frac{1}{2}, 1 + \frac{1}{2^2}, 1 - \frac{1}{2^3}, 1 + \frac{1}{2^4}, \ldots, 1 + \left(-\frac{1}{2}\right)^k, \ldots\right). \end{aligned}$$

There is no unique way to group the foregoing sequences – how you do so depends on the *criterion* you choose. For example, you can group the sequences on the basis of the *signs* of the numbers in the sequences. With this criterion, you obtain the following three groups:

$$\begin{array}{lll} \text{Group 1} = \{U, V, W, Z\} & \text{(all elements are } \geq 0), \\ \text{Group 2} = \{X\} & \text{(all elements are } < 0), \\ \text{Group 3} = \{Y\} & \text{(elements have mixed signs).} \end{array}$$

The way in which you decide to group sequences should be based on the properties you wish to isolate and to study. Accordingly, two other criterion for grouping sequences are described next.

Monotone Sequences

Another way to perform the grouping is on the basis of the relationship between successive numbers *within* the sequence. There are two different possibilities.

Decreasing Sequences. For example, each number in the sequence

$$U = \left(\frac{1}{1}, \frac{1}{2}, \frac{1}{3}, \frac{1}{4}, \ldots, \frac{1}{k}, \ldots\right)$$

is *smaller than* the preceding number. The same is *true* for the sequence

$$X = (-2, -4, -6, -8, \ldots, -2k, \ldots).$$

This is *not* the case for the remaining sequences in the foregoing list. On the basis of this criterion, can you create a symbolic definition that includes the sequences U and X (and all others like them) while *excluding* the sequences V, W, Y, and Z? The answer is given in the following definition.

Definition 6.14 *A sequence of real numbers $X = (x_1, x_2, \ldots)$ is* **decreasing** *(or* **monotonically decreasing***) if and only if for each $k \in \mathbb{N}$, $x_k \geq x_{k+1}$, that is, if*

$$x_1 \geq x_2 \geq \cdots \geq x_k \cdots.$$

In some cases, such as the foregoing examples of U and X, it is easy to see that the sequence is decreasing. In other cases, however, this property is not so obvious. For example, recall the sequence X given at the beginning of this section for finding the $\sqrt{2}$, in which $x_1 = 2$ and

$$x_{k+1} = \frac{1}{2}\left(x_k + \frac{2}{x_k}\right), \text{ for } k = 1, 2, \ldots. \tag{6.14}$$

In the next proposition, this sequence is shown to be decreasing.

Proposition 6.19 *The sequence $X = (x_k)$ in (6.14) is decreasing.*

Proof. First note that each $x_k > 0$ (which you can prove by induction on k). To see that this sequence X is decreasing, from (6.14) you have that

$$x_k - x_{k+1} = x_k - \frac{1}{2}\left(x_k + \frac{2}{x_k}\right) = \frac{x_k^2 - 2}{2x_k}.$$

Because the denominator $2x_k > 0$, the result will follow if $x_k^2 - 2 \geq 0$. However, this is the case because rewriting (6.14) results in the following:

$$x_{k-1}^2 - 2x_{k-1}x_k + 2 = 0. \tag{6.15}$$

Think of (6.15) as a quadratic equation in x_{k-1}. Because the discriminant is nonnegative, you have that

$$4x_k^2 - 8 \geq 0,$$

and so $x_k^2 - 2 \geq 0$, as required. Thus, $x_k \geq x_{k+1}$ and so X is a decreasing sequence.

QED

Increasing Sequences. You could equally well have grouped together those sequences whose numbers are getting progressively *larger*. In this case, you would have put the following sequences in the same group:

$$V = \left(1 - \frac{1}{1}, 1 - \frac{1}{2}, 1 - \frac{1}{3}, 1 - \frac{1}{4}, \ldots, 1 - \frac{1}{k}, \ldots\right)$$

and

$$W = (1, 2, 3, 4, \ldots, k, \ldots).$$

The appropriate definition for these types of sequences (analogous to Definition 6.14) follows.

> **Definition 6.15** *A sequence of real numbers $X = (x_1, x_2, \ldots)$ is* **increasing** *(or* **monotonically increasing***) if and only if for each $k \in \mathbb{N}$, $x_k \leq x_{k+1}$, that is, if*
>
> $$x_1 \leq x_2 \leq \cdots \leq x_k \cdots.$$

In some cases, such as the foregoing examples of V and W, it is easy to see that the sequence is increasing. In other cases, however, this property is not so obvious. For example, consider the sequence $X = (x_k)$ in which

$$x_k = \left(1 + \frac{1}{k}\right)^k, \text{ for } k = 1, 2, \ldots. \tag{6.16}$$

In the next proposition, this sequence is shown to be increasing.

Proposition 6.20 *The sequence $X = (x_k)$ in (6.16) is increasing.*

Proof. To establish that X is increasing, you must show that for each $k \in \mathbb{N}$, $x_k \le x_{k+1}$. This result follows by using the binomial theorem to expand both x_k and x_{k+1} in (6.16). In particular,

$$x_k = 1 + \frac{k}{1} \cdot \frac{1}{k} + \frac{k(k-1)}{2!} \cdot \frac{1}{k^2} + \cdots + \frac{k(k-1)\cdots 2 \cdot 1}{k!} \cdot \frac{1}{k^k}.$$

On dividing the power of k in the denominator of each term into the numerator, you obtain the following:

$$\begin{aligned} x_k = {} & 1 + 1 + \frac{1}{2!}\left(1 - \frac{1}{k}\right) + \cdots + \\ & \frac{1}{k!}\left(1 - \frac{1}{k}\right)\left(1 - \frac{2}{k}\right)\cdots\left(1 - \frac{k-1}{k}\right). \end{aligned} \tag{6.17}$$

Likewise,

$$\begin{aligned} x_{k+1} = {} & 1 + 1 + \frac{1}{2!}\left(1 - \frac{1}{k+1}\right) + \cdots + \\ & \frac{1}{(k+1)!}\left(1 - \frac{1}{k+1}\right)\left(1 - \frac{2}{k+1}\right)\cdots\left(1 - \frac{k}{k+1}\right). \end{aligned} \tag{6.18}$$

You can see that each of the terms in the expansion of x_k in (6.17) is less than or equal to the corresponding term in the expansion of x_{k+1} in (6.18) and, moreover, x_{k+1} has one more positive term than does x_k. It therefore follows that $x_k \le x_{k+1}$ and so the sequence X is increasing.

QED

Monotone and Oscillating Sequences. Sequences that are either increasing or decreasing are called **monotone**. For example, the sequences U, V, W, and X in the foregoing list are monotone.

Some sequences, however, are *not* monotone and are called **oscillating**. For example, the following sequences from the foregoing list are oscillating:

$$Y = (-1, 1, -1, 1, \ldots, (-1)^k, \ldots)$$

and

$$Z = \left(1 - \frac{1}{2}, 1 + \frac{1}{2^2}, 1 - \frac{1}{2^3}, 1 + \frac{1}{2^4}, \ldots, 1 + \left(-\frac{1}{2}\right)^k, \ldots\right).$$

Unbounded and Bounded Sequences

Another useful way to group sequences is on the basis of what is happening to the sequence *as a whole*, as described in what follows.

Unbounded Sequences. For example, the numbers in the sequence

$$W = (1, 2, 3, 4, \ldots, k, \ldots)$$

are *increasing without bound*. That is, not only are these numbers progressively larger, but also, they are getting *infinitely* large, as a whole. Likewise, the numbers in the sequence

$$X = (-2, -4, -6, -8, \ldots, -2k, \ldots)$$

are *decreasing without bound* – not only are these numbers progressively smaller, but also, they are getting *infinitely* small (that is, infinitely negative), as a whole. Even oscillating sequences can have no bounds, as in the following examples:

$$S = (1, -2, 3, -4, 5, -6, \ldots) \text{ and } T = (1, 2, 1, 3, 1, 4, \ldots).$$

On the basis of these examples, can you create a symbolic definition that describes the property of a sequence being "unbounded"?

As a first attempt, you might start with the property that,

> no matter how large the real number M is, you can always find a number in the sequence that is *larger* than M.

By using the quantifiers *for all* and *there is*, you can translate this property to the following symbolic definition:

> A sequence $X = (x_1, x_2, \ldots)$ is *unbounded* if and only if for all real numbers M, there is an $k \in \mathbb{N}$ such that $x_k > M$.

As you learned in Section 2.2.2, you must *verify* the definition to make sure that all desirable objects *do* satisfy the property (that for all real numbers M, there is a $k \in \mathbb{N}$ such that $x_k > M$, in this case). For example, the unbounded sequence $W = (1, 2, 3, \ldots)$ *does* satisfy this property. However, the unbounded sequence $X = (-2, -4, \ldots, -2k, \ldots)$ does *not* satisfy this property because, for example, for the real number $M = 1$, there is no $k \in \mathbb{N}$ such that $-2k > 1$. This error indicates that you must modify the definition so as to include this unbounded sequence X.

The reason X does not satisfy the foregoing property is because the numbers in this sequence are *negative*. On the basis of this observation, one way to modify the definition is as follows:

> A sequence $X = (x_1, x_2, \ldots)$ is *unbounded* if and only if for all real numbers M, there is a $k \in \mathbb{N}$ such that $x_k > M$ or $x_k < -M$.

An equivalent but more concise way to state this property is given in the following definition.

> **Definition 6.16** *A sequence of real numbers* $X = (x_1, x_2, \ldots)$ *is* **unbounded** *if and only if for every real number* $M > 0$*, there is a* $k \in \mathbb{N}$ *such that* $|x_k| > M$.

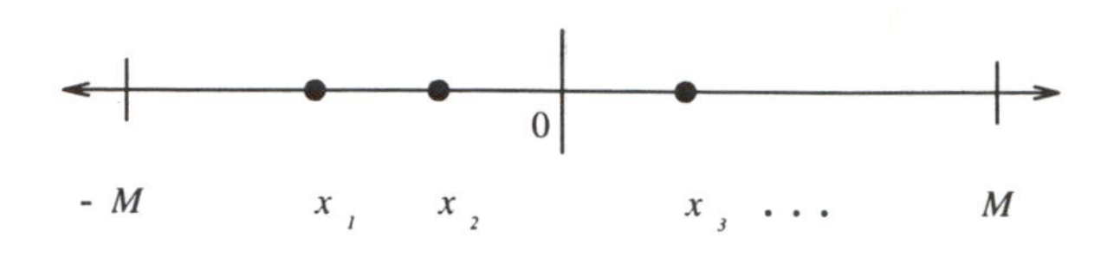

Figure 6.7: An Example of a Bounded Sequence of Real Numbers

You can verify for yourself that all of the foregoing special cases of unbounded sequences (S, T, W, and X) satisfy the property in Definition 6.16.

Bounded Sequences. As mentioned in Section 2.2, one of the advantages of a symbolic definition is that you can express the negation more easily. Applying the rules for negating the quantifiers in Definition 6.16 results in the following definition.

> **Definition 6.17** *A sequence of real numbers* $X = (x_1, x_2, \ldots)$ *is* **bounded** *if and only if there is a real number* $M > 0$ *such that for all* $k \in \mathbb{N}$, $|x_k| \leq M$.

An appropriate visual image associated with Definition 6.17 is shown in Figure 6.7, where you can see that all numbers in a bounded sequence lie on the real line somewhere between $-M$ and $+M$. Verify for yourself that the foregoing special cases of U, V, Y, and Z satisfy the property in Definition 6.17 and are therefore examples of bounded sequences. In the next two propositions, it is shown that the sequences in Proposition 6.19 and Proposition 6.20, respectively, are bounded.

> **Proposition 6.21** *The sequence* $X = (x_k)$ *in which* $x_1 = 2$ *and*
>
> $$x_{k+1} = \frac{1}{2}\left(x_k + \frac{2}{x_k}\right), \text{ for } k = 1, 2, \ldots$$
>
> *is bounded.*

Proof. To see that X is bounded, first note that, by Proposition 6.19, X is decreasing and because $x_1 = 2$, for each $k \in \mathbb{N}$, $x_k \leq 2$. Also, you can see by induction that $x_k \geq 0$. Thus, $|x_k| = x_k \leq 2$, and so X is bounded.

QED

Proposition 6.22 *The sequence $X = (x_k)$ in which*

$$x_k = \left(1 + \frac{1}{k}\right)^k, \text{ for } k = 1, 2, \ldots$$

is bounded.

Proof. First note that, by Proposition 6.20, X is increasing and because $x_1 = 2$, for each $k \in \mathbb{N}$, $2 \leq x_k$. To establish that X is bounded, recall (6.17) in the proof of Proposition 6.20, which establishes that

$$x_k = 1 + 1 + \frac{1}{2!}\left(1 - \frac{1}{k}\right) + \cdots +$$
$$\frac{1}{k!}\left(1 - \frac{1}{k}\right)\left(1 - \frac{2}{k}\right)\cdots\left(1 - \frac{k-1}{k}\right).$$

You can use this expression to show that $x_k \leq 3$ by noting that for each $i = 1, \ldots, k$, $(1 - i/k) < 1$, so you have the following inequality:

$$x_k < 1 + 1 + \frac{1}{2!} + \cdots + \frac{1}{k!}. \tag{6.19}$$

The next step is to use induction to show that for each $i = 1, \ldots, k$, $1/i! \leq 1/2^{i-1}$, as done in Proposition 1.4 in Section 1.6.5. Thus, (6.19) becomes

$$x_k < 1 + 1 + \frac{1}{2} + \cdots + \frac{1}{2^{k-1}}. \tag{6.20}$$

The final step is to note that

$$\frac{1}{2} + \cdots + \frac{1}{2^{k-1}} = 1 - \frac{1}{2^k - 1} < 1.$$

This combined with (6.20) yields that $x_k \leq 1 + 1 + 1 = 3$. You now know that the sequence X is bounded because, for each $k \in \mathbb{N}$, $|x_k| = x_k \leq 3$, thus completing the proof.

QED

6.3.4 Comparisons of Sequences and Sets

It is important to note the differences between a sequence and a set. For instance, the sequence

$$X = (1, 2, 1, 2, \ldots)$$

is not the same as the set

$$X = \{1, 2\}.$$

This is because a sequence *always* has an infinite number of elements, some of which can repeat. In contrast, a set can have a *finite* number of elements, none of which repeat.

As another example of the difference between a sequence and a set, you know that the *order* of the elements in a set is irrelevant, so the following sets are the same:

$$\{1, 2\} = \{2, 1\}.$$

In contrast, the order of the elements in a sequence *is* important. For that reason,

$$(1, 2, 1, 2, \ldots) \neq (2, 1, 2, 1, \ldots).$$

In summary, sequences are different from sets because

1. A sequence has an infinite number of elements but a set can have a finite number of elements.
2. Two or more elements of a sequence can be the same whereas the elements of a set are all different.
3. The *order* of the elements in a sequence is important whereas the order of the elements in a set is irrelevant.

Although you cannot always create a sequence from a set (because the set may have a finite number of elements, for example), you *can* create a set from a sequence. For example, from the sequence

$$X = (1, 2, 3, \ldots),$$

you can create the associated set

$$X' = \{1, 2, 3, \ldots\}.$$

As another example, from the sequence

$$X = (1, 1, 1, \ldots),$$

you can create the associated set

$$X' = \{1\}.$$

In general, associated with each sequence of real numbers $X = (x_1, x_2, \ldots)$ is the following set:

$$X' = \{x_k : k \in \mathbb{N}\}.$$

You can therefore compare properties of sequences with properties of the corresponding sets, as shown in the following proposition.

Proposition 6.23 *Suppose that $X = (x_1, x_2, \ldots)$ is a sequence of real numbers and that $X' = \{x_k : k \in \mathbb{N}\}$ is the associated set. The sequence X is bounded if and only if the set X' is bounded.*

Proof. Suppose first that $X = (x_1, x_2, \ldots)$ is a bounded sequence which, according to Definition 6.17, means that there is a real number $M > 0$ such that for all $k \in \mathbb{N}$, $|x_k| \leq M$. This same value of M shows that the set X' is also a bounded set because, for any element $x \in X'$, there is an $k \in \mathbb{N}$ such that $x = x_k$. The sequence X is bounded, so it follows that

$$|x| = |x_k| \leq M,$$

and so the set X' is bounded by Definition 6.9 in Section 6.2.3.

Suppose now that the set X' is bounded which, by definition, means that there is a real number $M > 0$ such that for all elements $x \in X'$, $|x| \leq M$. This same value of M shows that the sequence X is bounded. This is because, for each $k \in \mathbb{N}$, the element x_k of the sequence is also an element of the set X', so $|x_k| \leq M$. Thus, X is a bounded sequence by Definition 6.17 and this completes the proof.

QED

As another example of related properties between sets and sequences, recall the infimum and supremum properties of sets of real numbers from Section 6.2:

> A nonempty set of real numbers that is bounded below (above) has an infimum (supremum).

As a result, if a sequence of real numbers $X = (x_1, x_2, \ldots)$ is bounded, then, by Proposition 6.23, the corresponding set X' is bounded and hence has an infimum and supremum. In Section 6.4, you will see how the infimum and supremum of the set X' play a critical role in the study of the sequence X.

6.4 Convergence of Sequences of Real Numbers

In Section 6.3, you learned about sequences of real numbers and various operations you can perform on them. To motivate the topic of this section, look again at the sequence of numbers generated by the algorithm for finding the value of $\sqrt{2}$ given at the beginning of Section 6.3. From the first five values in the list, it appears that those numbers are getting "closer and closer" to the desired value of $\sqrt{2}$. The issue of whether or not a sequence is getting closer and closer to some number is called **convergence**. In that regard, there are two separate questions pertaining to a sequence $X = (x_1, x_2, \ldots)$:

1. *Given* a specific real number x, does the sequence X converge to x – *yes* or *no*?

2. Is there *any* real number to which the sequence X does converge and, if so, to *which* real number?

Both of these issue are examined in this section.

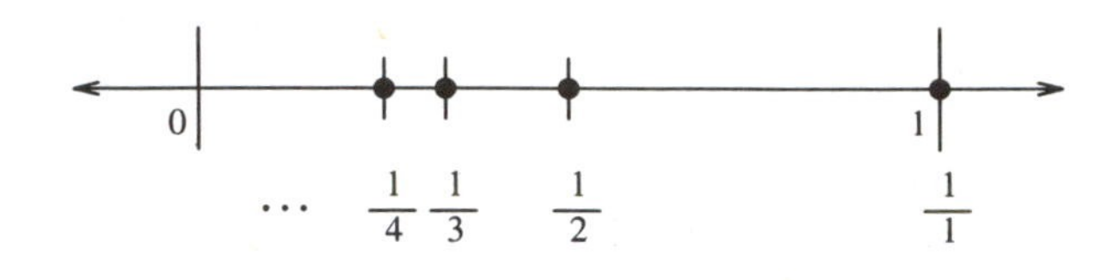

Figure 6.8: The Numbers in the Sequence $X = (1/k)$ Getting Closer to 0

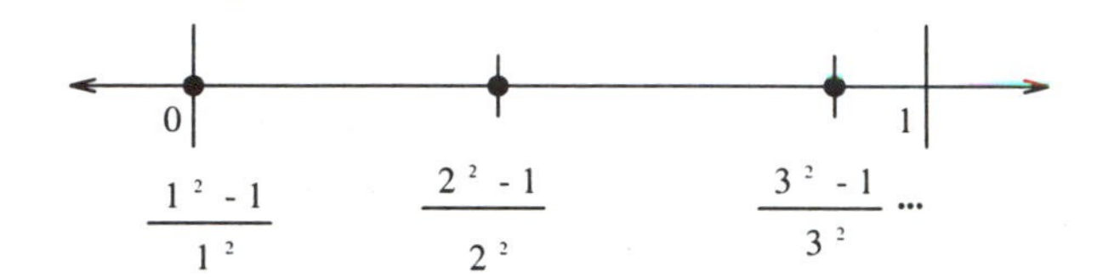

Figure 6.9: The Numbers in the Sequence $X = \left(\frac{k^2-1}{k^2}\right)$ Getting Closer to 1

6.4.1 Convergence of a Sequence to a Given Real Number

Suppose that $X = (x_1, x_2, \ldots)$ is a sequence of real numbers and that x is a specific real number to which you suspect that the numbers in X are getting closer. The objective now is to develop a symbolic definition for the concept of "the sequence X of real numbers is getting closer and closer to x," that is, for the sequence X to *converge* to x.

Developing the Definition

As described in Section 2.2.2, to develop a definition, you should use special cases to compare and contrast desirable objects – sequences that *do* converge to a specific real number – with undesirable objects – sequences that do *not* converge to a specific real number. For example, as seen in Figure 6.8, the numbers in the sequence

$$X = \left(\frac{1}{k} : k \in \mathbb{N}\right) \tag{6.21}$$

appear to get closer and closer to 0 – in other words, this sequence X converges to 0. Analogously, but perhaps not so obvious, is the fact that the sequence

$$X = \left(\frac{k^2 - 1}{k^2} : k \in \mathbb{N}\right) \tag{6.22}$$

illustrated in Figure 6.9 converges to 1.

In contrast, the sequence in (6.21) does *not* converge to -1. As another example, the sequence

$$X = (1, 2, 3, \ldots)$$

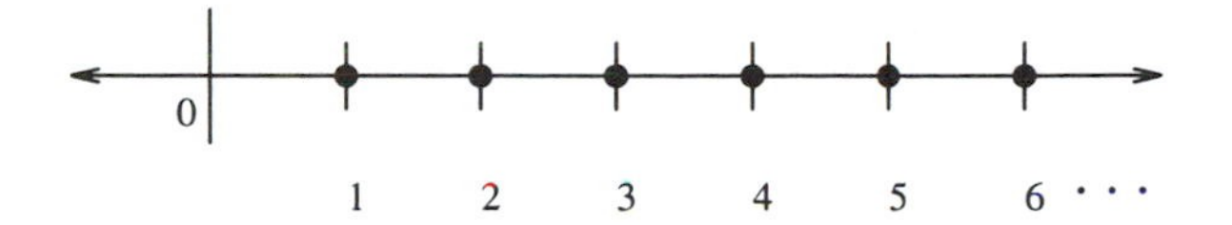

Figure 6.10: The Numbers in the Sequence $X = (1, 2, 3, \ldots)$ Are Not Getting Closer to 5

in Figure 6.10 does not converge to 5, or to any other real number, for that matter. On the basis of these examples, the objective is create a symbolic definition for the desirable sequences and given numbers – such as the ones in Figure 6.8 and Figure 6.9 – while excluding the undesirable sequences and given numbers – such as the one in Figure 6.10. So, let $X = (x_1, x_2, \ldots)$ be a sequence of real numbers and x be a given real number. What does it mean to say that X converges to x?

Intuitively, the further along you are in the sequence, the closer the numbers should be to the value x. That is, the larger the value of the subscript k, the closer x_k should be to x. So, how do you measure how close x_k is to x? One answer is to use the *absolute value* (see Section 6.1.2). That is,

$$|x_k - x|$$

is a measure of how close x_k is to x. You might now translate the property that "as k gets larger, x_k gets closer to x" to the following symbolic form:

> The sequence X *converges* to x if and only if for all $j, k \in \mathbb{N}$ with $k \geq j$, it follows that $|x_k - x| \leq |x_j - x|$.

Recall from Section 2.2.2 that you must *verify* this definition to make sure that all sequences X that converge to x satisfy the foregoing property. For example, you can see that the converging sequences in Figure 6.8 and Figure 6.9 *do* satisfy this property. However, is it also *true* that all sequences X and real numbers x that satisfy the foregoing property are what you think of as X converging to x? By trial and error, you should discover that the answer is *no*. For example, the sequence

$$X = \left(\frac{1}{1}, \frac{1}{2}, \frac{1}{3}, \ldots\right)$$

does *not* converge to $x = -1$, yet, this sequence and real number satisfy the property in the foregoing definition. This error means that you must modify the definition to exclude this undesirable sequence and real number. Doing so is a challenging task.

The key observation is that, not only do you want the values of x_k in the sequence to be getting closer to x, *but you also want* $|x_k - x|$ *to be getting closer to 0*. Mathematicians have discovered a clever method for expressing this property in symbolic form using the quantifiers *for all* and *there is*. The

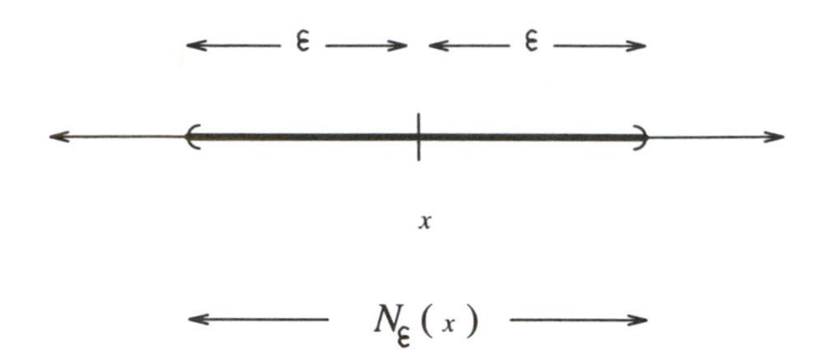

Figure 6.11: An ε-Neighborhood of the Real Number x

idea is to choose a fixed positive real number – say, $\varepsilon > 0$ – and to consider all real numbers y whose distance from x is less than ε, as illustrated in Figure 6.11. This set of points is so important to the concept of convergence that the following definition is warranted.

> **Definition 6.18** *Suppose that $x \in \mathbb{R}$ and that $\varepsilon > 0$ is a given positive real number. The ε-***neighborhood of** *x, denoted by $N_\varepsilon(x)$, is the following set:*
>
> $$N_\varepsilon(x) = \{y \in \mathbb{R} : |x - y| < \varepsilon\}.$$

Observe from Figure 6.11 that as ε gets closer to 0, the ε-neighborhoods of x are "shrinking to x." You can use this observation to create a symbolic definition of the sequence X converging to x by requiring that,

> no matter how close the value of ε is to 0, "most" of the points in the sequence lie inside the ε-neighborhood of x.

Translating this statement to symbolic form is still a challenge. For example, the property that

> no matter how close the value of ε is to 0, *something happens*

translates to

> for *every* $\varepsilon > 0$, *something happens*.

In this case, the *something that happens* is that "most of the points in the sequence lie inside the ε-neighborhood of x." But what, exactly, does "most" mean? Mathematicians have determined that "most" means "all the points in the sequence *after some point*, say, x_j." For example, "most of the sequence X" could mean all the points after x_5, that is, $x_6, x_7, \ldots$ (here, $j = 5$). As another example, "most of the sequence X" could mean all the points after x_{1000}, that is, $x_{1001}, x_{1002}, \ldots$ (here, $j = 1000$). More generally, the statement that

> for most of the sequence X, *something happens*,

is expressed symbolically as follows:

> *There is* a subscript $j \in \mathbb{N}$ such that *for all* $k \in \mathbb{N}$ with $k > j$, the *something happens* for each element x_k in X.

For the concept of convergence, the *something that happens* is that the points in the sequence after x_j are in the ε-neighborhood of x. By putting together all of the pieces, you obtain the following definition.

Definition 6.19 *Suppose that* $X = (x_1, x_2, \ldots)$ *is a sequence of real numbers and* $x \in \mathbb{R}$. *The* **sequence** X **converges to** x , *written* $\lim X = x$, *or* $\lim (x_k) = x$, *or* $(x_k) \to x$, *if and only if for every real number* $\varepsilon > 0$, *there is a* $j \in \mathbb{N}$ *such that for all* $k \in \mathbb{N}$ *with* $k > j$, $x_k \in N_\varepsilon(x)$.

Recalling Definition 6.18 for an ε-neighborhood, the following proposition provides several equivalent ways to say that the sequence X converges to x.

Proposition 6.24 *Suppose that* $X = (x_1, x_2, \ldots)$ *is a sequence of real numbers and that* $x \in \mathbb{R}$. *The following are equivalent:*

(a) X *converges to* x.

(b) For every real number $\varepsilon > 0$, *there is a* $j \in \mathbb{N}$ *such that for all* $k \in \mathbb{N}$ *with* $k > j$, $x_k \in N_\varepsilon(x)$.

(c) For every real number $\varepsilon > 0$, *there is a* $j \in \mathbb{N}$ *such that for all* $k \in \mathbb{N}$ *with* $k > j$, $|x_k - x| < \varepsilon$.

(d) For every real number $\varepsilon > 0$, *there is a* $j \in \mathbb{N}$ *such that for all* $k \in \mathbb{N}$ *with* $k > j$, $x - \varepsilon < x_k < x + \varepsilon$.

Proof. The equivalence of (a) and (b) is directly from Definition 6.19. The equivalence of (b) and (c) is the definition of $N_\varepsilon(x)$, that is,

$$x_k \in N_\varepsilon(x) \text{ if and only if } |x_k - x| < \varepsilon.$$

The equivalence of (c) and (d) follows from Proposition 6.6(d) in Section 6.1.2 because

$$|x_k - x| < \varepsilon \text{ if and only if } x - \varepsilon < x_k < x + \varepsilon.$$

This now establishes that (a) through (d) are all equivalent and completes the proof. QED

Proving that a Sequence Converges to a Given Real Number

With the use of Definition 6.19 and Proposition 6.24, you can now show that a sequence converges to a given point, as done in the following proposition.

> **Proposition 6.25** *If* $X = \left(\frac{1}{k} : k \in \mathbb{N}\right)$*, then* X *converges to 0.*

Proof. According to Proposition 6.24(c), it must be shown that for every real number $\varepsilon > 0$, there is a $j \in \mathbb{N}$ such that for all $k \in \mathbb{N}$ with $k > j$, $|x_k - 0| < \varepsilon$. The choose method is appropriate (see Section 1.6.4) because of the appearance of the first quantifier *for all*. Thus, let $\varepsilon > 0$, for which you must show that there is a $j \in \mathbb{N}$ such that for all $k > j$, $|x_k - 0| < \varepsilon$.

The appearance of the quantifier *there is* now means that the construction method (see Section 1.6.3) is used to construct the natural number j. The value you construct for j must satisfy the property that

$$\text{for all } k \in \mathbb{N} \text{ with } k > j,\ |x_k - 0| = |x_k| = \frac{1}{k} < \varepsilon. \tag{6.23}$$

So the question is, Given a value for $\varepsilon > 0$, what choice of j allows you to prove the statement in (6.23)?

To answer this question, suppose that you have *already* constructed the value for j. To show that (6.23) is *true*, the appearance of the quantifier *for all* suggests that you use the choose method to choose a value for $k \in \mathbb{N}$ with $k > j$. You will then need to show that

$$\frac{1}{k} < \varepsilon. \tag{6.24}$$

The result in (6.24) follows provided that you have chosen a value for $j \in \mathbb{N}$ that satisfies

$$\frac{1}{j} < \varepsilon \tag{6.25}$$

because then, from the fact that $k > j$ and (6.25), you have

$$\frac{1}{k} < \frac{1}{j} < \varepsilon,$$

which is the desired result in (6.24). Note finally that by Proposition 6.11(b) in Section 6.2.2, you can find a $j \in \mathbb{N}$ such that (6.25) is satisfied.

In summary, having chosen a value for the real number $\varepsilon > 0$, by Proposition 6.11(b), you can find a natural number j such that

$$\frac{1}{j} < \varepsilon. \tag{6.26}$$

To see that the value of j in (6.26) is correct, let $k > j$, for which you must show that

$$|x_k - 0| < \varepsilon.$$

But it now follows that

$$\begin{aligned}
|x_k - 0| &= |x_k| \quad \text{(property of 0)} \\
&= \left|\frac{1}{k}\right| \quad \text{(definition of } x_k\text{)} \\
&= \frac{1}{k} \quad (k > 0) \\
&< \frac{1}{j} \quad (k > j) \\
&< \varepsilon \quad \text{[from (6.26)].}
\end{aligned}$$

This shows that X converges to 0 and completes the proof. QED

The technique used in Proposition 6.25 is typical of the way in which you can prove that a sequence $X = (x_1, x_2, \ldots)$ converges to a given real number x. That is, you choose a value of $\varepsilon > 0$. You then construct a value for the natural number j in such a way that you can prove that for all $k \in \mathbb{N}$ with $k > j$, $|x_k - x| < \varepsilon$. This process is illustrated again in the next proposition.

Proposition 6.26 *The sequence X from (6.22) in which*

$$x_k = \frac{k^2 - 1}{k^2}, \text{ for } k = 1, 2, \ldots$$

converges to $x = 1$.

Proof. According to the discussion preceding this propsition, you should choose a value of $\varepsilon > 0$. Now you must construct a value for $j \in \mathbb{N}$ for which you can prove that for all $k \in \mathbb{N}$ with $k > j$, $|x_k - 1| < \varepsilon$. So, suppose you have *already* found this j. To prove that the value for j is correct, you need to choose a value for $k \in \mathbb{N}$ with $k > j$ and then show that

$$\begin{aligned}
|x_k - 1| &= \left|\frac{k^2 - 1}{k^2} - 1\right| \\
&= \left|\frac{k^2 - 1 - k^2}{k^2}\right| \\
&= \frac{1}{k^2} \\
&< \varepsilon.
\end{aligned}$$

This last inequality will follow provided that

$$k^2 > \frac{1}{\varepsilon}$$

or equivalently, if

$$k > \frac{1}{\sqrt{\varepsilon}}.$$

But, because $k > j$, you should choose $j \in \mathbb{N}$ so that

$$j > \frac{1}{\sqrt{\varepsilon}},$$

and this you can do by Proposition 6.10 in Section 6.2.2.

In summary, having chosen a value for the real number $\varepsilon > 0$, by Proposition 6.10, you can find a natural number j such that

$$j > \frac{1}{\sqrt{\varepsilon}}. \tag{6.27}$$

To see that the value j in (6.27) is correct, let $k \in \mathbb{N}$ with $k > j$, for which you must show that

$$|x_k - 1| < \varepsilon.$$

But it now follows that

$$\begin{aligned} |x_k - 1| &= \left|\frac{k^2-1}{k^2} - 1\right| && \text{(definition of } x_k\text{)} \\ &= \frac{1}{k^2} && \text{(algebra)} \\ &< \frac{1}{j^2} && (k > j) \\ &< \varepsilon && \text{[from (6.27)].} \end{aligned}$$

This shows that X converges to 1 and completes the proof. QED

Proving that a Sequence Does Not Converge to a Given Real Number

As discussed in Section 2.2, one of the advantages of a symbolic definition is that you can write the negation more easily. For example, applying the rules for negating the quantifiers in Definition 6.19 results in the following:

> The sequence X does *not* converge to the real number x if and only if there is a real number $\varepsilon > 0$ such that for all $j \in \mathbb{N}$, there is a $k \in \mathbb{N}$ with $k > j$ such that $x_k \notin N_\varepsilon(x)$, that is, $|x_k - x| \geq \varepsilon$.

You can use this statement to prove that the sequence in (6.21) does not converge to -1, as shown in the following proposition.

> **Proposition 6.27** *If* $X = \left(\frac{1}{k} : k \in \mathbb{N}\right)$, *then* X *does not converge to* -1.

Proof. To show that X does not converge to -1, you must construct a value for the real number $\varepsilon > 0$ such that for all $j \in \mathbb{N}$, there is a $k \in \mathbb{N}$ with $k > j$ such that $x_k \notin N_\varepsilon(-1)$. To determine what the value of ε should be, suppose you have *already* constructed

$$\varepsilon = \bar{\varepsilon}. \tag{6.28}$$

To see that the value of $\bar{\varepsilon}$ in (6.28) is correct, you must show that for all $j \in \mathbb{N}$, there is a $k \in \mathbb{N}$ with $k > j$ such that $x_k \notin N_{\bar{\varepsilon}}(-1)$. The appearance of the first quantifier *for all* now means you should proceed by the choose method so, choose $j \in \mathbb{N}$, for which you must show that there is a $k \in \mathbb{N}$ with $k > j$ such that $x_k \notin N_{\bar{\varepsilon}}(-1)$. That is, you must construct a value of $k > j$ such that

$$|x_k - (-1)| = \left|\frac{1}{k} + 1\right| = \frac{1}{k} + 1 \geq \bar{\varepsilon}. \tag{6.29}$$

The inequality in (6.29) indicates not only how you should construct $k \in \mathbb{N}$ but also what the value of $\varepsilon = \bar{\varepsilon}$ should be. In particular, if you construct $\varepsilon = \bar{\varepsilon} = 1/2$ (any positive real number less than 1 will do), then *any* $k \in \mathbb{N}$ satisfies (6.29).

In summary, then, to show that X does not converge to -1, construct

$$\varepsilon = \frac{1}{2} > 0. \tag{6.30}$$

To see that the value for ε in (6.30) is correct, choose $j \in \mathbb{N}$, for which you must show that there is a $k \in \mathbb{N}$ with $k > j$ such that $x_k \notin N_{\frac{1}{2}}(-1)$. Now construct k as any natural number for which $k > j$. You can see that $x_k \notin N_{\frac{1}{2}}(-1)$ because

$$|x_k - (-1)| = \left|\frac{1}{k} + 1\right| = \frac{1}{k} + 1 > \frac{1}{2} = \varepsilon.$$

Thus, X does not converge to -1. QED

A similar idea is used to show that this same sequence does not converge to 1, as shown in the next proposition.

> **Proposition 6.28** *If* $X = \left(\frac{1}{k} : k \in \mathbb{N}\right)$, *then* X *does not converge to* 1.

Proof. To show that X does not converge to 1, you must construct a value for the real number $\varepsilon > 0$ such that for all $j \in \mathbb{N}$, there is a $k \in \mathbb{N}$ with $k > j$ such that $x_k \notin N_\varepsilon(1)$. In this case, you can construct

$$\varepsilon = \frac{3}{4} > 0. \tag{6.31}$$

To see that the value for ε in (6.31) is correct, you must show that for all $j \in \mathbb{N}$, there is a $k \in \mathbb{N}$ with $k > j$ such that $x_k \notin N_{\frac{3}{4}}(1)$. The appearance of the first quantifier *for all* now means you should proceed by the choose method so, choose $j \in \mathbb{N}$, for which you must show that there is a $k \in \mathbb{N}$ with $k > j$ such that $x_k \notin N_{\frac{3}{4}}(1)$.

The appearance of the quantifier *there is* means that you must now construct a $k \in \mathbb{N}$ with $k > j$ such that $x_k \notin N_{\frac{3}{4}}(1)$. You can construct this value for k as any natural number greater than j that is also > 4, for then

$$\begin{aligned}
|x_k - 1| &= \left|\frac{1}{k} - 1\right| && \text{(definition of } x_k\text{)} \\
&= 1 - \frac{1}{k} && (k > 1) \\
&\geq \frac{3}{4} && (k > 4).
\end{aligned}$$

Thus, $x_k \notin N_{\frac{3}{4}}(1)$ and so X does not converge to 1. QED

6.4.2 Convergence Results

In many cases, you can think of the sequence you are interested in as a *combination* of two other sequences. For example, the sequence

$$Z = \left(3 + \frac{1}{1}, 3 + \frac{1}{2}, 3 + \frac{1}{3}, \ldots\right)$$

is the sum of the two sequences

$$X = (3, 3, 3, \ldots) \text{ and } Y = \left(\frac{1}{1}, \frac{1}{2}, \frac{1}{3}, \ldots\right).$$

The advantage of recognizing that $Z = X + Y$, as shown in the following proposition, is that you can conclude that Z converges to the sum of what X converges to and what Y converges to. In this case, because you know that X converges to 3 and that Y converges to 0 (see Proposition 6.25), it follows that

$$\lim Z = \lim X + \lim Y = 3 + 0 = 3.$$

Proposition 6.29 *Suppose that $x, y \in \mathbb{R}$ and that $X = (x_1, x_2, \ldots)$ converges to x and $Y = (y_1, y_2, \ldots)$ converges to y.*

(a) The sequence $X + Y$ converges to $x + y$.

(b) The sequence $X - Y$ converges to $x - y$.

(c) The sequence $X \cdot Y$ converges to xy.

(d) If $c \in \mathbb{R}$, then the sequence cX converges to cx.

(e) If all elements of Y are nonzero and if $y \neq 0$, then the sequence X/Y converges to x/y.

Proof. In each part, Definition 6.19 is used to prove that the sequence converges to the given real number. To that end, the techniques in Proposition 6.25 and Proposition 6.26 are used.

(a) To see that $X + Y$ converges to $x + y$, choose $\varepsilon > 0$ to be a positive real number. Now you must construct a value for $j \in \mathbb{N}$ for which you can prove that for all $k \in \mathbb{N}$ with $k > j$, $|x_k + y_k - (x + y)| < \varepsilon$. So, suppose you have *already* found a value for j. To prove that the value for j is correct, you need to choose a $k \in \mathbb{N}$ with $k > j$ and show that

$$\begin{aligned} |x_k + y_k - (x + y)| &= |(x_k - x) + (y_k - y)| \\ &\leq |x_k - x| + |y_k - y| \\ &< \varepsilon. \end{aligned}$$

This last inequality will follow from the previous inequality provided that both of the following conditions hold:

$$|x_k - x| < \frac{\varepsilon}{2}, \tag{6.32}$$

$$|y_k - y| < \frac{\varepsilon}{2}. \tag{6.33}$$

So, the issue is how to choose $j \in \mathbb{N}$ so that, when $k > j$, *both* (6.32) and (6.33) hold.

The answer is to use the hypotheses that X converges to x and that Y converges to y. In particular, specialize (see Section 1.6.6) the *for-all* statement in Definition 6.19 for X converging to x to conclude that

there is a $j_1 \in \mathbb{N}$ such that for all $k \in \mathbb{N}$ with $k > j_1$, $|x_k - x| < \varepsilon/2$.

Likewise, from the fact that Y converges to y, you know that

there is a $j_2 \in \mathbb{N}$ such that for all $k \in \mathbb{N}$ with $k > j_2$, $|y_k - y| < \varepsilon/2$.

The desired value for j is $j = \max\{j_1, j_2\}$, as now shown.

To see that the value of $j = \max\{j_1, j_2\}$ is correct, let $k > j$, so $k > j_1$ and $k > j_2$ and therefore (6.32) and (6.33) hold. It now follows that

$$\begin{aligned} |x_k + y_k - (x+y)| &= |(x_k - x) + (y_k - y)| && \text{(algebra)} \\ &\le |x_k - x| + |y_k - y| && \text{(triangle inequality)} \\ &< \frac{\varepsilon}{2} + \frac{\varepsilon}{2} && \text{[(6.32) and (6.33)]} \\ &= \varepsilon && \text{(algebra).} \end{aligned}$$

This shows that $X + Y$ converges to $x + y$ and completes the proof of this part.

(b) This proof is similar to that in part (a) and is left to the exercises.

(c) To see that $X \cdot Y$ converges to xy, choose $\varepsilon > 0$ to be a positive real number. Now you must construct a value for $j \in \mathbb{N}$ for which you can prove that for all $k \in \mathbb{N}$ with $k > j$, $|x_k y_k - xy| < \varepsilon$. So, suppose you have *already* found the value for j. To prove that the value for j is correct, you need to choose a $k \in \mathbb{N}$ with $k > j$ and show that

$$\begin{aligned} |x_k y_k - xy| &= |(x_k y_k - x_k y) + (x_k y - xy)| \\ &\le |x_k(y_k - y)| + |(x_k - x)y| \\ &= |x_k||y_k - y| + |x_k - x||y| \\ &= |x_k - x + x||y_k - y| + |x_k - x||y| \\ &\le |x_k - x||y_k - y| + |x||y_k - y| + |x_k - x||y| \\ &< \varepsilon. \end{aligned}$$

This last inequality will follow from the previous inequality provided that each of the three terms in that previous inequality is $< \varepsilon/3$, which in turn is *true* if *all* of the following conditions hold:

$$|x_k - x| < \sqrt{\frac{\varepsilon}{3}}, \tag{6.34}$$

$$|y_k - y| < \sqrt{\frac{\varepsilon}{3}}, \tag{6.35}$$

$$|x||y_k - y| < \frac{\varepsilon}{3}, \tag{6.36}$$

$$|x_k - x||y| < \frac{\varepsilon}{3}. \tag{6.37}$$

So, the issue is how to choose $j \in \mathbb{N}$ so that, when $k > j$, (6.34) through (6.37) hold.

The answer is to use the hypotheses that X converges to x and that Y converges to y. In particular, specialize the *for-all* statement in Definition 6.19 for X converging to x to conclude that both of the following are *true*:

> There is a $j_1 \in \mathbb{N}$ such that for all $k \in \mathbb{N}$ with $k > j_1$, $|x_k - x| < \sqrt{\varepsilon/3}$.
>
> There is a $j_2 \in \mathbb{N}$ such that for all $k \in \mathbb{N}$ with $k > j_2$, $|x_k - x||y| < \varepsilon/3$.

There is such a value for j_2 because, if $y = 0$, then *any* value for j_2 satisfies this property and if $y \neq 0$, then, by Definition 6.19, there is a $j_2 \in \mathbb{N}$ such that for all $k \in \mathbb{N}$ with $k > j_2$, $|x_k - x| < \varepsilon/(3|y|)$, or, equivalently, $|x_k - x||y| < \varepsilon/3$.

Likewise, from the fact that Y converges to y,

> There is a $j_3 \in \mathbb{N}$ such that for all $k \in \mathbb{N}$ with $k > j_3$, $|y_k - y| < \sqrt{\varepsilon/3}$.
>
> There is a $j_4 \in \mathbb{N}$ such that for all $k \in \mathbb{N}$ with $k > j_4$, $|x||y_k - y| < \varepsilon/3$ (using the same reasoning as for j_2).

The desired value for j is $\max\{j_1, j_2, j_3, j_4\}$, as now shown.

To see that the value of $j = \max\{j_1, j_2, j_3, j_4\}$ is correct, let $k > j$ so $k > j_1$, $k > j_2$, $k > j_3$, and $k > j_4$ and thus (6.34) through (6.37) hold. It now follows that

$$\begin{aligned}
|x_k y_k - xy| &= |(x_k y_k - x_k y) + (x_k y - xy)| && \text{(adding 0)} \\
&\leq |x_k(y_k - y)| + |(x_k - x)y| && \text{(triangle inequality)} \\
&= |x_k||y_k - y| + |x_k - x||y| && \text{(algebra)} \\
&= |x_k - x + x||y_k - y| + \\
&\quad\ |x_k - x||y| && \text{(adding 0)} \\
&\leq |x_k - x||y_k - y| + \\
&\quad\ |x||y_k - y| + |x_k - x||y| && \text{(triangle inequality)} \\
&< \sqrt{\frac{\varepsilon}{3}\left(\frac{\varepsilon}{3}\right)} + \frac{\varepsilon}{3} + \frac{\varepsilon}{3} && \text{[(6.34) through (6.37)]} \\
&= \varepsilon && \text{(algebra).}
\end{aligned}$$

This shows that $X \cdot Y$ converges to xy and completes the proof of this part.

(d) The proof is left to the exercises.

(e) Because all the elements of $Y = (y_k)$ are nonzero, you can define the sequence $Z = (z_k)$ in which

$$z_k = \frac{1}{y_k}, \text{ for } k = 1, 2, \ldots. \tag{6.38}$$

It suffices to show that the sequence Z in (6.38) converges to $1/y$ because then, by part (c), $X/Y = X \cdot Z$ converges to $x(1/y) = x/y$.

To see that the sequence Z defined by (6.38) converges to $1/y$, choose $\varepsilon > 0$ to be a positive real number. Now you must construct a value for $j \in \mathbb{N}$ for which you can prove that for all $k \in \mathbb{N}$ with

$k > j$, $|(1/y_k) - (1/y)| < \varepsilon$. So, suppose you have *already* found the value for j. To prove that the value for j is correct, you need to choose a $k \in \mathbb{N}$ with $k > j$ and show that

$$\begin{aligned}\left|\frac{1}{y_k} - \frac{1}{y}\right| &= \left|\frac{y - y_k}{y y_k}\right| \\ &= \frac{1}{|y y_k|}|y - y_k| \\ &< \varepsilon.\end{aligned}$$

This last inequality will follow from the previous equality provided that both of the following conditions hold:

$$\frac{1}{|y y_k|} < \frac{2}{|y|^2}, \tag{6.39}$$

$$|y - y_k| < \frac{1}{2}|y|^2\varepsilon. \tag{6.40}$$

So, the issue is how to choose $j \in \mathbb{N}$ so that when $k > j$, *both* (6.39) and (6.40) hold.

The answer is to use the hypothesis that Y converges to $y \neq 0$. For the inequality in (6.39), specialize the *for-all* statement in Definition 6.19 for Y converging to y to conclude that

there is a $j_1 \in \mathbb{N}$ such that for all $k \in \mathbb{N}$ with $k > j_1$,

$$|y - y_k| < \frac{1}{2}|y|.$$

Then, by the triangle inequality, you have that for $k > j_1$,

$$|y| - |y_k| \leq |y - y_k| < \frac{1}{2}|y|,$$

or equivalently, that

$$|y_k| > \frac{1}{2}|y|.$$

Now you can multiply both sides by $|y|$ and then take the reciprocals to obtain (6.39).

For the inequality in (6.40), specialize the *for-all* statement in Definition 6.19 for Y converging to y to conclude that

there is a $j_2 \in \mathbb{N}$ such that for all $k \in \mathbb{N}$ with $k > j_2$,

$$|y - y_k| < \frac{1}{2}|y|^2\varepsilon.$$

The desired value for j is $\max\{j_1, j_2\}$, as now shown.

To see that the value of $j = \max\{j_1, j_2\}$ is correct, let $k > j$, so $k > j_1$ and $k > j_2$ and thus (6.39) and (6.40) hold. It now follows that

$$\begin{aligned}
\left|\frac{1}{y_k} - \frac{1}{y}\right| &= \left|\frac{y - y_k}{yy_k}\right| && \text{(algebra)} \\
&= \frac{1}{|yy_k|}|y - y_k| && \text{(algebra)} \\
&< \left(\frac{2}{|y|^2}\right)\left(\frac{1}{2}|y|^2\varepsilon\right) && \text{[from (6.39) and (6.40)]} \\
&= \varepsilon && \text{(algebra).}
\end{aligned}$$

Consequently, the sequence $Z = (1/y_k)$ converges to $1/y$ and hence the sequence $X/Y = X \cdot Z$ converges to $x(1/y) = x/y$, thus completing the proof.

QED

6.4.3 Convergence of a Sequence to Some Real Number

Until now, in discussing the convergence of a sequence $X = (x_1, x_2, \ldots)$, it has been assumed that you *know* the specific value of $x \in \mathbb{R}$ to which you want to show that the sequence does (or does not) converge. In many cases, however, you may *not* know the value of x. For example, as you will soon learn, the sequence $X = (x_1, x_2, \ldots)$ in which

$$x_k = \left(1 + \frac{1}{k}\right)^k, \text{ for } k = 1, 2, \ldots \tag{6.41}$$

converges, even though you may not know the specific limit.

Definition of Convergence and Divergence of a Sequence

The first step is to translate the concept of "a sequence converging to some real number" to symbolic form, as given in the following definition.

> **Definition 6.20** *A sequence of real numbers $X = (x_1, x_2, \ldots)$* **converges** *if and only if there is a real number x such that X converges to x.*

For example, the sequence in (6.41) converges because, even though you do not know the specific value of x to which the sequence converges, *there is* a real number x such that X converges to x, as you will soon see.

As another example, the sequence $X = (x_1, x_2, \ldots)$ in which $x_k = 1/k$ also converges. In this case, by Proposition 6.25 in Section 6.4.1, you know that this sequence converges to 0.

In contrast, the sequence $X = (1, 2, 3, \ldots)$ does *not* converge. By negating Definition 6.20, you can state more precisely what this means, as follow.

Definition 6.21 *A sequence of real numbers $X = (x_1, x_2, \ldots)$* **diverges** *if and only if there is no real number x such that X converges to x.*

The next issue is to develop some methods for determining *when* a sequence converges.

Convergence of Bounded Monotone Sequences

The objective now is to develop a method for determining that *certain* sequences converge, *without having to know the specific value to which the sequence converges.* The approach is based on the infimum and supremum properties of real numbers (see Section 6.2), as stated in the following theorem.

Theorem 6.5 *If $X = (x_1, x_2, \ldots)$ is a sequence of real numbers that is both monotone and bounded, then X converges.*

Proof. Because a monotone sequence is one that is either increasing or decreasing, a proof by cases (see Section 1.6.11) is appropriate. But, without loss of generality, you can assume that the sequence is decreasing.

According to Definition 6.20, you must construct a real number x to which the sequence X converges. That number x is the infimum of the following set X' of real numbers that corresponds to X (see Section 6.3.4):

$$X' = \{x_k : k \in \mathbb{N}\}.$$

For the nonempty set X' to have an infimum, you must know that X' is bounded below, but this is proved in Proposition 6.23 in Section 6.3.4. In particular, because X is a bounded sequence, X' is a bounded set and so X' is bounded below. Therefore, by the infimum property of real numbers, X' has an infimum, say, x.

It remains to show that the sequence X converges to x. Accordingly, choose a real number $\varepsilon > 0$ for which you must construct a $j \in \mathbb{N}$ such that you can prove that for all $k \in \mathbb{N}$ with $k > j$, $|x_k - x| < \varepsilon$. Because x is the infimum of X', by Proposition 6.9 in Section 6.2.1, you can find a $j \in \mathbb{N}$ such that

$$x_j < x + \varepsilon. \tag{6.42}$$

To see that the value for j in (6.42) is correct, let $k \in \mathbb{N}$ with $k > j$. Because the sequence X is decreasing and $k > j$, you know that $x_k \leq x_j$. You therefore have from (6.42) that

$$x_k \leq x_j < x + \varepsilon. \tag{6.43}$$

On the other hand, x is a lower bound for X' and $\varepsilon > 0$ so

$$x - \varepsilon < x \leq x_k. \tag{6.44}$$

Combining (6.43) and (6.44) yields

$$x - \varepsilon < x \leq x_k \leq x_j < x + \varepsilon,$$

and therefore by subtracting x throughout, that

$$-\varepsilon < x_k - x < \varepsilon,$$

from which it follows by Proposition 6.6(d) in Section 6.1.2 that $|x_k - x| < \varepsilon$. This shows that X converges to x.

A similar proof shows that if X is increasing, then X converges to the supremum of X', thus completing the proof.

QED

You can often use Theorem 6.5 to show that a sequence X converges. Doing so requires that you show the sequence to be both monotone and bounded. Then, the real number x to which that sequence converges is the infimum (or supremum) of the correponding set X', which you must then find, if possible. This process is illustrated now with two examples.

Example 6.12 – Convergence of the Sequence $\left(1 + \frac{1}{k}\right)^k$ to $e = 2.718282$

Recall the sequence $X = (x_k)$ in which

$$x_k = \left(1 + \frac{1}{k}\right)^k, \text{ for } k = 1, 2, \ldots. \tag{6.45}$$

By Proposition 6.20 and Proposition 6.22 in Section 6.3.3, you know that X is both monotone and bounded. Consequently, by Theorem 6.5, X converges to some value, called the **Euler number** and denoted by e. Because the sequence is increasing, the number e is the supremum of the corresponding set X'. There is no closed-form expression for determining the exact value of e which, in fact, is irrational. There are, however, methods for finding the value of e to any desired degree of accuracy. The value of e to six decimal places is 2.718282.

Example 6.13 – Convergence of the Sequence $\frac{1}{2}\left(x^k + \frac{2}{x^k}\right)$ to $\sqrt{2}$

Recall the sequence $X = (x_k)$ for finding the $\sqrt{2}$, in which $x_1 = 2$ and

$$x_{k+1} = \frac{1}{2}\left(x_k + \frac{2}{x_k}\right), \text{ for } k = 1, 2, \ldots.$$

By Proposition 6.19 and Proposition 6.21 in Section 6.3.3, you know that X is both monotone and bounded. Consequently, by Theorem 6.5, X converges to some real number x, which is the infimum of the corresponding set X'. In this case, $x = \sqrt{2}$. This is because, from the definition of x_k and from Proposition 6.29 in Section 6.4.2, x satisfies the following property:

$$x = \frac{1}{2}\left(x + \frac{2}{x}\right),$$

or equivalently, $x^2 = 2$, so $x = \sqrt{2}$.

In this section, you have learned about convergence, which is the issue of whether a sequence of real numbers gets closer and closer to some real number. In Section 6.5, you will see how the techiques of generalization, abstraction, and developing axiomatic systems are used to extend the concept of sequences and convergence.

6.5 Generalizations of Sequences of Real Numbers

In Section 6.4, you learned about sequences of real numbers and what it means for such a sequence to converge. The objective of this section is to show how a sequential generalization, as described in Section 1.4, is used to extend these concepts to other mathematical objects. You will also see how abstraction is used and how to develop an axiomatic system for measuring the distance between two objects. You should review the material on abstraction and developing abstract and axiomatic systems in Chapters 1 and 2 and also the material on n-vectors in Chapter 4. Throughout this section, n is a fixed natural number. Also, *superscript* notation is used for the elements of a sequence instead of subscripts. Thus, a sequence of real numbers is written as $X = (x^1, x^2, \ldots)$, instead of $X = (x_1, x_2, \ldots)$.

6.5.1 Sequences of Vectors

One generalization of a sequence of real numbers $X = (x^1, x^2, \ldots)$ is a sequence in which each element is an n-vector. To formalize this concept, you can generalize Definition 6.13 in Section 6.3.1, as follows:

Definition 6.22 *A* **sequence of n-vectors** *is a function $X : \mathbb{N} \to \mathbb{R}^n$. For $k \in \mathbb{N}$, the notation $\mathbf{x}^k$ is used instead of $X(k)$ and the sequence is written as follows:*

$$X = (\mathbf{x}^1, \mathbf{x}^2, \mathbf{x}^3, \ldots),$$

or equivalently,

$$X = (\mathbf{x}^k : k \in \mathbb{N}) \text{ or } X = (\mathbf{x}^k).$$

A sequence of n-vectors is a generalization of a sequence of real numbers because, when you set the value of n to 1 in Definition 6.22, you obtain the definition of a sequence of real numbers. In other words, a sequence of real numbers is a special case of a sequence of n-vectors.

Each element $\mathbf{x}^k$ in the sequence is an n-vector and, as such, has n components – that is, $\mathbf{x}^k = (x_1^k, x_2^k, \ldots, x_n^k)$, as illustrated in the next example.

Example 6.14 – A Sequence of Vectors

The following is a sequence of vectors in two dimensions in which, for each $k = 1, 2, \ldots$, the two components of the vector $\mathbf{x}^k$ have the following values:

$$x_1^k = \frac{1}{k},$$

$$x_2^k = 1 - \frac{1}{k}.$$

An appropriate visual image of a sequence of vectors is illustrated in Figure 6.12, which corresponds to the sequence in Example 6.14.

The next step is to generalize all of the operations and concepts associated with sequences of real numbers to sequences of n-vectors. For example, the generalization of an ε-neighborhood to a ball of radius ε is discussed in Section 2.2.2. In the exercises, you are asked to generalize the unary and binary operations on sequences of real numbers to sequences of n-vectors. In what follows, the generalization of convergence is addressed.

6.5.2 Convergence of a Sequence of Vectors

Recall from Section 6.4 that when discussing the convergence of a sequence of *real numbers*, there are two issues:

1. *Given* a value for the real number x, does the sequence converge to x – *yes* or *no*?

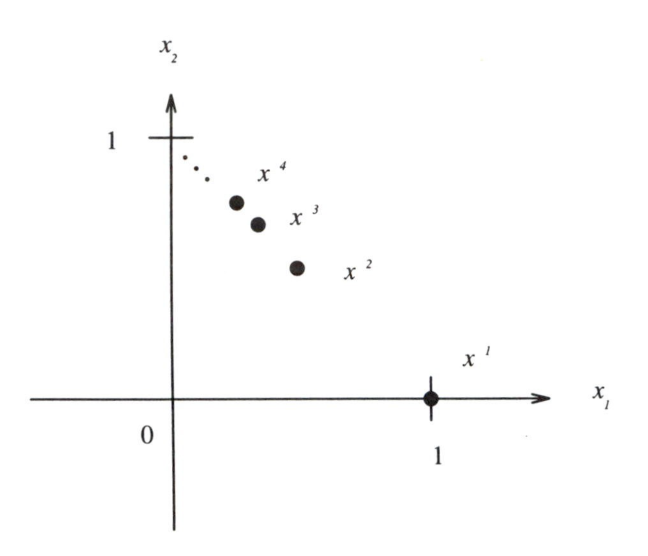

Figure 6.12: The Sequence of Vectors in Example 6.14

2. Does the sequence converge at *all*, that is, is there *any* real number to which the sequence converges?

These same two issues arise when discussing convergence of a sequence of n-vectors, both of which are addressed now.

Convergence of a Sequence of Vectors to a Given Vector

Consider now a sequence of n-vectors $X = (\mathbf{x}^1, \mathbf{x}^2, \ldots)$. Where a sequence of real numbers converges to a *real number*, a sequence of n-vectors should converge to an *n-vector*. The objective is to develop a symbolic definition of what it means for a sequence of n-vectors to *converge* to a given n-vector, say, $\mathbf{x} \in \mathbb{R}^n$.

There are several possible approaches for creating such a definition, all of which rely on Definition 6.19 in Section 6.4.1 for the convergence of a sequence of real numbers, the following version of which is used here:

> Suppose that $X = (x^1, x^2, \ldots)$ is a sequence of real numbers and $x \in \mathbb{R}$. The *sequence of real numbers X converges to x* if and only if for every real number $\varepsilon > 0$, there is a $j \in \mathbb{N}$ such that for all $k \in \mathbb{N}$ with $k > j$, $|x^k - x| < \varepsilon$.

One way to generalize the foregoing definition to n-vectors is based on the observation that a sequence of n-vectors $X = (\mathbf{x}^1, \mathbf{x}^2, \ldots)$ consists of n sequences of real numbers, one for each of the n coordinates. That is, from the sequence X of n-vectors, you can identify the following n sequences of real numbers:

$$\begin{aligned} X_1 &= (x_1^1, x_1^2, x_1^3, \ldots) \text{ (coordinate 1 of each vector)}, \\ X_2 &= (x_2^1, x_2^2, x_2^3, \ldots) \text{ (coordinate 2 of each vector)}, \\ \vdots \;\; &\vdots \qquad \vdots \qquad\qquad\qquad \vdots \\ X_n &= (x_n^1, x_n^2, x_n^3, \ldots) \text{ (coordinate } n \text{ of each vector)}. \end{aligned}$$

You can then define the sequence of n-vectors to converge to $\mathbf{x} \in \mathbb{R}^n$ if and only if *each* of the foregoing sequences of real numbers converges to the corresponding component of $\mathbf{x}$. You are asked to pursue this approach in the exercises.

A second approach to defining what it means for a sequence of n-vectors $X = (\mathbf{x}^1, \mathbf{x}^2, \ldots)$ to converge to a given n-vector $\mathbf{x}$ arises by trying directly to generalize the definition for sequences of real numbers. For example, a replacement of $\mathbb{R}$ with $\mathbb{R}^n$ and $x \in \mathbb{R}$ with $\mathbf{x} \in \mathbb{R}^n$ in the foregoing definition results in the following:

> Suppose that $X = (\mathbf{x}^1, \mathbf{x}^2, \ldots)$ is a sequence of n-vectors and $\mathbf{x} \in \mathbb{R}^n$. The *sequence of n-vectors X converges to* $\mathbf{x}$ if and only if for every real number $\varepsilon > 0$, there is a $j \in \mathbb{N}$ such that for all $k \in \mathbb{N}$ with $k > j$, $|\mathbf{x}^k - \mathbf{x}| < \varepsilon$.

However, a syntax error arises because $\mathbf{x}^k$ and $\mathbf{x}$ are n-vectors, and you cannot take the absolute value of the n-vector $\mathbf{x}^k - \mathbf{x}$. In other words, $|\mathbf{x}^k - \mathbf{x}|$ is *undefined*.

One solution to this problem is to recall that when x^k and x are *real numbers*, $|x^k - x|$ is a measure of how close x^k is to x. Thus, when $\mathbf{x}^k$ and $\mathbf{x}$ are *n-vectors*, what is needed is a method for measuring how close $\mathbf{x}^k$ is to $\mathbf{x}$. As described in Definition 4.5 in Section 4.1.3, one such measure is the *norm* of the difference of the two n-vectors. Recall that

$$\text{norm}(\mathbf{y}) = ||\mathbf{y}|| = \sqrt{(y_1)^2 + \cdots + (y_n)^2}.$$

Thus, when $\mathbf{x}^k$ and $\mathbf{x}$ are n-vectors, $||\mathbf{x}^k - \mathbf{x}||$ is a measure of how close $\mathbf{x}^k$ is to $\mathbf{x}$. With this idea, the definition of a sequence of n-vectors converging to a given n-vector $\mathbf{x}$ follows.

> **Definition 6.23** *Suppose that $X = (\mathbf{x}^1, \mathbf{x}^2, \ldots)$ is a sequence of n-vectors and $\mathbf{x} \in \mathbb{R}^n$. The* **sequence of n-vectors X converges to $\mathbf{x}$** *if and only if for every real number $\varepsilon > 0$, there is a $j \in \mathbb{N}$ such that for all $k \in \mathbb{N}$ with $k > j$, $||\mathbf{x}^k - \mathbf{x}|| < \varepsilon$.*

Convergence of a Sequence of Vectors to Some Vector

As mentioned earlier in this section, you might not know the specific n-vector $\mathbf{x}$ to which the sequence X converges. Nevertheless, the sequence may be converging to *some* n-vector. Analogous to Definition 6.20 in Section 6.4.3 is the following definition.

> **Definition 6.24** *A* **sequence of** n**-vectors** $X = (\mathbf{x}^1, \mathbf{x}^2, \ldots)$ **converges** *if and only if there is an n-vector* $\mathbf{x}$ *such that X converges to* $\mathbf{x}$.

6.5.3 An Abstract System for Sequences: Metric Spaces

The next step in the process of generalizing a sequence of real numbers is to use abstraction. That is, you should now think of each element x^k of the sequence $X = (x^1, x^2, \ldots)$ as an *object*, rather than as a specific item (such as a real number or an n-vector), in which each x^k belongs to some underlying set of objects, say, S. Now try to create an appropriate definition by generalizing Definition 6.22. You should arrive at the following.

> **Definition 6.25** *Let S be a given set. A* **sequence in** S *is a function $X : \mathbb{N} \to S$. For $k \in \mathbb{N}$, the notation x^k is used instead of $X(k)$ and the sequence is written as follows:*
>
> $$X = (x^1, x^2, x^3, \ldots),$$
>
> *or equivalently,*
>
> $$X = (x^k : k \in \mathbb{N}) \text{ or } X = (x^k).$$

Proceeding with this idea, the next step is to create a definition of what it means for the sequence $X = (x^1, x^2, \ldots)$ of objects in S to converge to a given object $x \in S$. What problems arise when you try to generalize Definition 6.23?

If you generalize Definition 6.23 by replacing $\mathbb{R}^n$ with S, you obtain the following:

> Suppose that $X = (x^1, x^2, \ldots)$ is a sequence in a set S and $x \in S$. The *sequence X in S converges to x* if and only if for every real number $\varepsilon > 0$, there is a $j \in \mathbb{N}$ such that for all $k \in \mathbb{N}$ with $k > j$, $||x^k - x|| < \varepsilon$.

A syntax error arises in that the expression $||x^k - x||$ does not make sense. This is because the binary operation of subtracting the objects x^k and x has not been defined – that is, $x^k - x$ is *undefined*. Observe that even if $x^k - x$ *is* an element of S, the concept of the *norm* of $x^k - x$ is also undefined.

You can overcome these syntax errors by recalling that when x^k and x are real numbers (or n-vectors, for that matter), $||x^k - x||$ is a measure of how close x^k is to x. To generalize this idea to the case where x^k and x are *objects* in the set S, you need a method for computing how close two such objects are.

Metric Spaces

So suppose that x and y are two objects in S. What is needed now is a method for computing a *real number* that represents the distance from x to y. Although you do not know specifically how to find this number, you can represent the computation symbolically, as follows:

Let $d(x, y)$ = a real number representing the distance from x to y.

In other words, you can think of d as a *function* that associates to each pair of elements in S, a real number representing the distance from x to y. In the notation of sets and functions:

$$d : S \times S \longrightarrow \mathbb{R}.$$

It is important to realize that not *every* function $d : S \times S \rightarrow \mathbb{R}$ is a *valid* distance function. For example, you would not consider a value of $d(x, y) = -1$ to be a measure of the distance from x to y because the distance between two objects should always be *nonnegative*. In other words, the function d must satisfy certain properties in order to be considered a *distance* function.

These properties come from working with many specific cases of distance functions, such as the absolute value of the difference of two real numbers and the norm of the difference of two n-vectors. Those properties then manifest themselves in the form of *axioms* for an associated axiomatic system, which is presented in the following definition.

Definition 6.26 *Let S be a set and $d : S \times S \rightarrow \mathbb{R}$. The pair (S, d) is a* **metric space** *if and only if the following conditions hold for all $x, y, z \in S$:*

1. $d(x, y) \geq 0$ (nonnegativity).
2. $d(x, y) = 0$ *if and only if* $x = y$ (definiteness).
3. $d(x, y) = d(y, x)$ (symmetry).
4. $d(x, z) \leq d(x, y) + d(y, z)$ (triangle inequality).

Several examples of metric spaces follow.

Example 6.15 – The Metric Space of Real Numbers

If $S = \mathbb{R}$, then (S, d) is a metric space when the function $d : \mathbb{R} \times \mathbb{R} \to \mathbb{R}$ is defined by

$$d(x, y) = |x - y|.$$

Properties (a), (b), and (c) of Definition 6.26 follow directly from the definition of the absolute value. Property (d) is a consequence of the triangle inequality of the absolute value because

$$\begin{aligned} d(x, z) &= |x - z| \\ &= |(x - y) + (y - z)| \\ &\leq |x - y| + |y - z| \\ &= d(x, y) + d(y, z). \end{aligned}$$

Example 6.16 – The Metric Space of n-Vectors

If $S = \mathbb{R}^n$, then (S, d) is a metric space when you define the function $d : \mathbb{R}^n \times \mathbb{R}^n \to \mathbb{R}$ by

$$d(\mathbf{x}, \mathbf{y}) = ||\mathbf{x} - \mathbf{y}||.$$

Properties (a), (b), and (c) of Definition 6.26 follow directly from the definition of the norm. Property (d) is a consequence of the property of the norm because

$$\begin{aligned} d(\mathbf{x}, \mathbf{z}) &= ||\mathbf{x} - \mathbf{z}|| \\ &= ||(\mathbf{x} - \mathbf{y}) + (\mathbf{y} - \mathbf{z})|| \\ &\leq ||\mathbf{x} - \mathbf{y}|| + ||\mathbf{y} - \mathbf{z}|| \\ &= d(\mathbf{x}, \mathbf{y}) + d(\mathbf{y}, \mathbf{z}). \end{aligned}$$

Example 6.17 – The Discrete Metric Space

As you are asked to verify in the exercises, if S is a set, then (S, d) is a metric space when the function $d : S \times S \to \mathbb{R}$ is defined as follows:

$$d(x, y) = \begin{cases} 1, & \text{if } x \neq y \\ 0, & \text{if } x = y. \end{cases}$$

This metric space is called the **discrete metric space on S**.

It is possible to define several *different* distance functions d on a set S, each of which makes (S, d) a metric space. For instance, if $S = \mathbb{R}^n$ and $\mathbf{x} = (x_1, \ldots, x_n)$ and $\mathbf{y} = (y_1, \ldots, y_n)$ are n-vectors, then the following distance function makes $(\mathbb{R}^n, d)$ a metric space:

$$d(\mathbf{x}, \mathbf{y}) = |x_1 - y_1| + \cdots + |x_n - y_n|.$$

Another distance function for which $(\mathbb{R}^n, d)$ is a metric space is

$$d(\mathbf{x}, \mathbf{y}) = \max\{|x_1 - y_1|, \ldots, |x_n - y_n|\}.$$

Convergence of a Sequence in a Set S

With the concept of a metric space, you now have the ability to measure the distance between two objects in a set S. This, in turn, allows you to generalize the definition of convergence of a sequence of real numbers and n-vectors to a sequence in a set S, as given in the following definition.

> **Definition 6.27** *Suppose that* $X = (x^1, x^2, \ldots)$ *is a sequence in a set* S, $d : S \times S \to \mathbb{R}$ *is a function with* (S, d) *being a metric space, and* $x \in S$. *The* **sequence** X **in** S **converges to** x *if and only if for every real number* $\varepsilon > 0$, *there is a* $j \in \mathbb{N}$ *such that for all* $k \in \mathbb{N}$ *with* $k > j$, $d(x^k, x) < \varepsilon$.

Definition 6.27 assumes that you *know* the value of $x \in S$ to which the sequence X converges. In the event that you do not know this value, the following definition (analogous to Definitions 6.20 and 6.24) is appropriate.

> **Definition 6.28** *The* **sequence** $X = (x^1, x^2, \ldots)$ **in a set** S **converges** *if and only if there is an element* $x \in S$ *such that* X *converges to* x.

You can use these definitions to prove many results pertaining to sequences in a set S but that material is beyond the scope of this book.

In this section, you have seen how generalization, abstraction, and axiomatic systems are used to create the abstract concept of a sequence in a set S. A summary of all the material in this chapter follows.

Chapter Summary

In this chapter, you have seen how the ideas of Chapters 1 and 2 are applied in the area of real analysis to differentiate the real numbers from the integers and the rationals. You have learned about sequences and convergence, which are useful in solving problems with algorithms that generate an infinite number of points. A summary of these topics is now provided.

The Algebraic and Order Properties of the Real Numbers

You have learned the following similarities and differences between the integers ($\mathbb{Z}$), the rationals ($\mathbb{Q}$), and the real numbers ($\mathbb{R}$).

Similarities Between the Sets $\mathbb{Z}$, $\mathbb{Q}$, and $\mathbb{R}$

1. You can apply the binary operations of addition, subtraction, multiplication, and division to the numbers in each of these sets. In each case, these operations satisfy certain properties, such as associativity, commutativity, and the like, as presented in Section 6.1.1.
2. You can compare two numbers in each of these sets with the binary relations $=, >, \geq, <,$ and $\leq$ to determine their relative order in relation to each other.
3. All three sets contain an infinite number of elements.

Differences Between the Sets $\mathbb{Z}$, $\mathbb{Q}$, and $\mathbb{R}$

1. The result of dividing two integers is not necessarily an integer, but the result of dividing two rationals (or two reals) *is* a rational (or a real), if the divisor is not zero.
2. You know that $\mathbb{Z} \subseteq \mathbb{Q} \subseteq \mathbb{R}$, but there are elements of $\mathbb{Q}$ (such as $1/2$) that are not in $\mathbb{Z}$. Similarly, there are elements of $\mathbb{R}$ (such as $\sqrt{2}$) that are not in $\mathbb{Q}$.
3. Between any two distinct real (or rational) numbers, you can find another real (or rational) number. In contrast, between two consecutive integers, you cannot find another integer.
4. Each set contains an infinite number of elements. By using bijective functions, you learned that the number of elements in $\mathbb{Z}$ and $\mathbb{Q}$ is the same as in $\mathbb{N}$. However, the number of elements in $\mathbb{Z}$ is *not* the same as the number of elements in $\mathbb{R}$, which has more elements than $\mathbb{Z}$.
5. The infimum and supremum properties distinguish the real numbers from the rationals. This is because a nonempty bounded set of real numbers has an infimum and a supremum that are real numbers (although neither the infimum nor the supremum need belong to the set). In contrast, the infimum and supremum of a nonempty bounded set of rational numbers need *not* be a rational number.

Sequences of Real Numbers

A sequence of real numbers is a generalization of an n-vector in that a sequence is a countably infinite ordered list of real numbers. The solution to many problems in mathematics, engineering, and the sciences are obtained by an algorithm that generates an infinite sequence.

In addition to the basic operations of adding, subtracting, multiplying, and dividing two sequences, you learned about several types of sequences with special properties, namely, those that are monotone – either increasing or decreasing – and those that are bounded – sequences all of whose numbers are within some fixed distance from 0. Keep in mind that sequences and sets, though related, are different because a sequence of real numbers is an infinite ordered list in which the elements can repeat, whereas, a set can be finite or infinite, but all elements are distinct and their order is irrelevant.

You then learned about the fundamental concept of convergence. Convergence is the issue of whether the numbers in a sequence are getting closer and closer to some limiting value (that is often the solution to the specific problem under investigation). There are two separate convergence questions pertaining to a sequence $X = (x_1, x_2, \ldots)$ of real numbers:

1. *Given* a specific real number x, does the sequence X converge to x – *yes* or *no*?

2. Is there *any* real number to which the sequence X *does* converge and, if so, to *which* real number?

You saw how ε-neighborhoods are used to develop a symbolic definition of what it means for a sequence of real numbers to converge to a given real number. A bounded monotone sequence of real numbers is shown to converge to the infimum or supremum of the set whose elements are those of the sequence.

Generalizations of Sequences of Real Numbers

In Section 6.5, generalization is used to create sequences of n-vectors that include sequences of real numbers as a special case. The concepts of convergence are generalized by using the norm as a measure of the distance between two n-vectors. Then, through the use of abstraction, you learned about a sequence of objects belonging to a set S. To generalize the notion of convergence to such a sequence, it is necessary to develop a method for determing how close two objects in S are. A metric space is an axiomatic system that enables you to do so and consequently, to develop appropriate definitions of convergence for a sequence in S.

Exercises

Exercise 6.1 Prove each of the following properties pertaining to the real numbers a and b. You may use any of the results in Section 6.1.

(a) If $a \geq 0$ and $b \leq 0$, then $ab \leq 0$.

(b) If $a \leq 0$ and $b \leq 0$, then $ab \geq 0$.

(c) If $0 \leq a \leq b$, then $a^2 \leq b^2$.

Exercise 6.2 Prove each of the following properties pertaining to the real numbers a and b. You may use any of the results in Section 6.1.

(a) If $a > 1$, then $\sqrt{a} < a$.

(b) $|a| - |b| \leq |a - b|$.

(c) If $a, b \geq 0$, then $(a + b)/2 \geq \sqrt{ab}$.

Exercise 6.3 Generalize Theorem 6.1 in Section 6.1.2 by proving that if n is a natural number and $a_1, \ldots, a_n \in \mathbb{R}$, then $|a_1 + \cdots + a_n| \leq |a_1| + \cdots + |a_n|$.

Exercise 6.4 Recall the assumption of the existence of the positive real numbers, that is, there is a subset P of $\mathbb{R}$ satisfying the following three axioms:

(i) For all $a, b \in P,\ a + b \in P$.

(ii) For all $a, b \in P,\ ab \in P$.

(iii) For all $a \in P$, exactly one of the following hold:
$a \in P,\ a \in \{0\}$, or $-a \in P$.

Consider applying abstraction to generalize this concept by replacing the set $\mathbb{R}$ with a general set S of objects (and so P is now a subset of S). Identify all syntax errors that now arise in Axioms (i), (ii), and (iii). What special element(s) in S and what operations on S are needed to eliminate all of these syntax errors?

Exercise 6.5 Suppose that a and b are real numbers. Prove that $a \leq b$ if and only if for every real number $\varepsilon > 0,\ a \leq b + \varepsilon$.

Exercise 6.6 Suppose that a and b are real numbers. Prove that $a \geq b$ if and only if for every real number $\varepsilon > 0,\ a \geq b - \varepsilon$.

Exercise 6.7 Consider generalizing Theorem 6.2 in Section 6.2.2 in which the $\sqrt{2}$ is proved to be the infimum of the following set:

$$T = \{s \in \mathbb{R} : s > 0 \text{ and } s^2 > 2\}.$$

(a) Suppose that $n > 1$ is a natural number. What set should you use instead of T if you want to prove the existence of the nth root of 2?

(b) Prove that your set in part (a) *has* an infimum.

Exercise 6.8 Look again at the set T in Exercise 6.7.

(a) Suppose that a is a positive real number. What set should you use instead of T if you want to prove the existence of the $\sqrt{a}$?

(b) Prove that your set in part (a) *has* an infimum.

Exercise 6.9 State and prove a proposition analogous to Proposition 6.8 in Section 6.2.1 for the supremum instead of the infimum.

Exercise 6.10 State and prove a proposition analogous to Proposition 6.9 in Section 6.2.1 for the supremum instead of the infimum.

Exercise 6.11 Prove Proposition 6.15 in Section 6.2.4, that is, show that if S and T are sets for which $S \subseteq T$ and S is infinite, then T is infinite.

Exercise 6.12 Generalize Proposition 6.17 in Section 6.2.4 by proving that if n is a natural number and $S_1, \ldots, S_n$ are nonempty, countably infinite, disjoint sets, then $S_1 \cup S_2 \cup \cdots \cup S_n$ is countably infinite.

Exercise 6.13 What problem might arise if you were to generalize the dot product of two n-vectors to the dot product of two *sequences*? In other words, suppose that $X = (x_1, x_2, \ldots)$ and $Y = (y_1, y_2, \ldots)$ are sequences of real numbers. What problem might arise if you were to define

$$X \cdot Y = x_1y_1 + x_2y_2 + \cdots ?$$

Exercise 6.14 Suppose that $c, x, y \in \mathbb{R}$, $X = (x_1, x_2, \ldots)$ is a sequence of real numbers that converges to x, and $Y = (y_1, y_2, \ldots)$ is a sequence of real numbers that converges to y. Complete Proposition 6.29 in Section 6.4.2 by proving each of the following:

(a) $X - Y$ converges to $x - y$.

(b) cX converges to cx.

Exercise 6.15 Recall the concept of a group from Definition 5.6 in Section 5.2.1.

(a) Do sequences of real numbers constitute a group under the operation of addition? Why or why not? Explain.

(b) Do sequences of real numbers constitute a group under the operation of multiplication? Why or why not? Explain.

Exercise 6.16 Do sequences of real numbers constitute a ring (see Definition 5.9 in Section 5.3.1) under the operations of addition and multiplication? Why or why not? Explain.

Exercise 6.17 Generalize the binary operations of adding, subtracting, multiplying, and dividing two sequences of real numbers to two sequences of n-vectors or explain why you cannot do so.

Exercise 6.18 What syntax errors arise if you try to generalize each of the following concepts from sequences of real numbers to sequences in a set S?

(a) A monotone sequence.

(b) A bounded sequence.

Exercise 6.19 Suppose that $X = (\mathbf{x}^1, \mathbf{x}^2, \ldots)$ is a sequence of n-vectors and that $\mathbf{x} = (x_1, \ldots, x_n)$ is an n-vector. Recall, from Section 6.5.2, that one approach to defining what it means for X to converge to $\mathbf{x}$ is for the n sequences of the components of the n-vectors in X to converge to the respective components of $\mathbf{x}$. Develop an appropriate symbolic definition.

Exercise 6.20 On the basis of your definition in Exercise 6.19, prove that a sequence X of n-vectors converges to an n-vector $\mathbf{x}$ (according to Definition 6.23 in Section 6.5.2) if and only if X has the property in Exercise 6.19.

Exercise 6.21 Complete Example 6.17 in Section 6.5.3 by proving that if S is a set and $d : S \times S \to \mathbb{R}$ is defined by

$$d(x, y) = \begin{cases} 1, & \text{if } x \neq y \\ 0, & \text{if } x = y, \end{cases}$$

then (S, d) is a metric space.

Exercise 6.22 Suppose that $\mathbf{x} = (x_1, \ldots, x_n)$ and $\mathbf{y} = (y_1, \ldots, y_n)$ are n-vectors. Prove that if $d : \mathbb{R}^n \times \mathbb{R}^n \to \mathbb{R}$ is defined by

$$d(\mathbf{x}, \mathbf{y}) = |x_1 - y_1| + \cdots + |x_n - y_n|,$$

then (S, d) is a metric space.

Exercise 6.23 Consider a sequence of $(m \times n)$ matrices as a special case of a sequence in a set S, that is, let S be the set of all $(m \times n)$ matrices. To work with the concept of convergence, you need to have a specific method for computing the distance d between two $(m \times n)$ matrices A and B. Create a distance function d in such a way that (S, d) satisfies the properties of a metric space. That is, prove that for your function d, (S, d) is a metric space. (Hint: Use the distances between corresponding elements of the matrices A and B to create a *single* real number that represents the distance from A to B.)

Appendix A

SUMMARY OF PROOF TECHNIQUES

The following material is a summary of the proof techniques as presented in Chapter 13 of *How to Read and Do Proofs*, second edition, by Daniel Solow copyright ©1990 John Wiley & Sons, Inc. and is reprinted with permission from John Wiley & Sons, Inc. Throughout this appendix, A and B are statements for which you want to prove that A *implies* B is true.

The Forward-Backward Method

With the forward-backward method, you can assume that A is true and your job is to prove that B is true. Through the forward process, you derive from A a sequence of statements, $A1, A2, \ldots$, that are necessarily true as a result of A being assumed true. This sequence is not random. It is guided by the backward process whereby, through asking and answering the key question, you derive from B a new statement, $B1$, with the property that if $B1$ is true, then so is B. This backward process can then be applied to $B1$, obtaining a new statement, $B2$, and so on. The objective is to link the forward sequence to the backward sequence by generating a statement in the forward sequence that is precisely the same as the last statement obtained in the backward sequence. Then, like a column of dominoes, you can do the proof by going forward along the sequence from A all the way to B.

The Construction Method

When obtaining the sequence of statements, watch for quantifiers to appear, for then the construction, choose, induction, and/or specialization methods may be useful in doing the proof. For instance, when the quantifier *there is* arises in the backward process in the standard form:

> There is an "object" with a "certain property" such that "something happens,"

consider using the construction method to produce the desired object. With the construction method, you work forward from the assumption that A is true to construct (produce, or devise an algorithm to produce, etc.) the object. However, the actual proof consists of showing that the object you constructed satisfies the certain property and also that the something happens.

The Choose Method

On the other hand, when the quantifier *for all* arises in the backward process in the standard form:

> For all "objects" with a "certain property," "something happens"

consider using the choose method. Here, your objective is to design a proof machine that is capable of taking any object with the certain property and proving that the something happens. To do so, you select (or choose) an object that does have the certain property. You must conclude that, for that object, the something happens. Once you have chosen the object, it is best to proceed by working forward from the fact that the chosen object does have the certain property (together with the information in A, if necessary) and backward from the something that happens.

The Induction Method

The induction method should be considered (even before the choose method) when the statement B has the form:

> For every integer n greater than or equal to some initial one, a statement $P(n)$ is true.

The first step of the induction method is to verify that the statement is true for the first possible value of n. The second step requires you to show that if $P(n)$ is true, then $P(n + 1)$ is true. Remember that the success of a proof by induction rests on your ability to relate the statement $P(n+1)$ to $P(n)$ so that you can make use of the assumption that $P(n)$ is true. In other words, to perform the second step of the induction proof, you should write down the statement $P(n)$, replace n everywhere by $(n+1)$ to obtain $P(n+1)$, and then see if you can express $P(n + 1)$ in terms of $P(n)$. Only then will you be able to use the assumption that $P(n)$ is true to reach the conclusion that $P(n + 1)$ is also true.

The Specialization Method

When the quantifier *for all* arises in the forward process in the standard form:

For all "objects" with a "certain property," "something happens"

you will probably want to use the specialization method. To do so, watch for one of these objects to arise, often in the backward process. By using specialization, you can then conclude, as a new statement in the forward process, that the something does happen for that particular object. That fact should then be helpful in reaching the conclusion that B is true. When using specialization, be sure to verify that the particular object does satisfy the certain property, for only then will the something happen.

If statements contain more than one quantifier, that is, they are nested, process them in the order in which they appear from left to right. As you read the first quantifier in the statement, identify its objects, certain property, and something that happens. Then apply an appropriate technique based on whether the statement is in the forward or backward process, and whether the quantifier is *for all* or *there is*. This process is repeated until all quantifiers have been dealt with.

The Contrapositive Method

When the original statement B contains the key word *no* or *not*, or when the forward-backward method fails, you should consider the contrapositive or contradiction method. To use the contrapositive approach, write down the statements $NOT\ B$ and $NOT\ A$ using the techniques in Section 1.6.10. Then, by beginning with the assumption that $NOT\ B$ is true, your job is to conclude that $NOT\ A$ is true. This is best accomplished by applying the forward-backward method, working forward from the statement $NOT\ B$ and backward from the statement $NOT\ A$. Remember to watch for quantifiers to appear in the forward or backward process for, if they do, then the corresponding construction, choose, induction, and/or specialization methods may be useful.

The Contradiction Method

In the event that the contrapositive method fails, there is still hope with the contradiction method. With this approach, you assume not only that A is true but also that B is false. This gives you two facts from which you must derive a contradiction to something that you know to be true. Where the contradiction arises is not always obvious, but it will be obtained by working forward from the statements A and $NOT\ B$.

The Uniqueness Methods

For problems containing the key words *unique, either/or,* and *max/min* you can use associated proof techniques. For example, when the conclusion of a proposition requires you to show that

> there is a unique "object" with a "certain property" such that "something happens,"

you can use the direct uniqueness method. With this method, you first establish the existence of the desired object, say X, using the construction (or contradiction) method. Then you assume that Y is a second object satisfying the certain property and for which the something happens. Using all this information (together with the hypothesis, if necessary), you must work forward to conclude that the two objects, X and Y, are the same, that is, that they are equal. You can also work backward from the fact that X is the same as Y.

With the indirect uniqueness method, you also begin by establishing the existence of the desired object X. To show uniqueness, however, you then assume that Y is a different object that also satisfies the certain property and the something that happens. You must use all this information, especially the fact that X and Y are different, to reach a contradiciton.

The Either/Or Methods

When the key words *either/or* arise, there are two proof techniques available depending, respectively, on whether those key words appear in the forward or backward processes. For example, you should use a proof by elimination when trying to prove a proposition of the form "If A then C OR D." To do so, you should assume that A is true and C is not true (that is, A and NOT C); you should then show that D is true. This can be accomplished best by using the forward-backward method. Alternatively, you can assume that A and NOT D are true; in this case you would have to show that C is true.

A proof by cases is used when trying to prove a proposition of the form "If C OR D then B." Two proofs are required. In the first case you assume that C is true and then prove that B is true; in the second case you assume that D is true and then prove that B is true.

The Max/Min Method

When the conclusion of a proposition requires you to show that the smallest (largest) element of a set of real numbers is less (greater) than or equal to a particular real number, say x, then you should use a max/min method. Doing so involves rewriting the statement in an equivalent form using the quantifier *for all* or *there is,* whichever is appropriate. Once in this form, you can then apply the choose or construction method to do the proof.

Conclusion

In trying to prove that *A implies B*, let the form of these statements guide you as much as possible. For example, you should scan the statement B for certain key words, as they will often indicate how to proceed. If you come across the quantifier *there is*, then consider the construction method, whereas the quantifier *for all* suggests using the choose or induction method. When the statement B contains the word *no* or *not*, you will probably want to use the contrapositive or contradiction method. Other key words to look for are *uniqueness*, *either/or*, and *maximum* and *minimum*, for then you would use the corresponding uniqueness, either/or, and max/min methods. If you are unable to choose an approach based on the form of B, then you should proceed with the forward-backward method. Tables A.1 and A.2 provide a complete summary.

You are now ready to "speak" mathematics. Your new "vocabulary" and "grammar" are complete. You have learned the three major proof techniques for proving propositions, theorems, lemmas, and corollaries: the forward-backward, contrapositive, and contradiction methods. You have come to know the quantifiers and the corresponding construction, choose, induction, and specialization methods. For special situations, your bag of proof techniques includes the uniqueness, either/or, and max/min methods. If all of these proof techniques fail, you may wish to stick to Greek – after all, it's all Greek to me.

Proof Technique	**When to Use It**	**What to Assume**
Forward-Backward	As a first attempt, or when B does not have a recognizable form	A
Contrapositive	When B has the word *no* or *not* in it	$NOT\ B$
Contradiction	When B has the word *not* in it, or when the first two methods fail	A and $NOT\ B$
Construction	When B has the term *there is, there exists*, etc.	A
Choose	When B has the term *for all, for each*, etc.	A, and choose an object with the certain property
Induction	When B is true for each integer beginning with an initial one, say n_0	The statement is true for n

Table A.1: Summary of Proof Techniques

What to Conclude	How to Do It
B	Work forward from A and apply the backward process to B.
$NOT\ A$	Work forward from $NOT\ B$ and backward from $NOT\ A$.
Some contradiction	Work forward from A and NOT B to reach a contradiction.
There is the desired object	Guess, construct, etc., the object. Then show that it has the certain property and that the something happens.
That the something happens	Work forward from A and the fact that the object has the certain property. Also work backward from the something that happens.
The statement is true for $n+1$. Also show it true for n_0.	First substitute n_0 everywhere and show it true. Then invoke the induction hypothesis for n to prove it true for $n+1$.

Proof Technique	**When to Use It**	**What to Assume**
Specialization	When A has the term *for all*, *for each*, etc.	A
Direct Uniqueness	When B has the word *unique* in it	There are two such objects, and A
Indirect Uniqueness	When B has the word *unique* in it	There are two different objects, and A
Proof by Elimination	When B has the form C OR D	A and NOT C or A and NOT D
Proof by Cases	When A has the form C OR D	Case 1: C Case 2: D
Max/min 1	When B has the form max $S \leq x$ or min $S \geq x$	Choose an s in S, and A
Max/min 2	When B has the form max $S \geq x$ or min $S \leq x$	A

Table A.2: Summary of Proof Techniques (Continued)

What to Conclude	How to Do It
B	Work forward by specializing A to one particular object having the certain property.
The two objects are equal	Work forward using A and the properties of the objects. Also work backward to show the objects are equal.
Some contradiction	Work forward from A using the properties of the two objects and the fact that they are different.
D	Work forward from A and NOT C, and backward from D
or	or
C	Work forward from A and NOT D, and backward from C.
B B	First prove that C implies B; then prove that D implies B.
$s \leq x$ or $s \geq x$	Work forward from A and the fact that s is in S. Also work backward.
Construct s in S so that $s \geq x$ or $s \leq x$	Use A and the construction method to produce the desired s in S.

SOLUTIONS TO ODD-NUMBERED EXERCISES

Solutions to Chapter 1 Exercises

1.1

(a) **Inputs:** A right triangle XYZ with legs whose length are x and y.

Outputs: The area of the triangle XYZ in terms of x and y.

(b) **Inputs:** n (the degree of the polynomial) and the coefficients $a_0, \ldots, a_n$ that make up the polynomial $p(x) = a_0 + a_1 x + \cdots + a_n x^n$.

Outputs: The n roots of the polynomial $p(x)$.

(c) **Inputs:** The 5 cities and the known nonstop fares between the appropriate pairs of cities.

Outputs: The fares between *all* pairs of cities.

1.3

(a) From the quadratic formula, the closed-form solution is:

$$x = \frac{3 \pm \sqrt{9-8}}{4} = \frac{3 \pm 1}{4} = 1 \text{ or } \frac{1}{2}.$$

(b) A numerical method is needed to identify the least price.

(c) Area = $b^2 - a^2$.

1.5

(a) The interchanges result in the following sequence of lists:

10, 13, 8, 9, 12
8, 13, 10, 9, 12
8, 9, 10, 13, 12
8, 9, 10, 13, 12
8, 9, 10, 12, 13

(b) The numerical method results in the following approximation for the slope:

$$\begin{aligned}\text{Slope} &\approx \frac{f(1+0.01)-f(1)}{0.01}\\ &= \frac{(2^{-1.01}-1.01)-(2^{-1}-1)}{0.01}\\ &\approx -1.345375.\end{aligned}$$

1.7 You can find x by computing $\log[(x_1^2)\cdots(x_n^2)]$ because

$$\begin{aligned}x &= 2[\log(x_1)+\cdots+\log(x_n)]\\ &= 2\log(x_1)+\cdots+2\log(x_n)\\ &= \log(x_1^2)+\cdots+\log(x_n^2)\\ &= \log[(x_1^2)\cdots(x_n^2)].\end{aligned}$$

1.9 For every positive integer n,

$$\frac{1}{n}-\frac{1}{n+1}<\frac{1}{n^2}.$$

1.11 Let $x_1,\ldots,x_n$ be n nonnegative real numbers. Then

$$(x_1x_2\cdots x_n)^{1/n}\leq\frac{x_1+\cdots+x_n}{n}.$$

1.13 (Answers other than those given here are possible.)

(a) Similarities:

(i) All numbers are real.

(ii) All numbers have infinite decimal expansions.

Differences:

(i) Both 1/3 and 0.0909 . . . have *repeating* decimal expansions, which is not the case for $\sqrt{2}$ and π.

(ii) Both 1/3 and 0.0909 . . . are rational numbers; neither $\sqrt{2}$ nor π is a rational number.

(b) Group 1 – $\{\sqrt{2}, \pi\}$ (both numbers in this group have infinite nonrepeating decimal expansions).

Group 2 – $\{1/3, 0.0909\ldots\}$ (both numbers in this group are rational).

1.15 (Answers other than those given here are possible.)

(a) Similarities:

(i) All sets are nonempty.

(ii) All sets contain an infinite number of elements.

Differences:

(i) The sets in (i), (iv), and (v) are intervals; the sets in (ii) and (iii) are not.

(ii) All elements in the sets in (iii) and (iv) are ≥ 0 but not those in the other sets.

(b) Group 1 – the sets in (i), (iv), and (v) (because they are intervals).

Group 2 – the sets in (ii) and (iii) (because they are not intervals).

or

Group 1 – the sets in (iii) and (iv) (because all elements are ≥ 0).

Group 2 – the sets in (i), (ii), and (v) (because they have some elements whose values are < 0).

1.17

(a) $\{$real numbers $x : a \leq x \leq b\}$ (where $a \leq b$ are real numbers).

(b) $\{(x, y) : a_1 \leq x \leq b_1$ and $a_2 \leq y \leq b_2\}$ (where $a_1 \leq b_1$ and $a_2 \leq b_2$ are real numbers).

(c) $\{(x_1, \ldots, x_n) :$ for each $i = 1, \ldots, n,\ a_i \leq x_i \leq b_i\}$ (where for each $i = 1, \ldots, n,\ a_i \leq b_i$ are real numbers).

1.19 One sequential generalization is from a triangle, to a polygon in the plane, to a polyhedron in n dimensions, to an arbitrary set in n dimensions.

1.21 The operation x^3 makes no sense when **x** is an n-vector. The operation $\sqrt{y}$ makes no sense when **y** is an n-vector.

1.23 Let x be an object belonging to some set, say, S. Then define an operation ν that associates to each object $x \in S$, another object, $\nu(x)$, belonging to the same set. The abstract system is (S, ν).

1.25

(a) The truth table for the converse statement q *implies* p is:

p	q	q *implies* p
T	T	T
T	F	T
F	T	F
F	F	T

(b) The truth table for the inverse statement (*not* p) *implies* (*not* q) is:

p	q	q *implies* p	*not* p	*not* q	(*not* p) *implies* (*not* q)
T	T	T	F	F	T
T	F	T	F	T	T
F	T	F	T	F	F
F	F	T	T	T	T

The truth of q *implies* p is the same as that of (*not* p) *implies* (*not* q), namely, these statements are *true* in all cases except when p is *false* and q is *true*.

1.27 The following are valid key questions and answers (others are also possible):

(a) **Key question:** How can I show that one real number is less than or equal to another real number?

Key answer: Show that the square of the first number is less than or equal to the square of the second number, that is,

$$ab \leq \frac{(a+b)^2}{4}.$$

Key answer: Show that the difference of the two real numbers is less than or equal to 0, that is,

$$\sqrt{ab} - \frac{a+b}{2} \leq 0.$$

Key question: How can I show that the square root of the product of two real numbers is less than or equal to the average of the two real numbers?

Key answer: Show that the product of the two numbers is less than or equal to the square of the average of the two numbers, that is,

$$ab \leq \frac{(a+b)^2}{4}.$$

Key answer: Show that the difference of the square root and the average of the two real numbers is less than or equal to 0, that is,

$$\sqrt{ab} - \frac{a+b}{2} \leq 0.$$

(b) **Key question:** How can I show that a set is not the empty set?

Key answer: Show that there is some element in the set, that is, show that

there is a point $(x, y) \in C^1 \cap C^2$.

Key answer: Show that the set contains a nonempty set, that is, show that

there is a nonempty set C such that $C \subseteq C^1 \cap C^2$.

Key question: How can I show that the intersection of two sets is not empty?

Key answer: Show that there is some element that is in both of the sets, that is, show that

there is a point $(x, y) \in C^1$ such that $(x, y) \in C^2$.

Key answer: Show that the intersection contains a nonempty set, that is, show that

there is a nonempty set C such that $C \subseteq C^1 \cap C^2$.

1.29 The construction method is used in each case as follows:

(a) Construct an angle t between 0 and π and then show that this angle t satisfies $\sin(t) = \cos(t)$.

(b) Construct a real number x and then show that this value of x satisfies $x^2 = y$.

1.31 To show that there is a point on the graph of the line and the parabola, it must be shown that

there is a real number x such that $ax^2 + bx + c = dx + e$,

or equivalently, that

there is a real number x such that $ax^2 + (b - d)x + (c - e) = 0$.

It is possible to find such a real number from the quadratic formula provided that $a \neq 0$ and the discriminant is nonnegative, that is, if

$$(b - d)^2 - 4a(c - e) \geq 0.$$

However, $(b - d)^2 \geq 0$ and from the hypothesis that $a(c - e) < 0$, it follows that $-4a(c - e) > 0$ and also that $a \neq 0$. Thus, the desired value for the real number x is

$$x = \frac{-(b - d) \pm \sqrt{(b - d)^2 - 4a(c - e)}}{2a}.$$

1.33

(a) The statement is *true* for $n = 1$ because

$$\sum_{k=1}^{1} k = 1 = \frac{1(1 + 1)}{2}.$$

Assume that the statement is *true* for n, that is, that

$$\sum_{k=1}^{n} k = \frac{n(n + 1)}{2}.$$

By using this induction hypothesis, for $n + 1$, you have that

$$\begin{aligned}
\sum_{k=1}^{n+1} k &= \left(\sum_{k=1}^{n} k\right) + (n + 1) \\
&= \frac{n(n + 1)}{2} + (n + 1) \\
&= (n + 1)\left(\frac{n}{2} + 1\right) \\
&= \frac{(n + 1)(n + 2)}{2}.
\end{aligned}$$

Thus it has been shown that the statement is *true* for $n + 1$, and so the result follows by induction.

(b) The statement is *true* for $n = 1$ because

$$1! = 1 \leq 1^1.$$

Assume that the statement is *true* for n, that is, that

$$n! \leq n^n.$$

By using this induction hypothesis, for $n + 1$, you have that

$$\begin{aligned} (n+1)! &= (n+1)(n!) \\ &\leq (n+1)n^n \\ &\leq (n+1)(n+1)^n \\ &= (n+1)^{n+1}. \end{aligned}$$

Thus it has been shown that the statement is *true* for $n + 1$ and so the result follows by induction.

1.35 In the following, the chosen *object* with the *certain property* becomes information in the forward process. The statement to be proved is that the chosen *object* satisfies the *something that happens*.

(a) Choose an angle t.
Show that $\sin^2(t) + \cos^2(t) = 1$.

(b) Choose real numbers t, x, and y with $0 \leq t \leq 1$.
Show that $f(tx + (1 - t)y) \leq tf(x) + (1 - t)f(y)$.

1.37 Let A and B be arbitrary sets (the word *let* indicates that the choose method is used). You must show that $A \cap B \subseteq A$. According to the definition of a subset, you must show that for all elements $x \in A \cap B$, $x \in A$. So now, let $x \in A \cap B$ (the word *let* again indicates that the choose method is used). You must show that $x \in A$. But this follows immediately from the definition of $A \cap B$ because if $x \in A \cap B$, then $x \in A$. Thus it has been shown that for any sets A and B, $A \cap B \subseteq A$.

1.39 The following indicate how specialization is used to produce a new statement in the forward process.

(a) You must first verify that A is an angle (which it is) and then you can state that $\sin^2(A) + \cos^2(A) = 1$.

(b) You must first show that $0 \leq \lambda \leq 1$ and then you can state that $f(\lambda x + (1 - \lambda)z) \leq \lambda f(x) + (1 - \lambda)f(z)$.

1.41 To show that $f \leq h$, by definition you must show that for all real numbers x, $f(x) \leq h(x)$. So, let $\bar{x}$ be a real number (the word *let* indicates that the choose method is used). You must show that $f(\bar{x}) \le h(\bar{x})$. From the hypothesis, you know that $f \le g$, which means that for all real numbers x, $f(x) \le g(x)$. In particular, because $\bar{x}$ is a real number, you can say that $f(\bar{x}) \le g(\bar{x})$ (here is where specialization is applied to the special object $\bar{x}$). Likewise, from the hypothesis that $g \le h$, you know that for all real numbers x, $g(x) \le h(x)$. In particular, for $\bar{x}$, you can state that $g(\bar{x}) \le h(\bar{x})$ (here again specialization is applied to $\bar{x}$). Thus, because $f(\bar{x}) \le g(\bar{x})$ and $g(\bar{x}) \le h(\bar{x})$, it follows that $f(\bar{x}) \le h(\bar{x})$. It has therefore been shown that $f \le h$ and so the proof is complete.

1.43

(a) $x < -1$ or $x > 1$.

(b) $x \notin A$ and $x \notin B$ (or $x \in A^c$ and $x \in B^c$).

(c) There are elements $x, y \in A$ with $x \neq y$ such that $f(x) = f(y)$.

(d) There is an integer k with $1 < k < n$ such that k divides n.

1.45 By the contrapositive method, you can assume that the conclusion is *false*, that is, that there is an integer i with $1 \le i \le n$ such that $x_i \neq 0$. Thus, $x_i > 0$ (because you are given that $x_i \ge 0$). You must now prove that the hypothesis is *false*, that is, that $x_1 + \cdots + x_n \neq 0$. However, because $x_i > 0$ and each other $x_j \ge 0$, it follows that $x_1 + \cdots + x_n > 0$ and so $x_1 + \cdots + x_n \neq 0$, thus completing the proof.

1.47 By contradiction, you can assume that the conclusion is *false*, that is, that *there is* an even integer x such that $x^2 + 2mx + 2n = 0$. You can now contradict the fact that n is odd by establishing that n is even, as follows. Because x is even, by definition, there is an integer k such that $x = 2k$, from which it follows that

$$x^2 + 2mx + 2n = (2k)^2 + 2m(2k) + 2n = 4k^2 + 4mk + 2n = 0,$$

or equivalently, that $n = -2(k^2 + 2mk)$, which means that n is even. This contradiction completes the proof.

1.49 You must first use the construction method to establish that there is a positve integer m such that m divides n and n divides m. One such positive integer is $m = n$ because n is positive and n divides n.

According to the direct uniqueness method, you should now assume that k is also a positive integer for which k divides n and n divides k. You must then show that $k = m$. However, because k divides n, by definition, there is an integer a such that $n = ak$. Likewise, because n divides k, there is an integer b such that $k = bn$. Combining these two yields that $k = bn = b(ak) = (ba)k$. Now $k \neq 0$ so $ba = 1$. Because a and b are integers, it must be that $a, b = \pm 1$. Thus, $k = bn = \pm n$. But $k > 0$, so it must be that $k = n$, and because $m = n$, it follows that $k = m$, thus completing the proof.

1.51 From the hypothesis that $x^2+x-6 \geq 0$, you know that $(x+3)(x-2) \geq 0$. Because you are also assuming that $x > -2$, you have that $x + 3 > 0$ and so you can divide through by $x + 3$ to obtain that $x - 2 \geq 0$, that is, $x \geq 2$, and so the proof is complete.

Solutions to Chapter 2 Exercises

2.1

(a) The vectors $t\mathbf{d}$ for all possible values of the real number t:

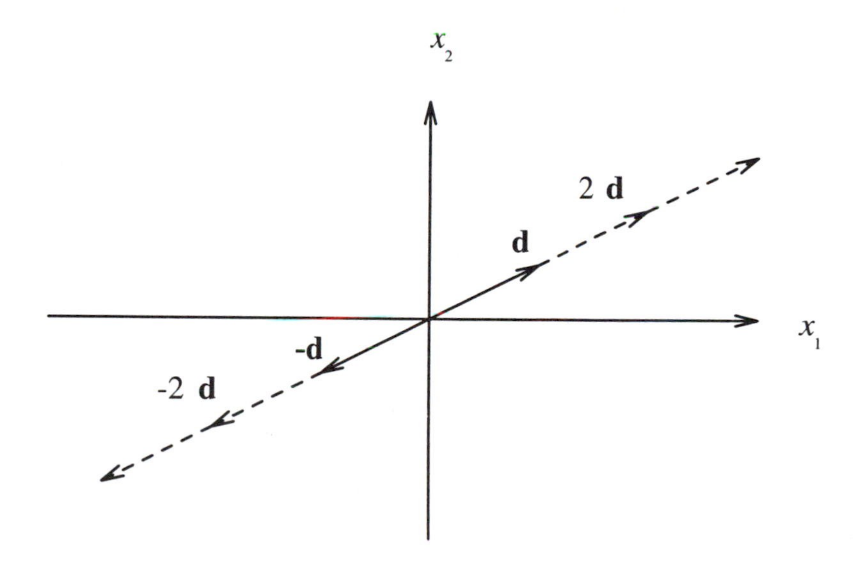

(b) The set of points in A that are not in B and the set of points in B that are not in A:

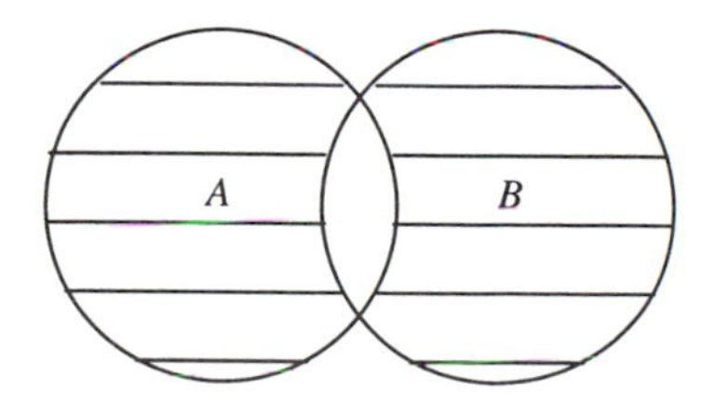

(c) A function f having several points x^* for which $f(x^*) = x^*$:

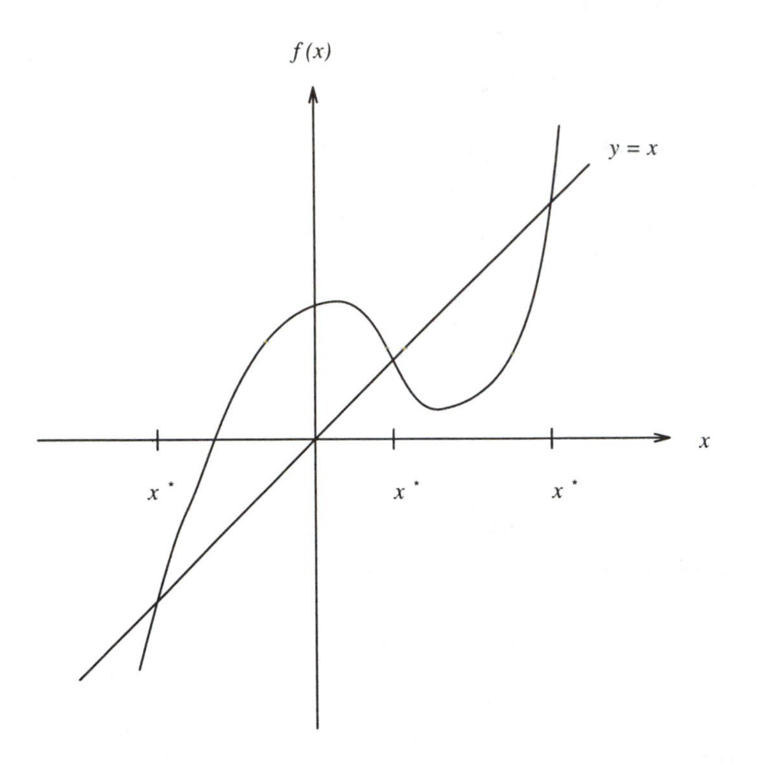

(d) In the following graph, there is one vertex corresponding to each of the first 6 integers. An edge from vertex i to vertex j indicates that the integer i divides the integer j evenly.

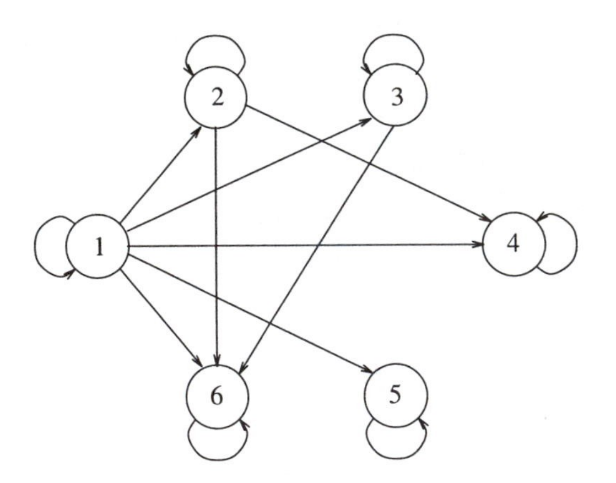

2.3

(a) Let a be the amount by which you want to shift the graph of the function f up. Then, create the new function g defined by

$$g(x) = f(x) + a.$$

(b) The vector that points from $\mathbf{x}$ to $\mathbf{y}$ is $\mathbf{y} - \mathbf{x} = (y_1 - x_1, y_2 - x_2)$, or, in n dimensions:

$$\mathbf{y} - \mathbf{x} = (y_1 - x_1, \ldots, y_n - x_n).$$

(c) For every element x in the set S, $x < u$.

2.5

(a) $y_i = 1 - x_i$.

(b) In this case,

$$\begin{aligned} \mathbf{x}' &= \mathbf{x} + t\mathbf{d} \\ &= (x_1, x_2) + t(d_1, d_2) \\ &= (x_1 + td_1, x_2 + td_2). \end{aligned}$$

(c) The equation for a plane in three dimensions is:

$$ax + by + cz = d,$$

where a, b, c, and d are given real numbers. Therefore, if you want all of the set A to be on one side of this plane and all of the set B to be on the other side, you need the following *two* statements:

(i) For all $(x, y, z) \in A,\ ax + by + cz \leq d$.

(ii) For all $(x, y, z) \in B,\ ax + by + cz \geq d$.

2.7

(a) For the definition of *divides*:

(i) Objects – integers a and b.

(ii) Name – divides.

(iii) Property – there is an integer k such that $b = ka$.

(b) For the definition of the *infimum of a set T of real numbers*:

(i) Objects – a real number t.

(ii) Name – the infimum of a set T of real numbers.

(iii) Property – (1) t is a lower bound for T and (2) for any lower bound s for T, $s \leq t$.

(c) For the definition of *linearly independent vectors*:

(i) Objects – a collection of n-vectors $\mathbf{x}^1, \ldots, \mathbf{x}^k$.

(ii) Name – linearly independent vectors.

(iii) Property – for all real numbers $t_1, \ldots, t_k$ with $t_1\mathbf{x}^1 + \cdots + t_k\mathbf{x}^k = \mathbf{0}$, it follows that $t_1 = \cdots = t_k = 0$.

2.9 (Answers other than those given here are possible.)

(a) (i) 3 divides 6 because 6 = 2 (3).

(ii) 3 does not divide 5.

(b) (i) 3 is an infimum of the set $\{x : 3 \le x \le 5\}$.

(ii) 2 is not an infimum of the set $\{x : 3 \le x \le 5\}$.

2.11

(a) A visual image associated with the statement that for all real numbers x, $f(x) \le g(x)$ is the following:

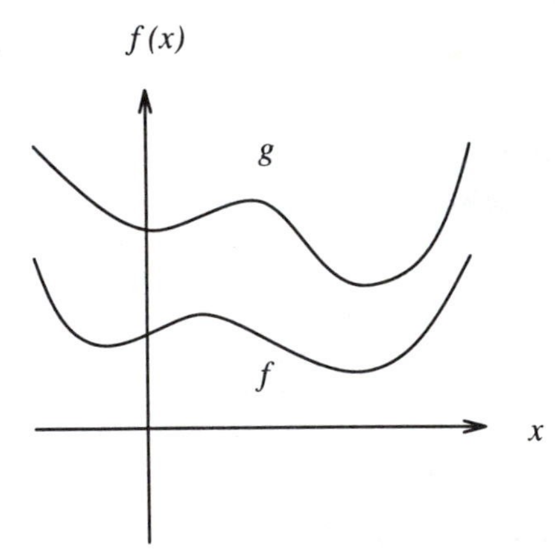

(b) A visual image associated with the statement that for all real numbers x, $f(x) \ge mx + b$ is the following:

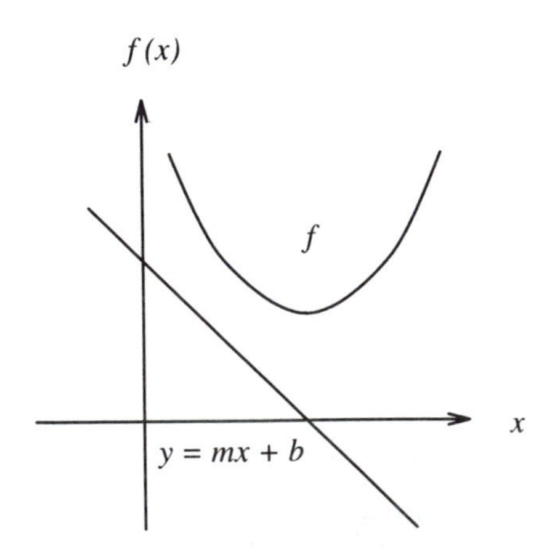

2.13

(a) The meaning of x is undefined.

(b) There are no syntax errors.

(c) You cannot compute the expressions $|\mathbf{x} - \mathbf{x}'|$ and $|\mathbf{y} - \mathbf{x}'|$ because $\mathbf{x}$, $\mathbf{x}'$, and $\mathbf{y}$ are points in the plane and not real numbers.

2.15

(a) Yes. If S satisfies the property in **A**, then S satisfies the property in **B**.

(b) No. The following set S satisfies the property in **B** but not the property in **A** because there is a point $\mathbf{y} \in S$ between the points $\mathbf{x} \in S$ and $\mathbf{z} \in S$ but the line segment connecting $\mathbf{x}$ to $\mathbf{z}$ is *not* contained in the set.

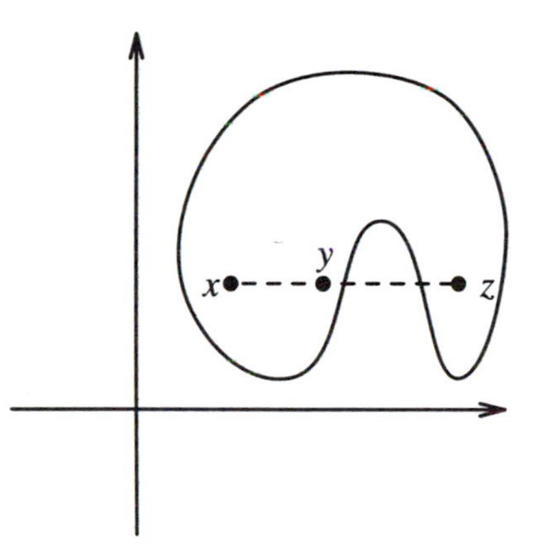

(c) A correct definition is:

> A set S of points in the plane is **convex** if and only if for all elements $\mathbf{x}, \mathbf{z} \in S$, and for all points $\mathbf{y}$ on the line segment connecting $\mathbf{x}$ to $\mathbf{z}$, $\mathbf{y} \in S$.

2.17 The parabola $ax^2 + bx + c$ is **U-shaped** if and only if $a > 0$.

2.19 An n-vector $\mathbf{x}$ is in the **boundary of a set** S if and only if for all real numbers $r > 0$, $B_r(\mathbf{x}) \cap S \neq \emptyset$ and $B_r(\mathbf{x}) \cap S^c \neq \emptyset$ (where $B_r(\mathbf{x})$ is the ball of radius r centered at $\mathbf{x}$ and S^c is the complement of S).

2.21 Yes, $\Leftrightarrow$ is an equivalence relation because:

(a) $p \Leftrightarrow p$ (since $p \Rightarrow p$ is always *true*).

(b) If $p \Leftrightarrow q$, then $q \Leftrightarrow p$ because, from $p \Leftrightarrow q$, you have that $p \Rightarrow q$ and $q \Rightarrow p$, thus, $q \Rightarrow p$ and $p \Rightarrow q$, hence, $q \Leftrightarrow p$.

(c) If $p \Leftrightarrow q$ and $q \Leftrightarrow r$, then $p \Leftrightarrow r$ because, from $p \Leftrightarrow q$, you have that $p \Rightarrow q$ and $q \Rightarrow p$. Similarly, $q \Leftrightarrow r$ means that $q \Rightarrow r$ and $r \Rightarrow q$. Combining these facts means that $p \Rightarrow q \Rightarrow r$ and $r \Rightarrow q \Rightarrow p$, so $p \Rightarrow r$ and $r \Rightarrow p$, thus, $p \Leftrightarrow r$.

2.23 The new axiomatic system consisting of Axioms 1 and 2 only is a *generalization* of the original system consisting of Axioms 1, 2, and 3. This is because one special case of the new system is the one in which Axiom 3 is added, which is the original system. Equivalently, the original system satisfies all of the axioms of the new system, but not vice versa.

2.25 Let S be a set of objects and $\prec$ be a binary relation on S. The pair $(S, \prec)$ constitutes a **strict inequality** if and only if the following three conditions hold:

(i) $\prec$ is transitive on S.

(ii) For all $x \in S$, it is *not true* that $x \prec x$.

(iii) If $x, y \in S$ with $x \prec y$, then it is *not true* that $y \prec x$.

Solutions to Chapter 3 Exercises

3.1

(a) The universal set U and two sets A and B for which $A \cap B = \emptyset$:

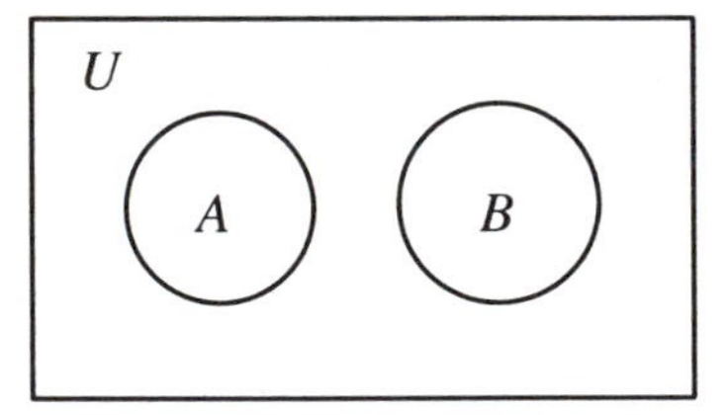

(b) The universal set U and two sets A and B for which $A \cap B$ has only one element:

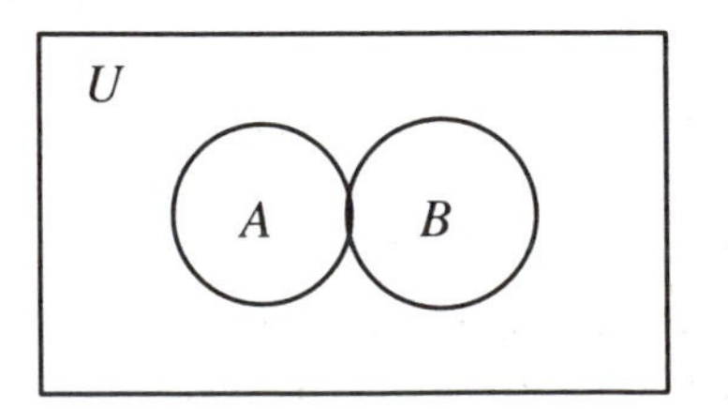

(c) The universal set U and two sets A and B for which $A \cap B$ has many elements:

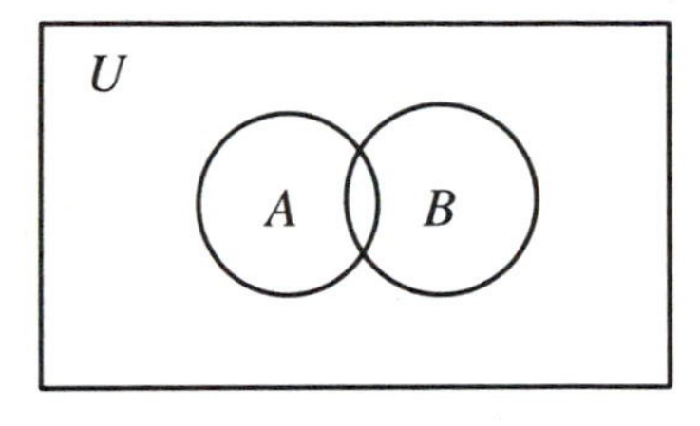

3.3

(a) $\{(x, y) : y \geq x^2 - 2x + 2\}$.

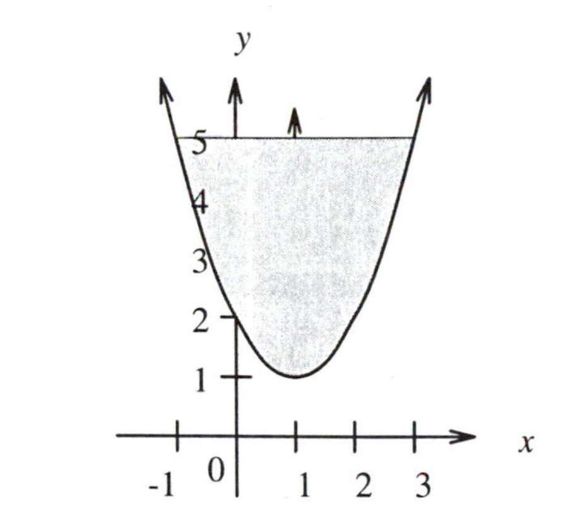

(b) $\{(x, y) : y \leq 4 - 2x\}$.

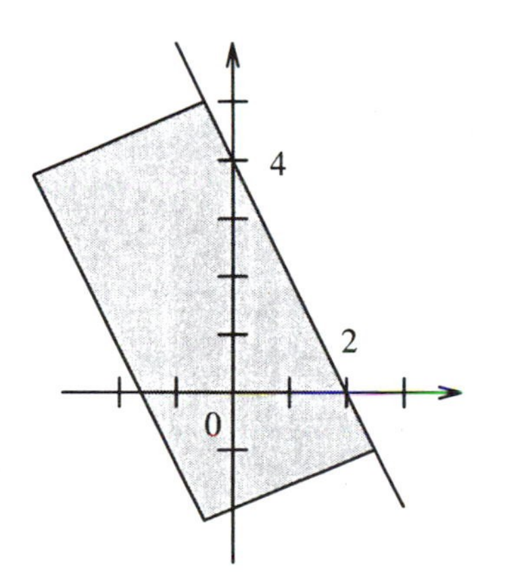

3.5

(a) European $\cup$ Asian.

(b) Leaders ∩ (Asian ∪ NorthAmerican).

(c) (Asian ∪ European) – Leaders.

3.7

(a) $\{(x, y) : a \le x \le b$ and $c \le y \le d\}$.

(b) $\{(x_1, \ldots, x_n)$: for each $i = 1, \ldots n,\ a_i \le x_i \le b_i\}$.

3.9 Let $x \in B$, so $x^2 - x - 2 = (x-2)(x+1) \le 0$. It must be shown that $-1 \le x \le 2$. But because $(x-2)(x+1) \le 0$, either

(i) $x - 2 \ge 0$ and $x + 1 \le 0$, or
(ii) $x - 2 \le 0$ and $x + 1 \ge 0$.

Now (i) cannot happen, for otherwise, $x \ge 2$ and $x \le -1$. Therefore, $x - 2 \le 0$ and $x + 1 \ge 0$, so $-1 \le x \le 2$ and hence $x \in A$. Thus, $B \subseteq A$ and the proof is complete.

3.11

(a) First it will be shown that $A \cup (B \cap C) \subseteq (A \cup B) \cap (A \cup C)$, so let $x \in A \cup (B \cap C)$. This means that $x \in A$ or $x \in B \cap C$. Proceeding by cases, if $x \in A$, then $x \in A \cup B$ and $x \in A \cup C$, whence $x \in (A \cup B) \cap (A \cup C)$. Alternatively, if $x \in B \cap C$, then $x \in B$ and $x \in C$, so $x \in A \cup B$ and $x \in A \cup C$, and again $x \in (A \cup B) \cap (A \cup C)$. Thus, in either case, $x \in (A \cup B) \cap (A \cup C)$ and so $A \cup (B \cap C) \subseteq (A \cup B) \cap (A \cup C)$.

It remains to show that $(A \cup B) \cap (A \cup C) \subseteq A \cup (B \cap C)$, so let $x \in (A \cup B) \cap (A \cup C)$. This means that $x \in A \cup B$ and $x \in A \cup C$. Proceeding by cases, if $x \in A$, then $x \in A \cup (B \cap C)$, as desired. Alternatively, if $x \notin A$, then $x \in B$ and $x \in C$, which means that $x \in B \cap C$ and so again, $x \in A \cup (B \cap C)$. Thus, in either case, $x \in A \cup (B \cap C)$ and so $(A \cup B) \cap (A \cup C) \subseteq A \cup (B \cap C)$.

It has now been shown that $A \cup (B \cap C) \subseteq (A \cup B) \cap (A \cup C)$ and that $(A \cup B) \cap (A \cup C) \subseteq A \cup (B \cap C)$, so $A \cup (B \cap C) = (A \cup B) \cap (A \cup C)$.

(b) First it will be shown that $(A \cup B)^c \subseteq A^c \cap B^c$, so let $x \in (A \cup B)^c$. This means that $x \notin A \cup B$, that is, $x \notin A$ and $x \notin B$. In other words, $x \in A^c \cap B^c$.

It remains to show that $A^c \cap B^c \subseteq (A \cup B)^c$, so let $x \in A^c \cap B^c$. This means that $x \in A^c$ and $x \in B^c$, that is, $x \notin A$ and $x \notin B$. In other words, x is neither in A nor in B, so $x \in (A \cup B)^c$.

It has now been shown that $(A \cup B)^c \subseteq A^c \cap B^c$ and $A^c \cap B^c \subseteq (A \cup B)^c$, so $(A \cup B)^c = A^c \cap B^c$.

3.13 The reason this is not a real-valued function is that there may be *two* solutions to the equation $ax^2 + bx + c = 0$ and a real-valued function must associate to the values $a, b,$ and c, a *unique* value.

3.15

(a) The following figure is appropriate for a function $f : \mathbb{R}^1 \to \mathbb{R}^1$:

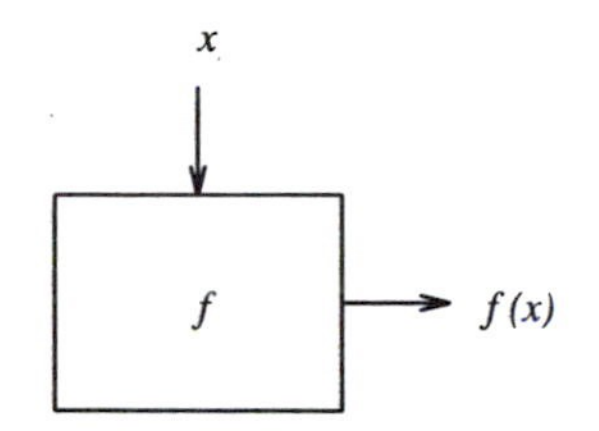

(b) The following figure is appropriate for a function $g : \mathbb{R}^n \to \mathbb{R}^1$:

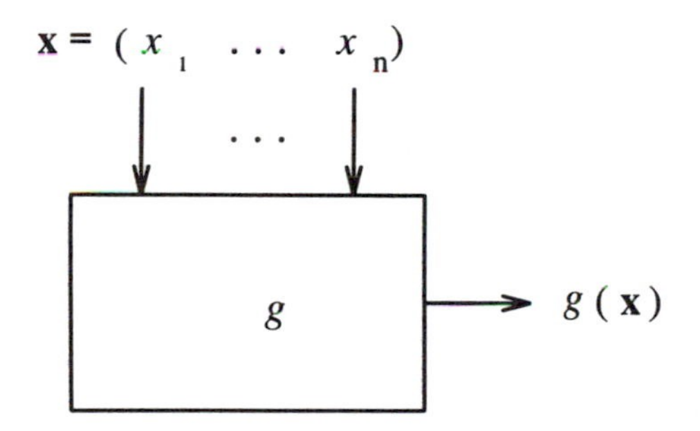

(c) The following figure is appropriate for a function $h : \mathbb{R}^n \to \mathbb{R}^p$:

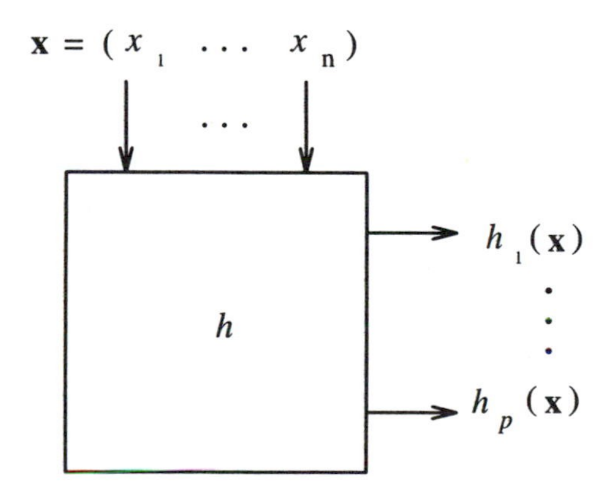

3.17

(a) $u : A \to A$.

(b) $f : \mathbb{R} - \{0\} \to \mathbb{R}$ (or $f : \mathbb{R} - \{0\} \to \mathbb{R} - \{0\}$).

(c) $f : \{\text{real numbers } a, b, c, d, e, f : ad - bc \neq 0\} \to \mathbb{R}^2$.

(d) $f : \{\text{real numbers } a, b, c : b^2 - 4ac \geq 0\} \to 2^{\mathbb{R}}$.

3.19

(a) A **fixed point of a function** $f : \mathbb{R} \to \mathbb{R}$ is a point $x \in \mathbb{R}$ such that $f(x) = x$.

(b) An **increasing function** is a function $f : \mathbb{R} \to \mathbb{R}$ such that for all $x, y \in \mathbb{R}$ with $x < y$, $f(x) < f(y)$.

3.21

(a) No syntax error arises because when $f, g : \mathbb{R}^n \to \mathbb{R}^1$, $(f + g)(\mathbf{x}) = f(\mathbf{x}) + g(\mathbf{x})$ makes sense and can be performed since both $f(\mathbf{x})$ and $g(\mathbf{x})$ are real numbers that you can add.

(b) No syntax error arises because when $f, g : \mathbb{R}^n \to \mathbb{R}^1$, $(f/g)(\mathbf{x}) = f(\mathbf{x})/g(\mathbf{x})$ makes sense and can be performed since both $f(\mathbf{x})$ and $g(\mathbf{x})$ are real numbers that you can divide, provided that $g(\mathbf{x}) \neq 0$.

3.23 By definition, it must be shown that for each $y \in B$, there is an element $x \in A$ such that $f(x) = y$, so let $y \in B$. Because $B =$ range f, it follows that $y \in$ range f. But now, from the definition of range f, there is an element $x \in A$ such that $f(x) = y$. Thus, f is surjective and the proof is complete.

3.25

(a) The function $\sin(x)$ is surjective because, for each $y \in [-1, 1]$, there is a value of $x \in \mathbb{R}$ such that $\sin(x) = y$. However, $\sin(x)$ is *not* injective because, for $y = 0$, both $x = 0$ and $x = 2\pi$ satisfy $\sin(x) = y$. Also, $\sin(x)$ is not bijective because $\sin(x)$ is not injective.

(b) The function $f : \{x \in \mathbb{R} : x \geq 0\} \to \mathbb{R}$ defined by $f(x) = \sqrt{x}$ is not surjective because, for $y = -1$, there is no value of $x \in \mathbb{R}$ such that $\sqrt{x} = -1$. However, f is injective because, if $x \geq 0$ and $y \geq 0$ and $\sqrt{x} = \sqrt{y}$, then $x = y$. This function is not bijective because f is not surjective.

(c) The function $f : \{x \in \mathbb{R} : x \geq 0\} \to \{x \in \mathbb{R} : x \geq 0\}$ defined by $f(x) = \sqrt{x}$ is surjective because, for any value of $y \geq 0$, there is an $x \geq 0$ such that $\sqrt{x} = y$, namely, $x = y^2$. This function is also injective because, if $x \geq 0$ and $y \geq 0$ and $\sqrt{x} = \sqrt{y}$, then $x = y$. This function is bijective because f is both surjective and injective.

3.27

(a) Create a graph with 6 vertices, one for each of the six people. Connect two vertices with an edge if the corresponding people know each other. The objective is to find three vertices that either form a cycle or three vertices no two of which are adjacent.

(b) Create a graph with two groups of 3 vertices each, corresponding to the two groups of microchips. You want to show that it is impossible to connect each vertex in one group to the three vertices in the other group in such a way that no two edges cross each other.

3.29

(a) Adding the edge from vertex i to vertex j translates to replacing a 0 with a 1 in row i and column j of the adjacency matrix A.

(b) Deleting the edge from vertex i to vertex j translates to replacing a 1 with a 0 in row i and column j of the adjacency matrix A.

(c) Adding a new vertex translates to adding a new row and column to the adjacency matrix A and filling in that row and column with zeroes.

(d) Deleting vertex i translates to removing row i and column i from the adjacency matrix A.

3.31 Each loop from vertex i to itself results in a value of 1 instead of 0 for the element A_{ii} in the adjacency matrix. To allow for multiple edges, the value of A_{ij} is now a nonnegative integer that represents the *number* of edges from vertex i to vertex j.

3.33 Let $G = (V, E)$ be a graph. The **complement of** G is the graph $G' = (V', E')$ in which $V' = V$ and $ij \in E'$ if and only if $ij \notin E$.

3.35 A graph $G = (V, E)$ is **bipartite** if and only if you can partition the vertices V into two groups, say, S and T, so that each edge in E connects a vertex in S to a vertex in T.

3.37 Yes, every component of G is a connected subgraph of G. In contrast, not every connected subgraph of G is a component. For example, the subgraph of G consisting of vertices 1 and 2 and the edge connecting those two vertices is a connected subgraph of G that is *not* a component.

3.39

(a) (i) No, the relation is not reflexive because a vertex is not adjacent to itself (recall that loops are not allowed).

(ii) Yes, the relation is symmetric because if a vertex i is adjacent to a vertex j, then that same edge indicates that j is adjacent to i (recall that the graph is undirected).

(iii) No, the relation is not transitive. Just because a vertex i is adjacent to a vertex j and j is adjacent to a vertex k does *not* mean that i is adjacent to k.

(b) (i) No, the relation is not reflexive because a vertex is not adjacent to itself (recall that loops are not allowed).

(ii) No, the relation is not symmetric. With a directed graph, just because a vertex i is adjacent to a vertex j does *not* mean that j is adjacent to i.

(iii) No, the relation is not transitive for the same reason as in part (a)(iii).

3.41

(a) If the proposition is *true*, then, to show that a graph is connected, you need only find a walk from vertex 1 to each other vertex, rather than from *every* vertex to every other vertex. In other words, if the proposition is *true*, then less computational effort is required to determine whether a graph is connected or not.

(b) If the graph G is connected then, by definition, there is a walk from every vertex i to every other vertex. In particular, there is a walk from vertex 1 to every other vertex.

For the converse, suppose there is a walk from vertex 1 to every other vertex, and let i and j be two vertices in G. You can create a walk from i to j by using the walk, W_{1i}, from vertex 1 to vertex i and the walk, W_{1j}, from vertex 1 to vertex j, as follows. Start at i and *reverse* W_{1i} to get to vertex 1. Then use W_{1j} to go from vertex 1 to vertex j.

Solutions to Chapter 4 Exercises

4.1 For $\mathbf{x} = (-1, 1, 0)$ and $\mathbf{y} = (0, 2, 3)$,

(a) $\mathbf{x} + 2\mathbf{y} = (-1, 1, 0) + 2(0, 2, 3) = (-1 + 2(0), 1 + 2(2), 0 + 2(3)) = (-1, 5, 6)$.

(b) $||\mathbf{x}|| = \sqrt{(-1)^2 + 1^2 + 0^2} = \sqrt{2}$.

(c) $\mathbf{x} \cdot \mathbf{y} = (-1, 1, 0) \cdot (0, 2, 3) = (-1)0 + (1)2 + (0)3 = 2$.

(d) Yes, $\mathbf{x} \leq \mathbf{y}$ because $-1 \leq 0,\ 1 \leq 2$, and $0 \leq 3$.

4.3 As the real number t varies, $t\mathbf{x}$ results in a straight line through the origin in the direction of the vector $\mathbf{x}$. When $\mathbf{x} = \mathbf{0}$, $t\mathbf{x} = \mathbf{0}$ for all t so, when $\mathbf{x} = \mathbf{0}$, $\{t\mathbf{x} : t \text{ is a real number}\} = \{\mathbf{0}\}$.

4.5

(a) This line is $\{\mathbf{y} = (y_1, y_2) : \mathbf{x} \cdot \mathbf{y} = x_1 y_1 + x_2 y_2 = 0\}$.

(b) One half space is $\{\mathbf{y} = (y_1, y_2) : x_1 y_1 + x_2 y_2 \leq 0\}$ and the other half space is $\{\mathbf{y} = (y_1, y_2) : x_1 y_1 + x_2 y_2 \geq 0\}$.

(c) For a given n-vector $\mathbf{x}$,

$$\begin{aligned} \text{the plane} &= \{\mathbf{y} \in \mathbb{R}^n : \mathbf{x} \cdot \mathbf{y} = 0\}, \\ \text{one half space} &= \{\mathbf{y} \in \mathbb{R}^n : \mathbf{x} \cdot \mathbf{y} \leq 0\}, \\ \text{other half space} &= \{\mathbf{y} \in \mathbb{R}^n : \mathbf{x} \cdot \mathbf{y} \geq 0\}. \end{aligned}$$

4.7

(a) One syntax error is that you cannot subtract the n-vector $z\mathbf{y}$ from the real number x. Thus, you cannot compute $|x - z\mathbf{y}|$. Also, you cannot compute $|\mathbf{y}|$ because $\mathbf{y}$ is an n-vector nor can you compare $|x - z\mathbf{y}|$ with $t|\mathbf{y}|$ using $\leq$.

(b) For n-vectors $\mathbf{y}$ and $\mathbf{z}$, one generalization is $|x - \mathbf{y} \cdot \mathbf{z}| \leq t\|\mathbf{y}\|$.

4.9 In each part, it is necessary to show that the two n-vectors are equal so, according to Definition 4.2, you must show that all n components of the vectors are the same. So, let i be an integer with $1 \leq i \leq n$.

(a) In this case,

$$\begin{aligned} [(\mathbf{x} + \mathbf{y}) + \mathbf{z}]_i &= (\mathbf{x} + \mathbf{y})_i + z_i \\ &= (x_i + y_i) + z_i \\ &= x_i + (y_i + z_i) \\ &= x_i + (\mathbf{y} + \mathbf{z})_i \\ &= [\mathbf{x} + (\mathbf{y} + \mathbf{z})]_i. \end{aligned}$$

(b) In this case,

$$(\mathbf{x} + \mathbf{0})_i = x_i + 0 = x_i \text{ and } (\mathbf{0} + \mathbf{x})_i = 0 + x_i = x_i.$$

(c) $[\mathbf{x} + (-\mathbf{x})]_i = x_i + (-x_i) = x_i - x_i = 0$.

(d) $[t(\mathbf{x} + \mathbf{y})]_i = t(x_i + y_i) = tx_i + ty_i = (t\mathbf{x})_i + (t\mathbf{y})_i = (t\mathbf{x} + t\mathbf{y})_i$.

(e) $[(t + u)\mathbf{x}]_i = (t + u)x_i = tx_i + ux_i = (t\mathbf{x})_i + (u\mathbf{x})_i = [(t\mathbf{x}) + (u\mathbf{x})]_i$.

(f) $(1\mathbf{x})_i = 1x_i = x_i = (\mathbf{x})_i$.

4.11 It is given that

$$A = \begin{bmatrix} 2 & 1 \\ 0 & -1 \end{bmatrix}, \quad B = \begin{bmatrix} 1 & 3 \\ -1 & 2 \end{bmatrix}, \quad C = \begin{bmatrix} 1 & 0 & 0 \\ 0 & 2 & -1 \end{bmatrix}.$$

(a) It follows that

$$C^t = \begin{bmatrix} 1 & 0 \\ 0 & 2 \\ 0 & -1 \end{bmatrix}.$$

(b) It follows that

$$\begin{aligned} A + 2B &= \begin{bmatrix} 2 & 1 \\ 0 & -1 \end{bmatrix} + 2\begin{bmatrix} 1 & 3 \\ -1 & 2 \end{bmatrix} \\ &= \begin{bmatrix} 2 & 1 \\ 0 & -1 \end{bmatrix} + \begin{bmatrix} 2 & 6 \\ -2 & 4 \end{bmatrix} \\ &= \begin{bmatrix} 2+2 & 1+6 \\ 0+(-2) & -1+4 \end{bmatrix} \\ &= \begin{bmatrix} 4 & 7 \\ -2 & 3 \end{bmatrix}. \end{aligned}$$

(c) It follows that

$$\begin{aligned} AB &= \begin{bmatrix} 2 & 1 \\ 0 & -1 \end{bmatrix} \begin{bmatrix} 1 & 3 \\ -1 & 2 \end{bmatrix} \\ &= \begin{bmatrix} 2(1)+1(-1) & 2(3)+1(2) \\ 0(1)+(-1)(-1) & 0(3)+(-1)(2) \end{bmatrix} \\ &= \begin{bmatrix} 1 & 8 \\ 1 & -2 \end{bmatrix}. \end{aligned}$$

(d) It follows that

$$\begin{aligned} AC &= \begin{bmatrix} 2 & 1 \\ 0 & -1 \end{bmatrix} \begin{bmatrix} 1 & 0 & 0 \\ 0 & 2 & -1 \end{bmatrix} \\ &= \begin{bmatrix} 2(1)+1(0) & 2(0)+1(2) & 2(0)+1(-1) \\ 0(1)+(-1)(0) & 0(0)+(-1)(2) & 0(0)+(-1)(-1) \end{bmatrix} \\ &= \begin{bmatrix} 2 & 2 & -1 \\ 0 & -2 & 1 \end{bmatrix}. \end{aligned}$$

(e) You cannot compute CA because C has three columns and A has only two rows.

4.13

(a) To compute $A\mathbf{x}$, $\mathbf{x}$ must be an $(n \times 1)$ matrix. Then, $A\mathbf{x}$ is an $(m \times 1)$ matrix, that is, a *column* vector.

(b) In this case,

$$
\begin{aligned}
(A\mathbf{x})_1 &= A_{1.} \cdot \mathbf{x} = (1,-1,0)\cdot(-1,1,2)^t = -2,\\
(A\mathbf{x})_2 &= A_{2.} \cdot \mathbf{x} = (2,0,-2)\cdot(-1,1,2)^t = -6,\\
(A\mathbf{x})_3 &= A_{3.} \cdot \mathbf{x} = (-3,1,3)\cdot(-1,1,2)^t = 10,
\end{aligned}
$$

so

$$
A\mathbf{x} = \begin{bmatrix} -2 \\ -6 \\ 10 \end{bmatrix}.
$$

(c) In this case,

$$
\begin{aligned}
A\mathbf{x} &= A_{.1}x_1 + A_{.2}x_2 + A_{.3}x_3\\
&= \begin{bmatrix} 1 \\ 2 \\ -3 \end{bmatrix}(-1) + \begin{bmatrix} -1 \\ 0 \\ 1 \end{bmatrix}(1) + \begin{bmatrix} 0 \\ -2 \\ 3 \end{bmatrix}(2)\\
&= \begin{bmatrix} -1 \\ -2 \\ 3 \end{bmatrix} + \begin{bmatrix} -1 \\ 0 \\ 1 \end{bmatrix} + \begin{bmatrix} 0 \\ -4 \\ 6 \end{bmatrix}\\
&= \begin{bmatrix} -2 \\ -6 \\ 10 \end{bmatrix}.
\end{aligned}
$$

4.15

(a) If you prove that $(AB)_{.j} = AB_{.j}$, then you need only multiply A by *one* column of B as opposed to computing AB, which requires multiplying A by all n columns of B. In other words, it is more efficient to compute $AB_{.j}$ rather than $(AB)_{.j}$. Likewise, it is more efficient to compute $A_{i.}B$ rather than $(AB)_{i.}$.

(b) For each $i = 1, \ldots, n$, $(AB)_{ij} = A_{i.} \cdot B_{.j}$, while $(AB_{.j})_i$ = component i of the vector $AB_{.j} = A_{i.} \cdot B_{.j}$, so $(AB)_{.j} = AB_{.j}$. Likewise, because $(AB)_{ij} = A_{i.} \cdot B_{.j}$ and $(A_{i.}B)_j$ = component j of the vector $A_{i.}B = A_{i.} \cdot B_{.j}$, it follows that $(AB)_{i.} = A_{i.}B$.

4.17 In each case, it is necessary to show that two $(m \times n)$ matrices are equal, so, let i and j be integers with $1 \le i \le m$ and $1 \le j \le n$.

(a) In this case,

$$\begin{aligned}[(A+B)+C]_{ij} &= (A+B)_{ij} + C_{ij} \\ &= (A_{ij} + B_{ij}) + C_{ij} \\ &= A_{ij} + (B_{ij} + C_{ij}) \\ &= A_{ij} + (B+C)_{ij} \\ &= [A+(B+C)]_{ij}.\end{aligned}$$

(b) In this case,

$$\begin{aligned}&(A+0)_{ij} = A_{ij} + 0 = A_{ij} \text{ and} \\ &(0+A)_{ij} = 0 + A_{ij} = A_{ij}.\end{aligned}$$

(c) $(A-A)_{ij} = A_{ij} - A_{ij} = 0.$

(d) $(0-A)_{ij} = 0 - A_{ij} = -A_{ij} = (-A)_{ij}.$

4.19

(a) Let $A \in \mathbb{R}^{m\times n}$, $B, C \in \mathbb{R}^{n\times p}$, and i and j be integers with $1 \le i \le m$ and $1 \le j \le p$. Then

$$\begin{aligned}[A(B-C)]_{ij} &= A_{i.} \cdot (B-C)_{.j} \\ &= A_{i.} \cdot (B_{.j} - C_{.j}) \\ &= (A_{i.} \cdot B_{.j}) - (A_{i.} \cdot C_{.j}) \\ &= (AB)_{ij} - (AC)_{ij} \\ &= (AB - AC)_{ij}.\end{aligned}$$

(b) Let $A, B \in \mathbb{R}^{m\times n}$, $C \in \mathbb{R}^{n\times p}$, and i and j be integers with $1 \le i \le m$ and $1 \le j \le p$. Then

$$\begin{aligned}[(A+B)C]_{ij} &= (A+B)_{i.} \cdot C_{.j} \\ &= (A_{i.} + B_{i.}) \cdot C_{.j} \\ &= (A_{i.} \cdot C_{.j}) + (B_{i.} \cdot C_{.j}) \\ &= (AC)_{ij} + (BC)_{ij} \\ &= (AC + BC)_{ij}.\end{aligned}$$

(c) Let $A, B \in \mathbb{R}^{m\times n}$, $C \in \mathbb{R}^{n\times p}$, and i and j be integers with $1 \le i \le m$ and $1 \le j \le p$. Then

$$\begin{aligned}[(A-B)C]_{ij} &= (A-B)_{i.} \cdot C_{.j} \\ &= (A_{i.} - B_{i.}) \cdot C_{.j} \\ &= (A_{i.} \cdot C_{.j}) - (B_{i.} \cdot C_{.j}) \\ &= (AC)_{ij} - (BC)_{ij} \\ &= (AC - BC)_{ij}.\end{aligned}$$

4.21 To see what PA is, observe that for each row $k = 1, \ldots, m$,

$$P_{k.} = \begin{cases} I_{k.}, & \text{if } k \neq i, j \\ I_{j.}, & \text{if } k = i \\ I_{i.}, & \text{if } k = j. \end{cases}$$

So, by Exercise 4.15,

$$(PA)_{k.} = P_{k.}A = \begin{cases} I_{k.}A, & \text{if } k \neq i, j \\ I_{j.}A, & \text{if } k = i \\ I_{i.}A, & \text{if } k = j \end{cases} = \begin{cases} A_{k.}, & \text{if } k \neq i, j \\ A_{j.}, & \text{if } k = i \\ A_{i.}, & \text{if } k = j. \end{cases}$$

4.23

(a) It is shown that for a permutation matrix P, $PP^t = P^tP = I$. This is *true* because

$$(PP^t)_{ij} = P_{i.} \cdot (P^t)_{.j} = P_{i.} \cdot P_{j.} = \begin{cases} 1, & \text{if } i = j \\ 0, & \text{if } i \neq j \end{cases} \quad \text{and}$$

$$(P^tP)_{ij} = (P^t)_{i.} \cdot P_{.j} = P_{.i} \cdot P_{.j} = \begin{cases} 1, & \text{if } i = j \\ 0, & \text{if } i \neq j. \end{cases}$$

Thus, P is nonsingular and $P^{-1} = P^t$.

(b) For an eta matrix E in which column k differs from column k of the identity matrix I, you have that

$$E_{ip} = \begin{cases} I_{ip}, & \text{if } p \neq k \\ E_{ik}, & \text{if } p = k \end{cases} \quad \text{and } E^{-1}_{pj} = \begin{cases} I_{pj}, & \text{if } j \neq k \\ -E_{pk}/E_{kk}, & \text{if } j = k \text{ and } p \neq k \\ 1/E_{kk}, & \text{if } j = k \text{ and } p = k. \end{cases}$$

The next step is to show that $EE^{-1} = E^{-1}E = I$, so, let i and j be integers with $1 \leq i \leq n$ and $1 \leq j \leq n$. Then

$$\begin{aligned} E_{i.} \cdot E^{-1}_{.j} &= \sum_{p=1}^{n} E_{ip}E^{-1}_{pj} \\ &= \sum_{p \neq k} E_{ip}E^{-1}_{pj} + E_{ik}E^{-1}_{kj} \\ &= \sum_{p \neq k} I_{ip}E^{-1}_{pj} + E_{ik}E^{-1}_{kj}. \ (1) \end{aligned}$$

The proof now proceeds by cases.

Case 1: $i = j$.

In this case, it must be shown that $E_{i.} \cdot E^{-1}_{.j} = 1$. This is done in two subcases. When $i = k$, the expression in (1) becomes

$$E_{i.} \cdot E^{-1}_{.j} = \sum_{p \neq k} (0)E^{-1}_{pj} + E_{kk}E^{-1}_{kk} = 0 + E_{kk}(1/E_{kk}) = 1.$$

Alternatively, when $i \neq k$, the expression in (1) is also 1 because

$$\begin{aligned} E_{i.} \cdot E^{-1}_{.j} &= \sum_{p \neq k} I_{ip} E^{-1}_{pj} + E_{ik} E^{-1}_{kj} \\ &= (1) E^{-1}_{ij} + E_{ik} E^{-1}_{kj} \\ &= (1) I_{ii} + E_{ik} I_{ki} \\ &= (1) 1 + E_{ik}(0) \\ &= 1. \end{aligned}$$

Thus, when $i = j$, $E_{i.} \cdot E^{-1}_{.j} = 1$.

Case 2: $i \neq j$.

In this case, it must be shown that $E_{i.} \cdot E^{-1}_{.j} = 0$. This is done again in subcases. When $i = k$, the expression in (1) becomes

$$\begin{aligned} E_{i.} \cdot E^{-1}_{.j} &= \sum_{p \neq k} (0) E^{-1}_{pj} + E_{kk} E^{-1}_{kj} \\ &= E_{kk} I_{kj} \\ &= E_{kk}(0) \\ &= 0. \end{aligned}$$

Alternatively, when $i \neq k$ the expression in (1) becomes

$$(1) E^{-1}_{ij} + E_{ik} E^{-1}_{kj}.$$

In the event that $j \neq k$, you then have that

$$\begin{aligned} E_{i.} \cdot E^{-1}_{.j} &= (1) E^{-1}_{ij} + E_{ik} E^{-1}_{kj} \\ &= (1)(0) + E_{ik}(0) \\ &= 0. \end{aligned}$$

Finally, if $j = k$, then

$$\begin{aligned} E_{i.} \cdot E^{-1}_{.j} &= (1) E^{-1}_{ij} + E_{ik} E^{-1}_{kj} \\ &= (1)(-E_{ik}/E_{kk}) + E_{ik}(1/E_{kk}) \\ &= 0. \end{aligned}$$

Thus, when $i \neq j$, $E_{i.} \cdot E^{-1}_{.j} = 0$. This shows that $EE^{-1} = I$. A similar computation shows that $E^{-1}E = I$ and so the formula for the inverse of an eta matrix is correct.

4.25

(a) The columns of $A \in \mathbb{R}^{m \times n}$ are **linearly independent** if and only if for all real numbers $t_1, \ldots, t_n$ with

$$A_{.1} t_1 + \cdots + A_{.n} t_n = \mathbf{0},$$

it follows that

$$t_1 = \cdots = t_n = 0.$$

That is, if $A\mathbf{t} = \mathbf{0}$, then $\mathbf{t} = \mathbf{0}$, where $\mathbf{t} = (t_1, \ldots, t_n)^t$ is a column vector.

(b) You cannot generalize the concept of a nonsingular matrix in a straightforward manner because, if $A \in \mathbb{R}^{m \times n}$ and $m \neq n$, then you cannot find a matrix B such that $AB = BA = I$ since you cannot compute both AB and BA.

4.27 Suppose first that there are real numbers u, v, and w, at least one of which is 1, such that $u\mathbf{x} + v\mathbf{y} + w\mathbf{z} = \mathbf{0}$. Assume, without loss of generality, that $u = 1$. By letting $a = u = 1, b = v$, and $c = w$, it follows that there are real numbers a, b, and c, not all zero (because $a = 1$), such that $a\mathbf{x} + b\mathbf{y} + c\mathbf{z} = \mathbf{0}$.

For the converse, suppose that there are real numbers a, b, and c, not all zero, such that $a\mathbf{x} + b\mathbf{y} + c\mathbf{z} = \mathbf{0}$. Without loss of generality, assume that $a \neq 0$. Then, dividing through by a yields

$$\mathbf{x} + \frac{b}{a}\mathbf{y} + \frac{c}{a}\mathbf{z} = \mathbf{0}.$$

Letting $u = 1, v = b/a$, and $w = c/a$, it follows that there are real numbers u, v, and w, at least one of which (namely, u) is 1, such that $u\mathbf{x} + v\mathbf{y} + w\mathbf{z} = \mathbf{0}$.

4.29 The vectors $\mathbf{v}^1, \ldots, \mathbf{v}^k \in V$ are **linearly independent** if and only if for all scalars $t_1, \ldots, t_k \in F$ with

$$(t_1 \odot \mathbf{v}^1) \oplus \cdots \oplus (t_k \odot \mathbf{v}^k) = \mathbf{0},$$

it follows that $t_1 = \cdots = t_k = 0$.

4.31 To generalize the operation of the dot product to a vector space $(F, +, *, V, \oplus, \odot)$, as defined in Definition 4.13 in Section 4.5.2, let $\mathbf{x}, \mathbf{y}, \mathbf{z} \in V$ and $t \in F$. The four axioms for the operation $\langle\ \rangle$ that combines two vectors to create a scalar become the following:

(i) $\langle \mathbf{x}, \mathbf{y} \rangle = \langle \mathbf{y}, \mathbf{x} \rangle$.

(ii) $\langle \mathbf{x} \oplus \mathbf{y}, \mathbf{z} \rangle = \langle \mathbf{x}, \mathbf{z} \rangle + \langle \mathbf{y}, \mathbf{z} \rangle$.

(iii) $\langle t \odot \mathbf{x}, \mathbf{y} \rangle = t * \langle \mathbf{x}, \mathbf{y} \rangle$.

(iv) $\langle \mathbf{x}, \mathbf{x} \rangle \geq 0$ and $\langle \mathbf{x}, \mathbf{x} \rangle = 0$ if and only if $\mathbf{x} = \mathbf{0}$.

Solutions to Chapter 5 Exercises

5.1

(a) By definition, because $a \mid b$, there is an integer k such that $b = ka$, so $-b = (-k)a$, and thus $a \mid (-b)$.

(b) By definition, because $a \mid b$, there is an integer k such that $b = ka$. Likewise, because $b \mid c$, there is an integer m such that $c = mb$. Thus, $c = mb = m(ka) = (mk)a$ and so $a \mid c$.

(c) By definition, because $a \mid b$, there is an integer k such that $b = ka$. Likewise, because $a \mid c$, there is an integer m such that $c = ma$. Adding these two equations yields that $b + c = ka + ma = (k + m)a$, and so $a \mid (b + c)$.

(d) By definition, because $a \mid b$, there is an integer k such that $b = ka$. Likewise, because $b \mid a$, there is an integer m such that $a = mb$. If $b = 0$, then $a = mb = 0$, and so $a = b$, as desired. Otherwise, $b \neq 0$ and $b = ka = k(mb) = (km)b$, so $km = 1$. But because k and m are integers, it must be that $k, m = \pm 1$, so $b = ka = \pm a$.

5.3

(a) An integer d is a **common divisor of the integers** a, b, **and** c if and only if $d \mid a$, $d \mid b$, and $d \mid c$.

(b) An integer d is a **common divisor of the** n **integers** $a_1, \ldots, a_n$ if and only if for each $i = 1, \ldots, n$, $d \mid a_i$.

5.5 Let G be a set of objects and let $\odot$ be a closed binary operation on G that combines two elements $a, b \in G$ to produce the element $a \odot b \in G$. The following is an abstraction of Definition 5.1:

> Let $a, b \in G$. Then a **divides** b, written $a \mid b$, if and only if there is an element $k \in G$ such that $b = a \odot k$.

(Observe that the statement $b = k \odot a$ is *not* correct because the operator $\odot$ need not be commutative.)

5.7

(a) An integer m is a **least common multiple of the integers** a **and** b if and only if (1) m is a common multiple of a and b and (2) if k is a common multiple of a and b, then $m \mid k$.

(b) An integer m is a **common multiple of the** n **integers** $a_1, \ldots, a_n$ if and only if for each $i = 1, \ldots, n$, $a_i \mid m$.

(c) An integer m is a **least common multiple** of the n integers $a_1, \ldots, a_n$ if and only if (1) m is a common multiple of $a_1, \ldots, a_n$ and (2) if k is a common multiple of $a_1, \ldots, a_n$, then $m \mid k$.

5.9 To establish that $(p, b) = 1$, it is shown that the only positive common divisor of p and b is 1, so, let d be a positive common divisor of p and b. Now, because p is prime and $d \mid p$, it must be that $d = 1$ or $d = p$. But now $d \neq p$ because, if $d = p$, then p would also divide b, which does not happen according to the hypothesis. Therefore, $d = 1$ is the *only* positive common divisor of p and b, so $(p, b) = 1$.

5.11 From Proposition 5.3, because $p \mid (ab)$ and p is prime, either $p \mid a$ or $p \mid b$. Now proceed by cases. If $p \mid a$, then there is an integer k such that $a = pk$. But because $0 \leq a < p$, it follows that $k = 0$ and so $a = 0$. Alternatively, if $p \mid b$, then $b = 0$.

5.13 When you use Definition 5.5 to determine if n is prime, you must check whether $k \mid n$ for each value of $k = 2, \ldots, n - 1$. On the other hand, if you prove the stated proposition, then you only need to check whether $k \mid n$ for each value of $2 \leq k \leq \sqrt{n}$, which is less work.

5.15

(a) No, $\{0, 1\}$ is not a group under addition because addition is *not* a closed binary operation on $\{0, 1\}$ since $1 + 1 = 2 \notin \{0, 1\}$.

(b) Yes, $\mathbb{Q}$ is a group under addition with 0 being the identity element. The inverse of any rational number $r \in \mathbb{Q}$ under addition is $-r \in \mathbb{Q}$.

(c) No, $\mathbb{Q}$ is *not* a group under division because division is *not* a closed binary operation on $\mathbb{Q}$. For example, if $r \in \mathbb{Q}$, then r divided by 0 is not a rational number.

5.17 The identity element of $(\mathbb{Z}_n, \odot)$ is $[1]$ because, for any $[a] \in \mathbb{Z}_n$,

$$[1] \odot [a] = [1(a)] = [a] \text{ and } [a] \odot [1] = [a(1)] = [a].$$

Associativity holds because, for $[a], [b], [c] \in \mathbb{Z}_n$,

$$\begin{aligned} [a] \odot ([b] \odot [c]) &= [a] \odot ([bc]) && \text{(definition of } \odot) \\ &= [a(bc)] && \text{(definition of } \odot) \\ &= [(ab)c)] && \text{(associativity of } \cdot) \\ &= [ab] \odot [c] && \text{(definition of } \odot) \\ &= ([a] \odot [b]) \odot [c] && \text{(definition of } \odot). \end{aligned}$$

Axiom (2) does *not* hold because there is no inverse for the element $[0] \in \mathbb{Z}_n$. That is, there is no element $[a] \in \mathbb{Z}_n$ with $[0] \odot [a] = [1]$.

5.19 Suppose that for $a, b, c \in G$, $b \odot a = c \odot a$. To see that $b = c$, apply a^{-1} on the *right* of both sides and use the axioms of a group to obtain the following:

$$\begin{aligned} (b \odot a) \odot a^{-1} &= (c \odot a) \odot a^{-1} \\ b \odot (a \odot a^{-1}) &= c \odot (a \odot a^{-1}) && \text{(associativity)} \\ b \odot e &= c \odot e && \text{(property of the inverse)} \\ b &= c && \text{(property of the identity).} \end{aligned}$$

5.21 To apply Exercise 5.19 to this special case, suppose that A, B, and C are nonsingular $(n \times n)$ matrices. If $BA = CA$, then $B = C$.

To apply Exercise 5.20 to this special case, suppose that A and B are nonsingular $(n \times n)$ matrices. Then the unique nonsingular $(n \times n)$ matrix C such that $CA = B$ is $C = BA^{-1}$.

5.23 The difficulty is that function composition might not make sense when all functions have A as their domain and B as their codomain. For example, if $f, g : A \to B$, then for $x \in A$, $(g \circ f)(x) = g(f(x))$, but $f(x) \in B$ and since $g : A \to B$, $g(f(x))$ results in a syntax error unless $f(x) \in A$. One condition that avoids this syntax error is $A = B$.

5.25

(a) If $a, b \in H$, then, because $\odot$ is a closed binary operation on H, it must be that $a \odot b \in H$.

(b) H, being a subgroup, has an identity element, say, f, so

$$f \odot f = f.$$

It is shown that $f = e$ and therefore that $e \in H$. So now let f^{-1} be the inverse of f in G. Applying f^{-1} to the left of both sides of the foregoing equality and using the properties of G, it follows that

$$\begin{aligned} f^{-1} \odot (f \odot f) &= f^{-1} \odot f \\ (f^{-1} \odot f) \odot f &= f^{-1} \odot f && \text{(associativity)} \\ e \odot f &= e && (f^{-1} \text{ is the inverse of } f \text{ in } G) \\ f &= e && (e \text{ is the identity in } G). \end{aligned}$$

(c) For $a \in H$, let a^{-1} be the inverse of a in G and $b \in H$ be the inverse of a in H, so b satisfies $a \odot b = b \odot a = e$, which is the identity element of H, from part (b). It is shown that $b = a^{-1}$ and therefore that $a^{-1} \in H$. But now you have that

$$\begin{aligned} a \odot b &= e \\ a^{-1} \odot (a \odot b) &= a^{-1} \odot e && (\text{apply } a^{-1} \text{ on the left}) \\ (a^{-1} \odot a) \odot b &= a^{-1} \odot e && \text{(associativity)} \\ e \odot b &= a^{-1} \odot e && (\text{property of } a^{-1}) \\ b &= a^{-1} && (\text{property of } e). \end{aligned}$$

5.27 To see that $\mathbb{Z}[\sqrt{2}]$ is a ring, let $a + b\sqrt{2}, c + d\sqrt{2}, e + f\sqrt{2}$ be elements of $\mathbb{Z}[\sqrt{2}]$. Note first that the identity element under $\oplus$ is $0 + 0\sqrt{2}$ because

$$(a + b\sqrt{2}) \oplus (0 + 0\sqrt{2}) = (a + 0) + (b + 0)\sqrt{2} = a + b\sqrt{2}$$

and

$$(0 + 0\sqrt{2}) \oplus (a + b\sqrt{2}) = (0 + a) + (0 + b)\sqrt{2} = a + b\sqrt{2}.$$

The inverse of $a + b\sqrt{2}$ under $\oplus$ is $-a - b\sqrt{2}$ because

$$(a + b\sqrt{2}) \oplus (-a - b\sqrt{2}) = (a - a) + (b - b)\sqrt{2} = 0 + 0\sqrt{2}$$

and

$$(-a - b\sqrt{2}) \oplus (a + b\sqrt{2}) = (-a + a) + (-b + b)\sqrt{2} = 0 + 0\sqrt{2}.$$

Associativity holds under $\oplus$ because

$$\begin{aligned}
[(a + b\sqrt{2}) \oplus (c + d\sqrt{2})] \oplus (e + f\sqrt{2}) &= [(a + c) + (b + d)\sqrt{2}] \oplus (e + f\sqrt{2}) \\
&= [(a + c) + e] + [(b + d) + f]\sqrt{2} \\
&= [a + (c + e)] + [b + (d + f)]\sqrt{2} \\
&= (a + b\sqrt{2}) \oplus [(c + e) + (d + f)\sqrt{2}] \\
&= (a + b\sqrt{2}) \oplus [(c + d\sqrt{2}) \oplus (e + f\sqrt{2})].
\end{aligned}$$

Commutativity holds under $\oplus$ because

$$\begin{aligned}
(a + b\sqrt{2}) \oplus (c + d\sqrt{2}) &= (a + c) + (b + d)\sqrt{2} \\
&= (c + a) + (d + b)\sqrt{2} \\
&= (c + d\sqrt{2}) \oplus (a + b\sqrt{2}).
\end{aligned}$$

The operation $\odot$ is associative because

$$\begin{aligned}
[(a + b\sqrt{2}) \odot (c + d\sqrt{2})] \odot (e + f\sqrt{2}) &= [(ac + 2bd) + (ad + bc)\sqrt{2}] \odot (e + f\sqrt{2}) \\
&= (ac + 2bd)e + 2(ad + bc)f + \\
&\quad [(ac + 2bd)f + (ad + bc)e]\sqrt{2} \\
&= [a(ce + 2df) + 2b(de + cf)] + \\
&\quad a(de + cf) + b(ce + 2df)]\sqrt{2} \\
&= (a + b\sqrt{2}) \odot [(ce + 2df) + (de + cf)\sqrt{2}] \\
&= (a + b\sqrt{2}) \odot [(c + d\sqrt{2}) \odot (e + f\sqrt{2})].
\end{aligned}$$

The distributive axioms hold because

$$\begin{aligned}
(a + b\sqrt{2}) \odot [(c + d\sqrt{2}) \oplus (e + f\sqrt{2})] &= (a + b\sqrt{2}) \odot [(c + e) + (d + f)\sqrt{2}] \\
&= a(c + e) + 2b(d + f) + \\
&\quad [a(d + f) + b(c + e)]\sqrt{2} \\
&= [(ac + 2bd) + (ad + bc)\sqrt{2}] + \\
&\quad [(ae + 2bf) + (af + be)\sqrt{2}] \\
&= [(a + b\sqrt{2}) \odot (c + d\sqrt{2})] \oplus \\
&\quad [(a + b\sqrt{2}) \odot (e + f\sqrt{2})].
\end{aligned}$$

Similarly,

$$\begin{aligned}
&[(c + d\sqrt{2}) \oplus (e + f\sqrt{2})] \odot (a + b\sqrt{2}) = \\
&((c + d\sqrt{2}) \odot (a + b\sqrt{2})) \oplus [(e + f\sqrt{2}) \odot (a + b\sqrt{2})].
\end{aligned}$$

5.29 Suppose that $(R, \oplus, \odot)$ is a ring and let $a, b, c \in R$.

(a) To see that $0 \in R$ is the unique zero element of R, suppose that $z \in R$ also satifies the property that

$$\text{for all } a \in R,\ a \oplus z = z \oplus a = a.$$

Then, by replacing a with 0, it follows that

$$0 \oplus z = z \oplus 0 = 0.$$

However, by the property of 0, $0 \oplus z = z$, so $z = 0$.

(b) To see that $-a$ is the unique negative of a in R, suppose that $d \in R$ also satisfies the property that

$$a \oplus d = 0.$$

Then applying $-a$ to the left of both sides yields that

$$(-a) \oplus (a \oplus d) = [(-a) \oplus a] \oplus d = 0 \oplus d = d$$

and

$$(-a) \oplus 0 = -a,$$

so $d = -a$.

(c) If $a \oplus b = a \oplus c$, then

$$\begin{aligned} (-a) \oplus (a \oplus b) &= (-a) \oplus (a \oplus c) \\ [(-a) \oplus a] \oplus b &= [(-a) \oplus a] \oplus c \\ 0 \oplus b &= 0 \oplus c \\ b &= c. \end{aligned}$$

(d) If $b \oplus a = c \oplus a$, then

$$\begin{aligned} (b \oplus a) \oplus (-a) &= (c \oplus a) \oplus (-a) \\ b \oplus [a \oplus (-a)] &= c \oplus [a \oplus (-a)] \\ b \oplus 0 &= c \oplus 0 \\ b &= c. \end{aligned}$$

(e) $-(-a)$ is that unique element $d \in R$ such that $(-a) \oplus d = d \oplus (-a) = 0$. But $d = a$ satisfies this property, so $-(-a) = d = a$.

Also, $-(a \oplus b)$ is that unique element $d \in R$ such that $d \oplus (a \oplus b) = (a \oplus b) \oplus d = 0$, But $d = (-a) \oplus (-b)$ satisfies this property because

$$\begin{aligned} d \oplus (a \oplus b) &= [(-a) \oplus (-b)] \oplus (a \oplus b) \\ &= [(-b) \oplus (-a)] \oplus (a \oplus b) \\ &= \{(-b) \oplus [(-a) \oplus a]\} \oplus b \\ &= [(-b) \oplus 0] \oplus b \\ &= (-b) \oplus b \\ &= 0. \end{aligned}$$

Similarly, $(a \oplus b) \oplus d = 0$. Therefore, $d = (-a) \oplus (-b) = -(a \oplus b)$.

(f) First note that $x = b - a = b \oplus (-a)$ is a solution to $a \oplus x = b$ because

$$\begin{aligned} a \oplus x &= a \oplus [b \oplus (-a)] \\ &= a \oplus [(-a) \oplus b] \\ &= [a \oplus (-a)] \oplus b \\ &= 0 \oplus b \\ &= b. \end{aligned}$$

To see that $x = b - a = b \oplus (-a)$ is the unique solution, suppose that y also satisfies $a \oplus y = b$. Then,

$$\begin{aligned} a \oplus y &= b \\ (-a) \oplus [a \oplus y] &= (-a) \oplus b \\ [(-a) \oplus a] \oplus y &= (-a) \oplus b \\ 0 \oplus y &= (-a) \oplus b \\ y &= b \oplus (-a). \end{aligned}$$

Thus, $y = x$ and so $x = b - a$ is the unique value of x for which $a \oplus x = b$.

A similar proof applies for the solution to $x \oplus a = b$.

(g) The proofs are by induction and are done here only for positive integers m and n. To prove that $m(a \oplus b) = (ma) \oplus (mb)$, by induction, note that for $m = 1$, $1(a \oplus b) = a \oplus b = (1a) \oplus (1b)$. Now assume the induction hypothesis that $(m-1)(a \oplus b) = [(m-1)a] \oplus [(m-1)b]$. For $m \geq 2$, you have that

$$\begin{aligned} m(a \oplus b) &= [(m-1)(a \oplus b)] \oplus (a \oplus b) \\ &= \{[(m-1)a] \oplus [(m-1)b]\} \oplus (a \oplus b) \\ &= \{[(m-1)a] \oplus a\} \oplus \{[(m-1)b] \oplus b\} \\ &= (ma) \oplus (mb). \end{aligned}$$

To prove that $m(na) = (mn)a$, when $n = 1$, you have that $m(1a) = m(a) = [m(1)]a$. Assume the induction hypothesis that $m[(n-1)a] = [m(n-1)]a$. Then for $n \geq 2$, you have that

$$\begin{aligned} m(na) &= m[(n-1)a \oplus a] && \text{(definition of } na) \\ &= \{m[(n-1)a]\} \oplus (ma) && \text{(from the first part)} \\ &= \{[m(n-1)]a\} \oplus (ma) && \text{(induction hypothesis)} \\ &= [m(n-1) + m]a && \text{(definition of } ka) \\ &= (mn)a. \end{aligned}$$

To prove that $(m+n)a = (ma) \oplus (na)$, by induction on n, for $n = 1$, $(m+1)a = (ma) \oplus a = (ma) \oplus (1a)$. Now assume the induction hypothesis that $(m+n-1)a = (ma) \oplus [(n-1)a]$. Then, for $n \geq 2$, you have that

$$\begin{aligned} (m+n)a &= (m+n-1)a \oplus a \\ &= \{(ma) \oplus [(n-1)a]\} \oplus a \\ &= (ma) \oplus \{[(n-1)a] \oplus a\} \\ &= (ma) \oplus (na). \end{aligned}$$

5.31 Suppose first that $(K, \oplus, \odot)$ is a subfield of a field $(F, \oplus, \odot)$. Now K must have a zero element and a unity element that are also in F but, by the uniqueness property of these elements in F, it must be that these elements in K are the same as 0 and 1 in F. Next, because $\oplus$ and $\odot$ are closed operations on K, it must be that for all $a, b \in K$, $a \oplus b \in K$ and $a \odot b \in K$. Also, if $a \in K$, then because $(K, \oplus, \odot)$ is a field, there must be an element $b \in K$ such that $a \oplus b = b \oplus a = 0$. Now b is also in F and because $-a$ is the unique such element in F, $b = -a \in K$. Finally, if $a \in K$ with $a \neq 0$, then, because $(K, \oplus, \odot)$ is a field, there must be an element $b \in K$ such that $a \odot b = b \odot a = 1$. But a^{-1} is the unique such element in F, so in fact $b = a^{-1} \in K$.

For the converse, assume that (1) $0 \in K$ and $1 \in K$, (2) for all $a, b \in K$, $a \oplus b \in K$ and $a \odot b \in K$, (3) for all $a \in K$, $-a \in K$, and (4) for all $a \in K$ with $a \neq 0$, $a^{-1} \in K$. It is shown that $(K, \oplus, \odot)$ is an integral domain whose nonzero elements form an Abelian group under $\odot$. But $(K, \oplus, \odot)$ is a ring (see Exercise 5.30) and in K, $\odot$ is commutative because $(F, \oplus, \odot)$ is a field and $\odot$ is commutative over F. Also, if $a, b \in K$ with $a \odot b = 0$, then, from the fact that $(F, \oplus, \odot)$ is a field, $a = 0$ or $b = 0$. To see that the nonzero elements of K form an Abelian group under $\odot$, observe first that $1 \in K$ is the identity element. Also, if $a \in K$ and $a \neq 0$, then, by (4), $a^{-1} \in K$. The operation $\odot$ is associative and commutative in K because these same properties hold in F.

Solutions to Chapter 6 Exercises

6.1 Let P be the positive real numbers.

(a) You have that

$a \geq 0, b \leq 0$	(given)
$a \geq 0, -b \geq 0$	(multiply by -1)
$a(-b) \geq 0$	(Property (2) of P)
$-(ab) \geq 0$	[Proposition 6.2(b)]
$ab \leq 0$	(multiply by -1).

(b) You have that

$a \leq 0, b \leq 0$	(given)
$-a \geq 0, -b \geq 0$	(multiply by -1)
$(-a)(-b) \geq 0$	(Property (2) of P)
$ab \geq 0$	[Proposition 6.2(c)].

(c) You have that

$$\begin{array}{ll}
0 \le a \le b & \text{(given)} \\
b - a \ge 0 & \text{(definition of } \le\text{)} \\
b + a \ge 0 & \text{(Property (1) of } P\text{)} \\
(b-a)(b+a) \ge 0 & \text{(Property (2) of } P\text{)} \\
b^2 - a^2 \ge 0 & \text{(algebra)} \\
a^2 \le b^2 & \text{(definition of } \le\text{).}
\end{array}$$

6.3 By induction on n, when $n = 1$, the result is *true* because $|a_1| \le |a_1|$. When $n = 2$, the result is *true* by Theorem 6.1. Now assume the induction hypothesis that $|a_1 + \cdots + a_n| \le |a_1| + \cdots + |a_n|$. Then

$$\begin{aligned}
|a_1 + \cdots + a_n + a_{n+1}| &= |(a_1 + \cdots + a_n) + a_{n+1}| \\
&\le |a_1 + \cdots + a_n| + |a_{n+1}| \quad \text{(tri. ineq.)} \\
&\le |a_1| + \cdots + |a_n| + |a_{n+1}| \quad \text{(ind. hyp.).}
\end{aligned}$$

6.5 Suppose first that $a \le b$ and let $\varepsilon > 0$. Then $a \le b \le b + \varepsilon$ by Proposition 6.4(b).

Now assume that for all $\varepsilon > 0$, $a \le b + \varepsilon$. By contradiction, suppose that $a > b$ and let $0 < \varepsilon < a - b$. (You can choose ε in this way because a real number exists between any pair of distinct real numbers.) Then you have $a \le b + \varepsilon < b + (a - b) = a$, which is a contradiction.

6.7

(a) Use $T' = \{s \in \mathbb{R} : s > 0 \text{ and } s^n > 2\}$.

(b) To show that T' has an infimum, by the Infimum Property of $\mathbb{R}$, you must show that $T' \ne \emptyset$ and that T' has a lower bound. But $T' \ne \emptyset$ because $s = 2 > 0$ and $s^n = 2^n > 2$, for $n > 1$, so $2 \in T'$. Also, 0 is a lower bound for T' because, for all $s \in T$, $s > 0$. Thus, T' has an infimum.

6.9 Proposition: If u is the supremum of a set T of real numbers, then u is unique.

Proof. Assume that v is also a supremum for T. You can now show that $u \le v$ and $v \le u$. Specifically, because both u and v are suprema of T, by condition (1) of the definition, both u and v are upper bounds for T. Consequently, by condition (2) in the definition of u being a supremum for T, it follows that $u \le v$. Likewise, by condition (2) in the definition of v being a supremum for T, it follows that $v \le u$. Thus, $u = v$ and so the supremum of the set T is unique.

6.11 By contradiction, assume that T is finite. But then $S \subseteq T$ and T is finite so, by Proposition 6.13 in Section 6.2.4, S is finite, contradicting the hypothesis that S is infinite.

6.13 The problem with generalizing the dot product of two n-vectors to two sequences X and Y is that it might happen that $X \cdot Y$ is ∞, which is not a real number. For example, if $X = (1, 1, \ldots)$ and $Y = (2, 2, \ldots)$, then

$$X \cdot Y = 1(2) + 1(2) + \cdots = \infty.$$

6.15

(a) Yes, the set of sequences of real numbers is a group under addition. Observe that addition is a closed binary operation on sequences of real numbers. The identity element is the zero sequence $(0, 0, \ldots)$ and the remaining properties follow from the definition of addition of sequences of real numbers. For example, for every sequence X of real numbers, the sequence $-X$ satisfies $X + (-X) = (-X) + X = 0$.

(b) No, sequences of real numbers do *not* form a group under multiplication, even though multiplication is a closed binary operation and the sequence $(1, 1, \ldots)$ is the identity element. This is because a sequence such as $(1, 0, 1, 0, \ldots)$ has no inverse under multiplication.

6.17 Let $X = (\mathbf{x}^1, \mathbf{x}^2, \ldots)$ and $Y = (\mathbf{y}^1, \mathbf{y}^2, \ldots)$ be sequences of n-vectors.

(i) $X + Y = (\mathbf{x}^1 + \mathbf{y}^1, \mathbf{x}^2 + \mathbf{y}^2, \ldots)$.

(ii) $X - Y = (\mathbf{x}^1 - \mathbf{y}^1, \mathbf{x}^2 - \mathbf{y}^2, \ldots)$.

(iii) $X \cdot Y = (\mathbf{x}^1 \odot \mathbf{y}^1, \mathbf{x}^2 \odot \mathbf{y}^2, \ldots)$, where each $\mathbf{x}^k \odot \mathbf{y}^k$ is defined as follows:

$$\mathbf{x}^k \odot \mathbf{y}^k = (x_1^k y_1^k, \ldots, x_n^k y_n^k).$$

(iv) X/Y cannot be generalized because you cannot divide the n-vector $\mathbf{x}^k$ by the n-vector $\mathbf{y}^k$.

6.19 The sequence of n-vectors $X = (\mathbf{x}^1, \mathbf{x}^2, \ldots)$ **converges to the n-vector** $\mathbf{x} = (x_1, \ldots, x_n)$ if and only if for each $i = 1, \ldots, n$, the sequence of real numbers $X_i = (x_i^1, x_i^2, \ldots)$ converges to x_i, that is, if and only if for each $i = 1, \ldots, n$ and for every real number $\varepsilon > 0$, there is a $j_i \in \mathbb{N}$ such that for all $k \in \mathbb{N}$ with $k > j_i$, $|x_i^k - x_i| < \varepsilon$.

6.21 Let $x, y, z \in S$. The four axioms of a metric space in Definition 6.26 in Section 6.5.3 are now verified for (S, d).

(i) By definition of d, $d(x, y) = 0$ or 1, so $d(x, y) \geq 0$.

(ii) If $x = y$, then, by definition of d, $d(x, y) = 0$. Conversely, if $d(x, y) = 0$, then, by definition of d, $x = y$.

(iii) If $x = y$, then $d(x, y) = 0 = d(y, x)$.
If $x \neq y$, then $d(x, y) = 1 = d(y, x)$.
Thus, in either case, $d(x, y) = d(y, x)$.

(iv) If $x = z$, then $d(x, z) = d(x, x) = 0 \leq d(x, y) + d(y, z)$.
If $x \neq z$, then $d(x, z) = 1$. Now either $y \neq x$ or $y \neq z$ (otherwise, $x = z$). In the former case, $d(x, y) = 1$, so $d(x, z) = 1 \leq d(x, y) + d(y, z)$. In the latter case, $d(y, z) = 1$, so again, $d(x, z) = 1 \leq d(x, y) + d(y, z)$.
Thus, in all cases, $d(x, z) \leq d(x, y) + d(y, z)$.

6.23 (Answers other than the one given here are possible.)
Define $d : \mathbb{R}^{m \times n} \times \mathbb{R}^{m \times n} \to \mathbb{R}$ by

$$d(A, B) = \sum_{i=1}^{m} \sum_{j=1}^{n} |A_{ij} - B_{ij}|.$$

The four axioms of a metric space in Definition 6.26 in Section 6.5.3 are now verified for (S, d).

(i) By definition of absolute value, for all $i = 1, \ldots, m$ and $j = 1, \ldots, n$, $|A_{ij} - B_{ij}| \geq 0$, so $d(A, B) \geq 0$.

(ii) If $A = B$, then for all $i = 1, \ldots m$ and for all $j = 1, \ldots n$, $A_{ij} = B_{ij}$, so $|A_{ij} - B_{ij}| = 0$ and hence $d(A, B) = 0$.
Conversely, if $d(A, B) = 0$, then

$$\sum_{i=1}^{m} \sum_{j=1}^{n} |A_{ij} - B_{ij}| = 0.$$

But the only way the sum of the nonnegative numbers $|A_{ij} - B_{ij}|$ can be 0 is if *all* of them are 0, that is,

for all $i = 1, \ldots m$ and for all $j = 1, \ldots n$, $|A_{ij} - B_{ij}| = 0$,

or equivalently,

for all $i = 1, \ldots m$ and for all $j = 1, \ldots n$, $A_{ij} = B_{ij}$,

in which case $A = B$.

(iii) You have that

$$\begin{aligned} d(A, B) &= \sum_{i=1}^{m} \sum_{j=1}^{n} |A_{ij} - B_{ij}| \\ &= \sum_{i=1}^{m} \sum_{j=1}^{n} |B_{ij} - A_{ij}| \\ &= d(B, A). \end{aligned}$$

(iv) For the $(m \times n)$ matrices A, B, and C, you have that

$$\begin{aligned}
d(A, C) &= \sum_{i=1}^{m} \sum_{j=1}^{n} |A_{ij} - C_{ij}| \\
&= \sum_{i=1}^{m} \sum_{j=1}^{n} |A_{ij} - B_{ij} + B_{ij} - C_{ij}| \\
&\leq \sum_{i=1}^{m} \sum_{j=1}^{n} (|A_{ij} - B_{ij}| + |B_{ij} - C_{ij}|) \\
&= \sum_{i=1}^{m} \sum_{j=1}^{n} |A_{ij} - B_{ij}| + \sum_{i=1}^{m} \sum_{j=1}^{n} |B_{ij} - C_{ij}| \\
&= d(A, B) + d(B, C).
\end{aligned}$$

GLOSSARY

Abelian group A group $(G, \odot)$ in which the commutative law that for all $a, b \in G$, $a \odot b = b \odot a$ also holds.

absolute value The following number computed from the real number a:

$$|a| = \begin{cases} a, & \text{if } a \geq 0 \\ -a, & \text{if } a < 0. \end{cases}$$

abstract algebra The study of numbers and abstract systems that behave like numbers.

abstract system One or more sets of objects together with some unary and/or binary operations on those objects.

abstraction The process of getting farther and farther away from specific items by working more and more with general objects.

add a vertex The act of including a new vertex in a graph.

add an edge The act of including a new edge in a graph.

adjacency matrix A method for representing a graph using a matrix A of numbers consisting of one row and one column for each vertex. The value of A_{ij} is 1, if vertex i is adjacent to vertex j and 0, otherwise.

adjacent vertices Two vertices that are connected by an edge in a graph.

Archimedean property The property that for any real number z, there is a natural number $n > z$.

axiom A property of an abstract system that is assumed to hold.

axiomatic system An abstract system together with a list of axioms that are assumed to hold.

backward process That part of the forward-backward method in which you try to establish that the conclusion of an implication is *true* by asking and answering a key question.

ball of radius r centered at the point $\mathbf{c} = (c_1, \ldots, c_n)$ The set

$$B_r(\mathbf{c}) = \{(x_1, \ldots, x_n) : (x_1 - c_1)^2 + \cdots + (x_n - c_n)^2 \leq r^2\}.$$

bijective function A function that is both injective and surjective.

binary operator A method for combining two objects of the same type to produce another object.

binary relation A method for combining two objects of the same type to produce a result of *true* or *false*.

bounded above The property of a set having an upper bound.

bounded below The property of a set having a lower bound.

bounded sequence of real numbers A sequence $X = (x_1, x_2, \ldots)$ of real numbers for which there is a real number $M > 0$ such that for all $k \in \mathbb{N}$, $|x_k| \leq M$.

bounded set A set T in $\mathbb{R}^n$ for which there is a real number $M > 0$ such that for all elements $\mathbf{x} \in T$, $||\mathbf{x}|| \leq M$.

Cartesian product The following set created from two sets A and B:

$$A \times B = \{(x, y) : x \in A \text{ and } y \in B\}.$$

characteristic function of a set A function whose value at any point x is 1, if x is in the set and 0, otherwise.

choose method A technique for proving that for all objects with a certain property, something happens. Use this method when the key words *for all* arise in the backward process.

closed-form solution A solution to a problem that is obtained from the given data by using a formula or simple rule.

closed interval from l to u The set of real numbers between l and u, that is, $\{x \in \mathbb{R} : l \leq x \leq u\}$.

closed operator on a set An operator on a set of objects whose result is also an object in the set.

codomain A set that contains all values that result from applying the function to all elements in the domain.

column vector An n-vector that is thought of as an $(n \times 1)$ matrix.

common divisor An integer that divides two given integers.

commutative ring A ring $(R, \oplus, \odot)$ such that for all $a, b \in R$, $a \odot b = b \odot a$.

complement of a set The set of all elements that are not in the set.

complement of a set A in a set B The set $A - B$ consisting of those elements of B that are not in A, that is, $A - B = \{x \in B : x \notin A\}$.

complete graph A graph in which each pair of vertices is connected by an edge.

complex numbers The set $\mathbb{C} = \{a + bi : a \text{ and } b \text{ are reals and } i = \sqrt{-1}\}$.

component One of the individual numbers in an n-vector.

composition The binary operation of combining two functions $f : A \to B$ and $g : B \to C$ to create the function $h : A \to C$ defined, for each $x \in A$, by $h(x) = (g \circ f)(x) = g(f(x))$.

conclusion The statement q in the implication p *implies* q.

conditional statement A statement whose truth depends on the value of one or more variables.

congruent modulo n An integer a is *congruent modulo n* to an integer b if and only if n divides $(a - b)$.

conjunction The statement p *and* q associated with the two statements p and q. The statement p *and* q is *true*, when both p and q are *true* and *false*, otherwise.

connected graph A graph in which there is a walk between each pair of vertices.

constant sequence A sequence in which each element has the same value.

construction method A technique for proving that there is an object with a certain property such that something happens. Use this technique when the key words *there is* arise in the backward process.

contained in A set A is *contained in* a set B means that A is a subset of B.

contradiction method A technique for proving that p *implies* q in which you assume p is *true* and q is *not true* and work forward from this information to reach a contradiction to some statement that you *know* is *true*.

contrapositive method A technique for proving that p *implies* q in which you assume q is *not true* and prove that p is *not true*. This is accomplished by working forward from *not* q and backward from *not* p.

contrapositive statement The statement (*not* q) *implies* (*not* p) associated with the statement p *implies* q.

convergence The issue of whether or not the values in a sequence are getting closer and closer to some value.

convergence of a sequence in a set The *sequence X in the set S converges* if and only if there is an element $x \in S$ such that X converges to x.

convergence of a sequence in a set to an element in the set The *sequence $X = (x^1, x^2, \ldots)$ in the set S converges to an element $x \in S$* if and only if for every real number $\varepsilon > 0$, there is a $j \in \mathbb{N}$ such that for all $k \in \mathbb{N}$ with $k > j$, $d(x^k, x) < \varepsilon$ (where $d : S \times S \to \mathbb{R}$ is a distance function, that is, (S, d) is a metric space).

convergence of a sequence of n-vectors The *sequence X of n-vectors converges* if and only if there is an n-vector $\mathbf{x}$ such that X converges to $\mathbf{x}$.

convergence of a sequence of n-vectors to an n-vector The *sequence of n-vectors $X = (\mathbf{x}^1, \mathbf{x}^2, \ldots)$ converges to the n-vector* $\mathbf{x}$ if and only if for every real number $\varepsilon > 0$, there is a $j \in \mathbb{N}$ such that for all $k \in \mathbb{N}$ with $k > j$, $||\mathbf{x}^k - \mathbf{x}|| < \varepsilon$.

convergence of a sequence of real numbers The *sequence X of real numbers converges* if and only if there is a real number x such that X converges to x.

convergence of a sequence of real numbers to a real number The *sequence $X = (x_1, x_2, \ldots)$ of real numbers converges to the real number x* if and only if for every real number $\varepsilon > 0$, there is a $j \in \mathbb{N}$ such that for all $k \in \mathbb{N}$ with $k > j$, $|x_k - x| < \varepsilon$.

converse statement The statement q *implies* p associated with the statement p *implies* q.

coordinate function Any one of the p outputs of a function $f : \mathbb{R}^n \to \mathbb{R}^p$.

coordinate of an n-vector One of the individual numbers in an n-vector.

countable set A set that is finite or countably infinite.

countably infinite set A set S for which there is a bijective $f : \mathbb{N} \to S$.

cycle A connected graph in which each vertex has degree 2.

data Known information that you can use to solve a problem.

decreasing sequence of real numbers A sequence of real numbers in which each successive element has a value less than or equal to that of the preceding element.

defining property of a set The property that an object must satisfy in order to be an element of the set.

definition A meaning given to a word (or group of words) that describes those objects satisfying certain desirable properties and excludes those objects not satisfying the properties.

degree of a vertex The number of edges incident to the vertex.

delete a vertex The act of removing a vertex (and all edges incident to that vertex) from a graph.

delete an edge The act of removing an edge from a graph.

denumerable set A countably infinite set.

diagonal elements of a matrix The elements A_{ii} ($i = 1, \ldots, n$) of a square matrix A of order n.

diagonal matrix A square matrix of order n in which all elements, except possibly for the diagonal elements, are 0.

dimension of a matrix The number m of rows and n of columns of the matrix.

dimension of a vector The number n of elements in a vector.

direct uniqueness method The method for proving that there is a unique object with a certain property such that something happens in which you first establish that *there is* such an object, say, X. You then assume that Y is also an object with the certain property and for which the something happens. You then must prove that $X = Y$, that is, that X and Y are the same.

directed edge An edge of a graph in which there is a direction associated with the edge.

directed graph A graph in which there is a direction associated with each edge.

discrete mathematics The study of mathematical problems involving a finite number of items.

disjunction The statement p *or* q associated with the two statements p and q. The statement p *or* q is *true* in all cases except when p and q are both *false*.

distance function A function $d : S \times S \to \mathbb{R}$ that associates to each pair of objects x and y in a set S, a real number $d(x, y)$ representing the *distance* from x to y. To be considered a distance, this function must satisfy the following axioms, for all $x, y, z \in S$:

1. $d(x, y) \geq 0$.
2. $d(x, y) = 0$ if and only if $x = y$.
3. $d(x, y) = d(y, x)$.
4. $d(x, z) \leq d(x, y) + d(y, z)$.

diverge A sequence that does not converge.

divides An integer a *divides* an integer b if and only if there is an integer k such that $b = ak$.

division algorithm An algorithm for dividing two integers that results in a unique integer remainder.

domain The set of all values to which a function can be applied.

dot product An operation that combines two n-vectors $\mathbf{x} = (x_1, \ldots, x_n)$ and $\mathbf{y} = (y_1, \ldots, y_n)$ to produce the following real number:

$$\mathbf{x} \cdot \mathbf{y} = \sum_{i=1}^{n} x_i y_i = x_1 y_1 + \cdots + x_n y_n .$$

edge list A method for representing a graph consisting of the number of vertices and edges and then a list of all the edges.

edge of a graph A line connecting two vertices in a graph that represents a relation between the objects corresponding to the two connected vertices.

either/or method A technique for proving p *implies* q when p or q contains the key words *either/or*.

element of a matrix Any of the individual numbers in the matrix.

element of a set Any item belonging to the set.

empty set A set containing no elements.

endpoint of an edge A vertex connected by that edge in a graph.

epsilon-neighborhood of a real number x The set of points defined by $N_\varepsilon(x) = \{y \in \mathbb{R} : |x - y| < \varepsilon\}$.

equality of functions Two functions $f : A \to B$ and $g : C \to D$ are *equal* if and only if $A = C,\ B = D$, and, for each $x \in A,\ f(x) = g(x)$.

equality of matrices Two $(m \times n)$ matrices A and B are *equal* if and only if for all $i = 1, \ldots, m$ and for all $j = 1, \ldots, n,\ A_{ij} = B_{ij}$.

equality of n-vectors Two n-vectors $\mathbf{x} = (x_1, \ldots, x_n)$ and $\mathbf{y} = (y_1, \ldots, y_n)$ are *equal* if and only if for all $i = 1, \ldots, n,\ x_i = y_i$.

equality of sets A set A *equals* a set B if and only if A is a subset of B and B is a subset of A.

equivalence relation A binary relation that is reflexive, symmetric, and transitive.

equivalent elements Two elements x and y in a set S having a binary relation R and for which $x\ R\ y$ and $y\ R\ x$.

eta matrix A square matrix of order n that is equal to the identity matrix except possibly for one column.

Euclidean algorithm A particular algorithm for finding the greatest common divisor of two integers.

Euler number The real number $e = 2.718282$ to which the following sequence $X = (x_1, x_2, \ldots)$ converges:

$$x_k = \left(1 + \frac{1}{k}\right)^k, \text{ for } k = 1, 2, \ldots.$$

even integer An integer n for which there is an integer k such that $n = 2k$.

existential quantifier The words *there is, there are, there exist*, or equivalent words that arise in statements.

field An integral domain $(F, \oplus, \odot)$ in which the nonzero elements form an Abelian group with respect to $\odot$.

finite set A set S that is empty or for which there is an $n \in \mathbb{N}$ and a bijective function $f : \{1, 2, \ldots, n\} \to S$.

forward-backward method A technique for proving that the statement p *implies* q is *true* in which you assume that p is *true* and try to reach the conclusion that q is *true*.

forward process That part of the forward-backward method in which you derive new statements from previously *true* statements.

function A rule that associates to each element x of a set A, a unique element $f(x)$ belonging to a set B. More formally, a function is a set f of ordered pairs (x, y) with $x \in A$ and $y \in B$ such that (1) for all $x \in A$, there is a $y \in B$ such that $(x, y) \in f$ and (2) if $(x, y) \in f$ and $(x, z) \in f$, then $y = z$.

Gaussian elimination A systematic numerical method for finding the inverse of a matrix or for determining that no such inverse exists.

generalization The process of creating a general class of problems (concepts, theorems, definitions, and so on) from one or more specific problems (concepts, theorems, definitions, and so on). The new class contains all of the special cases as well as some different problems.

graph A nonempty finite collection of vertices together with a finite set of edges, each of which connects a pair of vertices.

greater than For n-vectors $\mathbf{x} = (x_1, \ldots, x_n)$ and $\mathbf{y} = (y_1, \ldots, y_n)$, $\mathbf{x} > \mathbf{y}$ if and only if for all $i = 1, \ldots, n$, $x_i > y_i$.

greater than or equal to For n-vectors $\mathbf{x} = (x_1, \ldots, x_n)$ and $\mathbf{y} = (y_1, \ldots, y_n)$, $\mathbf{x} \geq \mathbf{y}$ if and only if for all $i = 1, \ldots, n$, $x_i \geq y_i$.

greatest common divisor An integer d is the *greatest common divisor* of the integers a and b if and only if (1) d is a common divisor of a and b and (2) for any common divisor c of a and b, c divides d.

greatest lower bound See infimum of a set.

group An axiomatic system consisting of a set G and a closed binary operation $\odot$ on G that satisfies the following three axioms:

1. There is an element $e \in G$ such that for all $a \in G$, $a \odot e = e \odot a = a$.
2. For each $a \in G$, there is an element $b \in G$ such that $a \odot b = b \odot a = e$.
3. For all $a, b, c \in G$, $(a \odot b) \odot c = a \odot (b \odot c)$.

hypothesis The statement p in the implication p *implies* q.

identity element The element e of a group $(G, \odot)$ that satisfies the property that for all $a \in G$, $a \odot e = e \odot a = a$.

identity function A function whose value at any point in the domain is that same point.

identity matrix The square matrix I of order n in which each element is 0 except for the diagonal elements, all of which are 1.

implication A statement of the form p *implies* q in which p and q are statements. The implication is *true* in all cases except when p is *true* and q is *false*.

included in A set A is *included in* a set B means that A is a subset of B.

increasing sequence of real numbers A sequence of real numbers in which each successive element has a value greater than or equal to that of the preceding element.

indirect proof See contradiction method.

indirect uniqueness method The method for proving that there is a *unique* object with a certain property such that something happens in which you first establish that *there is* such an object, say, X. You then assume that Y is a *different* object from X with the certain property and for which the something happens. You then must reach a contradiction.

induction A technique for proving that for all integers greater than or equal to some initial one, something happens.

inductive form A method for describing a sequence in which the value of each element, after the initial ones, is given by a closed-form expression in terms of the values of one or more of the preceding elements.

infimum of a set T of real numbers A real number t such (1) t is a lower bound for T and (2) for any lower bound s for T, $s \leq t$.

infimum property of real numbers The property that every nonempty set of real numbers that is bounded below has an infimum.

infinite set A set that is not finite.

injective function A function $f : A \to B$ such that for all $x, y \in A$ with $x \neq y$, $f(x) \neq f(y)$.

inner product See dot product.

input Known information that you can use to solve a problem.

input value A value to which a function is applied.

integers The set $\mathbb{Z} = \{\ldots -2, -1, 0, 1, 2, \ldots\}$.

integral domain A commutative ring $(R, \oplus, \odot)$ with a unity element $1 \neq 0$ such that if $a, b \in R$ with $a \odot b = 0$, then $a = 0$ or $b = 0$.

intersection A binary operation that combines two sets A and B to create the set $A \cap B = \{x : x \in A \text{ and } x \in B\}$.

interval from l to u See a closed interval from l to u.

inverse element The element a^{-1} associated with an element a in a group $(G, \odot)$ such that $a \odot a^{-1} = a^{-1} \odot a = e$ (where e is the identity element of the group).

inverse function For a given function $f : A \to B$, a function $g : B \to A$ is an *inverse function* if and only if g composed with f is the identity function on A and f composed with g is the identity function on B, that is, $g \circ f = i_A$ and $f \circ g = i_B$.

inverse image of a set Given a function $f : A \to B$ and a set $Y \subseteq B$, the inverse image of Y under f is the following set:

$$\widehat{f}^{-1}(Y) = \{x \in A : f(x) \in Y\}.$$

inverse of a matrix An $(n \times n)$ matrix A^{-1} associated with an $(n \times n)$ matrix A such that $AA^{-1} = A^{-1}A = I$ (the $(n \times n)$ identity matrix).

inverse statement The statement $(not\ p)\ implies\ (not\ q)$ associated with the statement $p\ implies\ q$.

invertible matrix A matrix that has an inverse.

invertible operator A closed unary operator that, when applied twice to an object, results in the original object.

irrational numbers The set of real numbers that are not rational.

key question The question you should ask when working backward in a proof of the statement $p\ implies\ q$, which is, How can I show that q is *true*?

least-integer principle The property that every nonempty set of positive integers has a least element.

least upper bound See supremum of a set.

length of a walk The number of edges of a graph in the walk.

length of an n-vector The real number number $||\mathbf{x}||$ associated with the n-vector $\mathbf{x} = (x_1, \ldots, x_n)$ that is computed by the following formula:

$$||\mathbf{x}|| = \sqrt{x_1^2 + \cdots + x_n^2}.$$

(See also norm).

less than For n-vectors $\mathbf{x} = (x_1, \ldots, x_n)$ and $\mathbf{y} = (y_1, \ldots, y_n)$, $\mathbf{x} < \mathbf{y}$ if and only if for all $i = 1, \ldots, n,\ x_i < y_i$.

less than or equal to For n-vectors $\mathbf{x} = (x_1, \ldots, x_n)$ and $\mathbf{y} = (y_1, \ldots, y_n)$, $\mathbf{x} \leq \mathbf{y}$ if and only if for all $i = 1, \ldots, n,\ x_i \leq y_i$.

linear algebra The study of ordered lists of numbers.

linearly dependent vectors The n-vectors $\mathbf{x}^1, \ldots, \mathbf{x}^k$ are *linearly dependent* if and only if there are real numbers $t_1, \ldots, t_k$, not all zero, such that

$$t_1\mathbf{x}^1 + \cdots + t_k\mathbf{x}^k = \mathbf{0}.$$

linearly independent vectors The n-vectors $\mathbf{x}^1, \ldots, \mathbf{x}^k$ are *linearly independent* if and only if for all real numbers $t_1, \ldots, t_k$ with

$$t_1\mathbf{x}^1 + \cdots + t_k\mathbf{x}^k = \mathbf{0},$$

it follows that $t_1 = \cdots = t_k = 0$.

list notation A method for representing a set in which you enclose the members of the set, separated from each other by a comma, in braces.

loop An edge that connects a vertex to itself in a graph.

lower bound for a set T of real numbers A real number t such that for each $x \in T,\ t \leq x$.

matrix A rectangular table of numbers organized in m rows of n columns.

member of a set Any item belonging to the set.

metric space An axiomatic system for describing the distance between two objects that consists of a set S and a function $d : S \times S \to \mathbb{R}$ satisfying the following axioms, for all $x, y, z \in S$:

1. $d(x, y) \geq 0$.
2. $d(x, y) = 0$ if and only if $x = y$.
3. $d(x, y) = d(y, x)$.
4. $d(x, z) \leq d(x, y) + d(y, z)$.

monotone sequence of real numbers A sequence of real numbers that is either increasing or decreasing.

monotonically decreasing sequence of real numbers See a decreasing sequence of real numbers.

monotonically increasing sequence of real numbers See an increasing sequence of real numbers.

multiple edges Two or more edges connecting the same pair of vertices in a graph.

multiple of a sequence of real numbers The sequence obtained by multiplying each element of a given sequence by a fixed value.

n-vector An ordered list of n real numbers.

natural numbers The set $\mathbb{N} = \{1, 2, 3, \ldots\}$.

negation of a statement p The statement *not* p, which is *true*, if p is *false* and *false*, if p is *true*.

negative of a scalar A scalar $-x$ that when added to the scalar x results in the scalar zero.

negative of a vector The vector $-\mathbf{x}$ that when added to the vector $\mathbf{x}$ results in the zero vector.

negative real numbers The set $\{-a : a \in P\}$, where P is the set of positive real numbers.

neighbors Two vertices that are connected by an edge in a graph.

nested quantifier Two or more quantifiers that arise in a statement.

nonsingular matrix A matrix that has an inverse.

norm A unary operator that associates to each n-vector $\mathbf{x}$ a real number, denoted by $||\mathbf{x}||$, representing the *length* of $\mathbf{x}$. To be considered a norm, this operation must satisfy the following axioms, for all $\mathbf{x}, \mathbf{y} \in \mathbb{R}^n$:

1. $||\mathbf{x}|| \geq 0$.
2. $||\mathbf{x}|| = 0$ if and only if $\mathbf{x} = \mathbf{0}$.
3. $||-\mathbf{x}|| = ||\mathbf{x}||$.
4. $||\mathbf{x} + \mathbf{y}|| \leq ||\mathbf{x}|| + ||\mathbf{y}||$.

numerical method A procedure or algorithm that uses the problem data to produce a solution or an approximate solution.

odd integer An integer n for which there is an integer k such that $n = 2k+1$.

one-to-one function A function $f : A \to B$ such that for all $x, y \in A$ with $x \neq y$, $f(x) \neq f(y)$.

onto function A function $f : A \to B$ such that for all $y \in B$, there is an $x \in A$ with $f(x) = y$.

order The relationship of numbers to each other in terms of size.

orthogonal vectors Two n-vectors that are perpendicular to each other, that is, whose dot product is 0.

oscillating sequence of real numbers A sequence of real numbers that is not monotone.

output The solution to a problem obtained from the data.

output value The result of applying a function to an input value.

partition of a set A splitting up of a set A into a collection of nonempety subsets that are pairwise disjoint and whose union is A.

path A walk in which no vertex is repeated.

permutation A rearrangement of items.

permutation matrix A square matrix of order n in which each row and each column has exactly one 1 and every other entry is 0.

positive real numbers The subset P of real numbers that satisfies the following three axioms:

1. For all $a, b \in P$, $a + b \in P$.
2. For all $a, b \in P$, $ab \in P$.
3. For all $a \in \mathbb{R}$, exactly one of the following is *true*:

$$a \in P,\ a \in \{0\},\ \text{or}\ -a \in P.$$

power set of a set A The set of all subsets of A.

prime An integer greater than 1 that is divisible only by 1 and itself.

proof A convincing argument, expressed in the language of mathematics, for showing that statements of the form p *implies* q are *true*.

proof by cases An either/or method for proving that p *implies* q when p has the form *either* p_1 *or* p_2. With a proof by cases, you first prove that p_1 *implies* q and then that p_2 *implies* q.

proof by elimination An either/or method for proving that p *implies* q when q has the form *either* q_1 *or* q_2. With a proof by elimination, you assume that p and *not* q_1 are *true* and then work forward to establish that q_2 is *true*. Alternatively, you can assume that p and *not* q_2 are *true* and then work forward to establish that q_1 is *true*.

proof technique A method for doing a proof.

range of a function The set of all values that result from applying the function $f : A \to B$ to every element in the domain A. In other words, range $f = \{f(x) : x \in A\}$.

rationals The set $\mathbb{Q} = \left\{ \frac{p}{q} : p \text{ and } q \text{ are integers and } q \neq 0 \right\}$.

real analysis The study of the properties of real numbers and related concepts.

real line A horizontal line used to visualize all the real numbers (negative, zero, and positive).

reals The set $\mathbb{R} = \{r : r \text{ is a number expressible in a decimal form}\}$.

reflexive relation A binary relation R on a set S with the property that for all $x \in S$, $x\ R\ x$.

ring An axiomatic system consisting of a set R together with two closed binary operations on R, denoted by $\oplus$ and $\odot$, that satisfy the following axioms:

1. The pair $(R, \oplus)$ is an Abelian group.
2. For all $a, b, c \in R$, $a \odot (b \odot c) = (a \odot b) \odot c$.
3. For all $a, b, c \in R$, $(a \oplus b) \odot c = (a \odot c) \oplus (b \odot c)$ and $a \odot (b \oplus c) = (a \odot b) \oplus (a \odot c)$.

row vector An n-vector that is thought of as a $(1 \times n)$ matrix.

scalars Any collection of objects that satisfy the same algebraic properties as the real numbers.

sequence in a set S A function $X : \mathbb{N} \to S$.

sequence of n-vectors A function $X : \mathbb{N} \to \mathbb{R}^n$.

sequence of real numbers A function $X : \mathbb{N} \to \mathbb{R}$.

sequential generalization A list of generalizations, each of which includes the preceding one as a special case.

set A collection of objects.

set-builder notation A method for representing a set in which you use a verbal and mathematical description of the property that the elements of the set satisfy.

singular matrix A matrix that has no inverse.

special case A specific problem (concept, theorem, and so on) that provides the basis for applying unification, generalization, and/or abstraction.

specialization method A proof technique for deriving a new statement in the forward process when you *know that* for all objects with a certain property, something happens. Use this method when the key words *for all* arise in the forward process.

square matrix of order n A matrix with n rows and n columns.

standard unit vector An n-vector which is the zero vector except in one component, whose value is 1.

statement A mathematical expression that is either *true* or *false*.

strict subset A set A is a *strict subset* of a set B if and only if A is a subset of B and A is not equal to B.

subfield A subset of a field $(F, \oplus, \odot)$ that is itself a field under the operations $\oplus$ and $\odot$.

subgraph A portion of a graph that is itself a graph.

subgroup A subset of a group $(G, \odot)$ that is itself a group under the operation $\odot$.

subring A subset of a ring $(R, \oplus, \odot)$ that is itself a ring under the operations $\oplus$ and $\odot$.

subset A set A is a *subset* of a set B if and only if all the elements of A are in B.

supremum of a set T of real numbers A real number u such that (1) u is an upper bound for T and (2) for any upper bound v for T, $v \geq u$.

supremum property of real numbers The property that a nonempty set of real numbers that is bounded above has a supremum.

surjective function A function $f : A \to B$ such that for all $y \in B$, there is an $x \in A$ with $f(x) = y$.

symmetric group of a set S The group consisting of all bijective functions $f : S \to S$ under the operation of function composition.

symmetric relation A binary relation R on a set S such that for all $x, y \in S$, if $x \; R \; y$, then $y \; R \; x$.

syntax error An error that arises when the symbols and/or operations in a mathematical statement do not make sense or cannot be performed.

trail A walk in a graph in which no edge is repeated.

transitive relation A binary relation R on a set S with the property that for all $x, y, z \in S$, if $x \; R \; y$ and $y \; R \; z$, then $x \; R \; z$.

transpose of a matrix The matrix obtained from a matrix A by making each row of A into a column.

triangle inequality The property that for any $a, b \in \mathbb{R}$, $|a + b| \leq |a| + |b|$.

trichotomy property The property that every real number is either positive, negative, or zero.

trivial graph A graph consisting of one vertex and no edges.

truth table A table indicating the truth of a complex statement in terms of the the truths of constituent statements.

unary operator A method for creating a new object from a given object.

unbounded sequence of real numbers A sequence $X = (x_1, x_2, \ldots)$ of real numbers such that for every real number $M > 0$, there is a $k \in \mathbb{N}$ such that $|x_k| > M$.

uncountable set A set that is not countable.

undirected graph See a graph.

unification The process of combining two or more problems (concepts, theorems, and so on) into a single, comprehensive problem that contains the original problems as special cases.

union A binary operation that combines two sets A and B to create the set $A \cup B = \{x : x \in A \text{ or } x \in B\}$.

uniqueness method A technique for proving that statements of the following form are *true*: There is a *unique* object with a certain property such that something happens.

unity element An element 1 of a ring $(R, \oplus, \odot)$ with the property that for all $a \in R$, $a \odot 1 = 1 \odot a = a$.

universal quantifier The words *for all, for each, for every, for any*, or equivalent words that arise in statements.

universal set An underlying set containing all possible objects that might be elements of a given set.

upper bound for a set T of real numbers A real number u such that for each $x \in T$, $x \leq u$.

vector of dimension n An ordered list of n real numbers.

vector space An axiomatic system consisting of a collection of scalars and vectors together with appropriate unary and binary operations that satisfy all of the algebraic properties of real numbers and n-vectors. See Section 4.5.2 for a list of the specific axioms.

Venn diagram A method for visualizing the relationship of sets to each other in which each set is represented by a circle.

vertex of a graph A circle in a graph that represents a particular object in a problem under consideration.

walk from v_0 to v_k An alternating sequence of vertices and edges in a graph that begins at the vertex v_0 and ends at the vertex v_k.

zero matrix The matrix in which each element is 0.

zero sequence The sequence in which each element is 0.

zero vector The n-vector in which each component is 0.

Index